T0188223

Performance Analysis and Synthesis for Discrete-Time Stochastic Systems with Network-Enhanced Complexities

Performance Analysis and Synthesis for Discrete-Time Stochastic Systems with Network-Enhanced Complexities

Derui Ding
Zidong Wang
Guoliang Wei

CRC Press
Taylor & Francis Group
Boca Raton London New York

CRC Press is an imprint of the
Taylor & Francis Group, an **informa** business

CRC Press
Taylor & Francis Group
6000 Broken Sound Parkway NW, Suite 300
Boca Raton, FL 33487-2742

First issued in paperback 2020

© 2019 by Taylor & Francis Group, LLC
CRC Press is an imprint of Taylor & Francis Group, an Informa business

No claim to original U.S. Government works

Version Date: 20180917

ISBN 13: 978-0-367-57092-7 (pbk)
ISBN 13: 978-1-138-61001-9 (hbk)

This book contains information obtained from authentic and highly regarded sources. Reasonable efforts have been made to publish reliable data and information, but the author and publisher cannot assume responsibility for the validity of all materials or the consequences of their use. The authors and publishers have attempted to trace the copyright holders of all material reproduced in this publication and apologize to copyright holders if permission to publish in this form has not been obtained. If any copyright material has not been acknowledged please write and let us know so we may rectify in any future reprint.

Except as permitted under U.S. Copyright Law, no part of this book may be reprinted, reproduced, transmitted, or utilized in any form by any electronic, mechanical, or other means, now known or hereafter invented, including photocopying, microfilming, and recording, or in any information storage or retrieval system, without written permission from the publishers.

For permission to photocopy or use material electronically from this work, please access www.copyright.com (http://www.copyright.com/) or contact the Copyright Clearance Center, Inc. (CCC), 222 Rosewood Drive, Danvers, MA 01923, 978-750-8400. CCC is a not-for-profit organization that provides licenses and registration for a variety of users. For organizations that have been granted a photocopy license by the CCC, a separate system of payment has been arranged.

Trademark Notice: Product or corporate names may be trademarks or registered trademarks, and are used only for identification and explanation without intent to infringe.

Library of Congress Cataloging-in-Publication Data

Names: Ding, Derui, author. | Wang, Zidong, 1966- author. | Wei, Guoliang, author.
Title: Performance analysis and synthesis for discrete-time stochastic systems with network-enhanced complexities / Derui Ding, Zidong Wang and Guoliang Wei.
Description: First edition. | Boca Raton, FL : CRC Press/Taylor & Francis Group, 2019. | Includes bibliographical references and index.
Identifiers: LCCN 2018025090| ISBN 9781138610019 (hardback : acid-free paper) | ISBN 9780429465901 (ebook)
Subjects: LCSH: Automatic control--Mathematics. | Stochastic processes. | Discrete-time systems.
Classification: LCC TJ213 .D54135 2019 | DDC 629.801/176--dc23
LC record available at https://lccn.loc.gov/2018025090

Visit the Taylor & Francis Web site at
http://www.taylorandfrancis.com

and the CRC Press Web site at
http://www.crcpress.com

This book is dedicated to the Dream Dynasty, consisting of a group of bright people who have been to Brunel University London to enjoy memorable research with challenging network-enhanced complexities through performance analysis and synthesis in a discrete (success-driven) yet stochastic (laziness-perturbed) way ......

Contents

Preface

The real-world systems are usually subject to various sorts of complexities such as parameter uncertainties, fading measurements, randomly occurring nonlinear disturbances, cyber-attacks, event-triggered communication mechanism, and so forth. These kinds of complexities, which could exist in a non-networked environment, are becoming even more severe (in terms of their degrees or intensities) in networked systems due mainly to the limited bandwidth and open channel media. In other words, these kinds of complexities have been greatly enhanced because of the usage of the communication networks. It is worth mentioning that the usage of the network enhanced complexities inevitably affects the system performance and could even lead to the instability of the considered systems. Therefore, it is of both theoretical significance and practical importance to investigate the impact on the system performance from these network-enhanced complexities and to develop new approaches addressing the analysis and synthesis of various systems.

In this book, we focus on the performance analysis and synthesis for several classes of discrete-time stochastic systems with various network-enhanced complexities. The complexities under consideration mainly include fading measurements, stochastic nonlinearities, event-triggered communication protocol, cyber-attacks, and so forth. The addressed systems cover general stochastic nonlinear systems, stochastic parameters systems, state-saturated time-varying systems and some special networked systems with a given topology (i.e., sensor networks and multi-agent systems). Furthermore, some novel concepts, models of network-enhanced complexities, and performance criteria in probability are proposed to account for the real-world engineering requirements. They mainly involve the consensus in probability, randomly occurring deception attacks, the discrete version of input-to-state stability in probability, envelope-constrained filtering performance, and so forth.

The content of this book can be divided into two parts conceptually. In the first part, a series of finite-horizon state estimation and control issues are investigated for discrete time-varying stochastic systems with network-enhanced complexities. Some new controller and estimator design schemes are developed in terms of the solutions to backward recursive Riccati difference equations (RDEs), Kalman-type iterative algorithms or recursive linear matrix inequalities (RLMIs). In the second part, the impact on the control or estimation performance are deeply discussed from the event-triggered communication protocol for discrete-time stochastic systems subject to packet dropouts or cyber attacks. Some novel design frameworks of distributed controllers and filters are proposed to guarantee the prescribed performance requirements, and some typical applications are effectively implemented in the realm of sensor networks and multi-agent systems.

The compendious framework and description of this book are given as follows. In Chapter 1, the research motivation is discussed, the research problems to be addressed in this book are proposed, and the outline of this book is given. Chapter 2 is concerned with the finite-horizon $\mathcal{H}_\infty$ control problem for discrete time-varying nonlinear systems with simultaneous presence of fading measurements and randomly occurring nonlinearities. A novel $\mathcal{H}_\infty$ control scheme based on RDEs is proposed by utilizing model transformation techniques combined with a certain $\mathcal{H}_2$-type criterion. In light of such a framework, the $\mathcal{H}_\infty$ consensus control and the distributed $\mathcal{H}_\infty$ state estimation are, respectively, discussed in Chapter 3 and Chapter 4 for discrete time-varying multi-agent systems with both missing measurements and parameter uncertainties, and discrete time-varying systems with both stochastic parameters and stochastic nonlinearities over sensor networks. In what follows, the dissipative control and $\mathcal{H}_\infty$ filtering focusing on similar network-enhanced complexities are, respectively, addressed in Chapter 5 and Chapter 6 for state-saturated time-varying systems. In addition, the envelope-constrained $\mathcal{H}_\infty$ filtering problem is investigated in Chapter 7 for a class of discrete time-varying stochastic systems with fading measurements and randomly occurring nonlinearities via ellipsoid description approaches.

The complexities of communication protocols or cyber-attacks have become another focus of research, and some interesting research is done in this book. First, the distributed recursive filter dependent on the solution of two RDEs is designed in Chapter 8 for discrete time-delayed stochastic systems subject to both uniform quantization and deception attack effects. By utilizing mathematical induction, a sufficient condition is established to ensure the asymptotic boundedness of the sequence of the error covariance. For event-triggered communication protocols, Chapter 9 is concerned with the event-based distributed $\mathcal{H}_\infty$ state estimation subject to packet dropouts over sensor networks. In this chapter, a novel distributed filter is put forward to effectively merge the available innovations from not only itself but also its neighboring sensors under the event-triggered communication protocol. In addition, event-triggered consensus control is studied in Chapter 10 for discrete-time stochastic multi-agent systems with state-dependent noise. The desired controller is designed in a critical theoretical framework established for analyzing the so-called input-to-state stability in probability (ISSiP) for general discrete-time nonlinear stochastic systems. When both cyber-attacks and communication protocols are taken into account, the event-based security control and the event-based consensus control are, respectively, investigated in Chapter 11 and Chapter 12 for discrete-time stochastic systems with multiplicative noise, and discrete-time multi-agent systems with lossy sensors.

For consensus control issues, an easy-to-design approach is developed by making use of eigenvalues and eigenvectors of the Laplacian matrix.

This book is a research monograph whose intended audience is graduate and postgraduate students as well as researchers.

Acknowledgements

We would like to acknowledge the help of many people who have been directly involved in various aspects of the research leading to this book. Special thanks go to Professor Daniel W. C. Ho from City University of Hong Kong, Professor James Lam from The University of Hong Kong, Professor Bo Shen from Donghua University, Shanghai, Professor Hongli Dong from Northeast Petroleum University, Daqing, China, Professor Huisheng Shu from Donghua University, Shanghai, and Professor Jun Hu from Harbin University of Science and Technology, Harbin, China.

Symbols

Symbol Description

$\mathbb{R}^n$ The n-dimensional Euclidean space.

$\mathbb{R}^{n \times m}$ The set of all $n \times m$ real matrices.

$\mathbb{R}^+$ The set of all nonnegative real numbers.

$\mathbb{Z}^+$ The set of all nonnegative integers.

A^T The transpose of the matrix A.

$|A|$ The determinant of the square matrix A.

$tr(A)$ The trace of a matrix A.

$\|A\|$ The norm of matrix A defined by $\|A\| = \sqrt{\text{trace}(A^T A)}$.

$\|A\|_F$ The Frobenius norm of matrix A.

$A^\dagger \in \mathbb{R}^{n \times m}$ The Moore-Penrose pseudo inverse of $A \in \mathbb{R}^{m \times n}$.

$\lambda_{\min}(A)$ The smallest eigenvalue of a square matrix A.

$\lambda_{\max}(A)$ The largest eigenvalue of a square matrix A.

I The identity matrix of compatible dimension.

0 The zero matrix of compatible dimension.

$\mathbf{1}$ The compatible dimensional column vector with all ones.

$\text{Prob}(\cdot)$ The occurrence probability of the event ".".

$\mathbb{E}\{x\}$ The expectation of the stochastic variable x.

$\mathbb{E}\{x \,|\, y\}$ The expectation of the stochastic variable x conditional on y.

$(\Omega, \mathscr{F}, \text{Prob})$ The complete probability space.

$(\Omega, \mathscr{F}, \{\mathscr{F}_k\}_{k \in \mathbb{Z}^+}, \text{Prob})$ The filtered probability space with a filtration $\{\mathscr{F}_k\}_{k \in \mathbb{Z}^+}$ satisfying the usual conditions.

$\mathbb{I}_{\mathscr{A}}$ The indicator function of set $\mathscr{A}$.

$\gamma^{-1}(\cdot)$ The inverse function of the monotone function $\gamma(\cdot)$.

$\mathscr{K}$ The class functions $\gamma : \mathbb{R}^+ \to \mathbb{R}^+$ that is a continuous strictly increasing function with $\gamma(0) = 0$.

$\mathscr{K}_\infty$ The class functions $\gamma : \mathbb{R}^+ \to \mathbb{R}^+$ that belongs to $\mathscr{K}$ with $\gamma(r) \to \infty$ as $r \to \infty$.

$\mathscr{K}\mathscr{L}$ The class functions $\gamma(s, k) : \mathbb{R}^+ \times \mathbb{R}^+ \to \mathbb{R}^+$ that is of class $\mathscr{K}$ for each fixed k, and is decreasing to zero as $k \to \infty$ for each fixed s.

$a \wedge b$ The minimum of a and b for any $a, b \in \mathbb{R}$.

$a \vee b$ The maximum of a and b for any a, $b \in \mathbb{R}$.

$\varphi \circ \psi : M \to S$

The composition of two functions $\varphi : M \to N$ and $\psi : N \to S$.

$\otimes$ The Kronecker product.

$\| \cdot \|_2$ The usual l_2 norm.

$l_2[0, \infty)$ The space of square summable sequences.

$\mathcal{L}_{[0,N]}$ The space of vector functions over $[0, N]$.

$X > Y$ The $X - Y$ is positive definite, where X and Y are real symmetric matrices.

$X \geq Y$ The $X - Y$ is positive semi-definite, where X and Y are real symmetric matrices.

$\text{diag}\{\cdots\}$

The block-diagonal matrix.

$*$ The ellipsis for terms induced by symmetry, in symmetric block matrices.

1

Introduction

Real-world systems are usually subject to various sorts of complexities such as parameter uncertainties, fading measurements, randomly occurring nonlinear disturbances, cyber-attacks, event-triggered communication mechanisms, and so forth. These kinds of complexities, which do exist in a non-networked environment, are becoming even more severe (in terms of their degree or intensity) in networked systems due mainly to the limited bandwidth and fluctuation of network load. In other words, complexities have been greatly enhanced because of the usage of communication networks. It is worth mentioning that the presence of network-enhanced complexities inevitably affects the system performance and could even lead to instability of the considered systems. In this case, it would be interesting to examine 1) how the networks have substantial impacts on the complexity, and 2) how to analyze/reduce such network-enhanced complexity. Therefore, the necessity of designing the controller/estimator arises naturally in situations where network-enhanced complexities are inevitable. As such, the performance analysis and synthesis of discrete-time stochastic systems with network-enhanced complexities serve as imperative yet challenging topics. In this chapter, we will reveal the research motivation of this book from the three aspects: addressed discrete-time stochastic systems, network-enhanced complexities, and performance analysis and engineering design synthesis.

1.1 Discrete-Time Stochastic Systems

With the rapid development of network technologies, signal process and system control are generally executed over communication networks since they have the benefits of decreasing the hardwiring, the installation cost and the implementation difficulties. Due to the discretization coming from sampling or digitized transmission, the analysis and its real-world application of discrete-time stochastic systems have been attracting increasing research attention. As such, one of the objectives of this book is to deal with the performance analysis and synthesis of discrete-time stochastic systems, which include, but are not limited to, stochastic nonlinear systems with predetermined constraint conditions, state-saturated time-varying systems, stochastic parameter systems, and some special networked stochastic systems with topologies (sensor networks, multi-agent systems).

Stochastic Nonlinear Systems

The interaction of stochastic fluctuations and nonlinearity play an important role for understanding the corresponding dynamic phenomena in electronic generators, lasers, mechanical, chemical, and biological systems [11]. In the past few years, the stability, $\mathcal{H}_\infty$ performance analysis and synthesis have been an active branch within the general research area of stochastic nonlinear problems, and a great number of techniques have been developed in order to meet the needs of practical engineering, see [27, 77, 138, 142, 154] and the references therein. Additionally, the majority of available results are highly dependent on some predetermined assumptions on nonlinear functions, such as sector-bounded conditions, Lipschitz conditions, or monotonicity conditions. It is not difficult to find that, up to now, there is a lack of systematic theory on these topics of general nonlinear discrete-time stochastic systems, and some essential difficulties still need to be overcome. For instance, in order to discuss the relation between the disturbance attenuation level and network-enhanced complexities, the corresponding Hamilton-Jacobi inequalities should be provided and the existence of solutions should be discussed. In addition, the effective application of backstepping approaches still has some insurmountable difficulties for stochastic nonlinear systems without the help of neural networks.

State-Saturated Stochastic Systems

In reality, the obstacles in delivering the high performance promises of traditional control/filter theories are often due to the physical limitations of system components, of which the most commonly encountered one stems from the saturation that occurs in any actuators, sensors, or certain system components. As a special case of nonlinear systems, saturation systems have attracted considerable research interest. In most existing literature, it has been implicitly assumed that the system state is free of saturation. However, in practice, state saturation is frequently encountered due to the physical limitation of the devices or due to protection equipment. Examples include mechanical systems with position and speed limits [58, 74], digital filters implemented in finite word-length format [73] and neural networks with saturation-type transfer function [122]. In these cases, all or part of the states are constrained to stay within a bounded set. If the state saturations are not considered in the analysis and synthesis procedure, the desired performance of the closed-loop system or the estimation error dynamics cannot be ensured. In other words, this kind of saturation characteristics can severely restrict the application of traditional design schemes concerning controllers or estimators. Up to now, for some *linear time-invariant systems* with state saturations, the stability and control problems have been investigated by many researchers (see, e.g., [36, 45, 58–61, 74, 91, 114]). However, the corresponding problems for *time-varying systems* have been largely overlooked due probably to the difficulty in finding appropriate methodologies, and this constitutes the main motivation for our research.

Stochastic Parameters Systems

It is quite common in practical engineering that the system parameters might exhibit unavoidable stochastic fluctuations around some deterministic nominal values. For instance, in network environments, some system parameters might be randomly perturbed within certain intervals due probably to abrupt phenomena such as random failures and repairs of the components, changes in the interconnections of subsystems, sudden environment changes, modification of the operating point of a linearized model of nonlinear systems, etc. This kind of system, with stochastic parameters, is customarily referred to as a stochastic parameter system [80, 145], state-dependent or multiplicative noise system with white parameters [17, 42], which can find extensive application in modeling a variety of real-world systems for, e.g., radar control, missile track estimation, satellite navigation, and digital control of chemical processes [80]. Recently, the control and state estimation problems for such systems have already stirred some research interest, and some recent results can be found in [49, 50, 80] and the references therein. On the other hand, all models for real-time systems should be time-varying since the system parameters could be changeable with time, temperature, operating point, and so forth. As such, the finite-horizon control or state estimation problem is of practical significance. Unfortunately, there have been very few results in the literature regarding this issue for time-varying stochastic parameter systems with network-enhanced complexities.

Multi-Agent Systems with Networked-Induced Randomness

During the last decade, multi-agent systems (MASs) have received compelling attention in the systems and control community due to their applications in various fields ranging from cooperative unmanned air vehicles, automated highway systems scheduling, air traffic control, to sensor networks. In the context of networked multi-agent systems, much research effort has been made on consensus control issues with stochastic communication noise and/or time-delays, see e.g., [44,54,68,82]. The representative methodologies mainly involve algebraic graph methods [67, 81, 148], stochastic analysis (e.g., stochastic matrix analysis, convergence of Markov processes, martingale theory) [54, 68, 82], system theory such as Lyapunov stability analysis [1, 48, 93, 121, 149], optimization theory [82, 144] and game theory [44, 68, 105]. It should be noted that the available approach might suffer from considerable computation burden when considering a multi-agent system with large dimensions and possibly network-enhanced complexities (e.g., missing measurements, parameter uncertainties and event-triggering protocol). Unfortunately, the consensus control problem with network-enhanced complexities has not yet received adequate attention despite its clear engineering significance in a networked environment.

1.2 Network-Enhanced Complexities

Since the network cable is of limited capacity, the system complexities have been inevitably enhanced, which results in some nontrivial challenges in system performance analysis and synthesis. In the past few years, some preliminary work has been done on network-enhanced complexities such as missing measurements, channel fadings, event-triggered communication mechanisms and randomly launched cyber-attacks. The latest research developments are summarized as follows.

Network-Induced Phenomena

As it is well known, the traditional control and state estimation schemes rely on the assumption that the measurement signals are perfectly transmitted. Such an assumption, however, is fairly conservative in many engineering practices when presented with unreliable communication links. In practical engineering within networked links, the measurement signals are usually subject to probabilistic information missing (data dropouts or packet losses) from a variety of reasons such as the high maneuverability of the tracked target, a fault in the measurement, intermittent sensor failures, network congestion, etc. As such, in the past decade, the control and state estimation problems with missing measurements have gained considerable research attention and a wealth of literature has appeared on this topic, see e.g., [30, 39,103,109,113,119,139] and the references therein. It is well recognized that most existing results regarding the missing measurements have concentrated on linear time-invariant systems, and the corresponding state estimation or control issues for time-varying stochastic systems, especially for time-varying stochastic nonlinear systems, have not been thoroughly investigated yet.

Compared with missing measurements, the network-induced channel fading problem has received much less research interest in the context of estimator/controller designs for systems connected via wireless and shared links. It should be pointed out that wireless channels are known to be susceptible to fading effects [34] which constitute one of the most dominant features of wireless communication links. To be more specific, when a signal is transmitted over a wireless channel, it is inevitably subject to some special physical phenomena such as reflection, refraction and diffraction, which lead to multi-path induced fading or shadow fading. Fading is often modeled by a time-varying stochastic mathematical model representing the transmitted signal's change in both amplitude and phase. Some representative models have been investigated, of which the analog erasure channel model, Rice fading channel model, and Rayleigh channel model are arguably the most popular ones, see [15, 16] for more details. In the case that the system measurements are transmitted via fading channels, the corresponding stability analysis, LQG control and Kalman filtering problems have attracted some initial research attention, see [34, 40, 86, 96, 141] and the references therein. Not that, in reality, the majority of practical systems exhibit a time-varying nature and the system dynamics is better quantified over a finite horizon. So far, however, the finite-horizon control and state estimation problems have not been properly investigated for time-varying nonlinear systems subject to fading measurements.

On the other hand, in reality, many real-world systems are often influenced by additive nonlinear disturbances and much research effort has been directed towards the rejection or attenuation of external disturbances. In a networked environment, the nonlinear disturbances might be caused by the random fluctuation of the network load and the unreliability of the communication links and, in such a case, the disturbances themselves could experience random abrupt changes in their type or intensity [31, 70]. Such network-induced nonlinear disturbances are customarily referred to as randomly occurring nonlinearities (RONs). Up to now, two typical approaches [51, 70] have been widely adopted to describe RONs: one way is to introduce a random variable obeying the Bernoulli binary distribution taking values of either 1 or 0, and the other is to characterize them by utilizing statistical means. Although RONs have received some initial research attention, they should also be considered with other phenomena that appear simultaneously in order to reflect network-enhanced complexities in a more realistic way.

Event-Triggered Communication Protocols

In most of the literature, a common assumption is that the controllers or estimators are time-triggered rather than event-triggered. Nevertheless, from a technical viewpoint, the event-triggered control or state estimation strategy is more attractive, especially in distributed real-time sensing and control, due mainly to the need for reducing the communication burden [85]. On the other hand, in order to prolong the lifetime of digital control circuits, an effective method is to reduce the triggering frequency of electronic components. Therefore, event-triggered scheduling seems to be more suitable for distributed control and state estimation of networked systems. As such, the event-triggered control and state estimation issues have received particular research attention in the past few years and some preliminary results have been reported in [35, 85, 106, 123] and the references therein.

Compared with the time-triggered control or state estimation scheme, the distinct feature of the event-triggered control or state estimation strategy is that the controller or estimator input is updated only when a certain specific event happens. Between updates of the controller or estimator inputs, the signal remains unchanged, and an appropriately generated updating time would guarantee the stability and other performances of the closed-loop system or the estimate error dynamic. It is worth noting that the updating time is typically unknown *a priori* since it is determined by event-triggered mechanisms dependent on time, sampling data, system states [35]. A naturally challenging problem is how to examine the effect on stability from the event-triggered condition. In order to overcome such a difficulty, two arguably representative theoretical frameworks, namely, switch system theory [33,137] and input-to-state stability (ISS) theory [9,84] have been widely employed to analyze the event-triggered control and state estimation issues. Up to now, most published results concerning event-triggered control and state estimation problems have been obtained for continuous-time systems, and the corresponding issues for discrete-time systems with stochastic disturbances have not gained adequate research attention due primarily to the difficulty in both the mathematical analysis and the design of suitable event-triggered mechanisms. For example, the theory of ISS in probability has not been taken into consideration for discrete-time stochastic nonlinear systems.

Randomly Launched Cyber-Attacks

As is well known, sensors, controllers and controlled plants are often connected over a common network medium for networked control systems. In such an engineering setting, exchanged data without security protection can be easily exploited by attackers (or adversaries). A cyber-attack can be regarded as methods, processes, or means used to maliciously attempt to reduce network reliability or steer the plant to the adversaries' desired operating point [24]. According to their implementation types, the cyber-attacks can be generally classified into Denial-of-Service (DoS) attacks [38], replay attacks [157] and deception attacks [22,92], see Tab. 1.1. Up to now, by employing the techniques of dynamic programming or Lyapunov stability theory, some preliminary results concerning security control problems have been reported in the literature, see [6,13] for the case of DoS attacks and [5,125] for the case of deception attacks.

Note that, from the viewpoint of defenders (for plants), the attacks possess *random nature* since the successes of the attacks largely depend on the *detection ability* of protection equipment or software, the *communication protocols* and the *network conditions* (e.g., network load, network congestion, network transmission rate) that are typically randomly fluctuated. For instance, the false data sent by deception attackers could be identified by using some hardware, software tools, or algorithms (e.g., χ^2 detectors) which leads to a failed attack. In addition, in a multipath routing protocol with a (T, N) secret sharing scheme, the secure message is divided into N shares such that it can be easily recovered from any T or more shares [78]. Based on this fact, a Bernoulli process or Markov process with known statistical information has been employed to control DoS attacks [6]. However, randomly occurring deception attacks have not received proper research attention for the following two reasons: 1) it is non-trivial to describe such attacks; and 2) it is challenging to develop appropriate methodology to analyze the transient dynamics of closed-loop systems due to both the stochastic nature and the interference signals transmitted by attackers.

TABLE 1.1
Mathematical models of cyber-attacks.

Types	Mathematical models
DoS attacks	$\bar{y}_k \in \emptyset$ which means the transmission of information y_k is unsuccessful. Here, $\bar{y}_k$ and y_k stand for the received data and the measurement data, respectively.
Replay attacks	$\bar{y}_k \in Y_k$ where Y_k is the set of past information.
Deception attacks	$\bar{y}_k = y_k + y_k^a$ where y_k^a is the injected information by attackers.

1.3 Performance Analysis and Engineering Design Synthesis

Combined with the introduction, we are now in a position to show the performance criterion with practical research motivation.

Input-to-State Stability in Probability

As we all know, stability is the most fundamental performance requirement for almost all real-world systems. By employing a Lyapunov-like function (functional) method, comparison principle, or Razumikhin techniques, some representative results have been established for time-varying delayed systems, switching systems, impulsive systems and so forth. For example, internally mean-square stability has been discussed in [156] for a continuous Markovian process by introducing multi-step state transition conditional probability functions. Stability and stabilization for a class of discrete-time semi-Markov jump linear systems has been intensively investigated in [153] by adopting a semi-Markov kernel, of which the probability density function of sojourn-time is dependent on both current and next system mode. In [75], some easily verifiable conditions have been obtained which ensure almost sure asymptotic stability and p-th moment asymptotic stability of nonlinear stochastic differential systems with polynomial growth.

Among the various stability properties about disturbances, the input-to-state stability (ISS) theory has been playing an important role in the performance analysis and synthesis of a large class of nonlinear systems since its inception by Sontag [116]. In the past decade, ISS has been successfully applied in various fields such as robotic systems [7], signal processing [8], tracking control [10], swarm formation [87], model predictive control [71], networked control systems, and so forth. Furthermore, the ISS concept has been widely adopted to deal with *stochastic* nonlinear systems, which leads to several new stability definitions including ISSiP [76], γ-ISS [129], p-th moment ISS [53], noise-to-state stability [25] and exponential ISS [130]. It is worth emphasizing that these ISS-related concepts mentioned above have been exclusively studied for continuous-time systems, and the corresponding investigation for the discrete-time cases remains an open problem despite its crucial importance in analyzing the dynamic behaviors of discrete-time stochastic nonlinear systems.

Envelope-Constrained Performance and $\mathcal{H}_\infty$ Performance

The design of many filters can often be cast as a constrained optimization problem where the constraints are defined by the specifications of the filter in signal processing and communications. It should be pointed out that such specifications can arise either from practical considerations [20] or from the standards set by certain regulatory bodies. Examples include the communication channel equalization problems, the radar and sonar detection problems, and the robust antenna and filter design problems [20, 94]. As a topical representative, envelope-constrained filtering (ECF) is to find a filter such that the filtering error output stimulated by a specified input signal lies within a desired envelope and the effect from the input noises is also minimized [127, 135].

In the time domain, the ECF issue is usually cast as a finite-dimensional constrained quadratic optimization problem which can be effectively handled by using linear matrix inequality approaches [124, 128, 150, 155]. Almost all ECF-relevant literature has been concerned with linear time-invariant systems, and the corresponding investigation of nonlinear time-varying systems has not received proper research attention for the following two reasons: 1) it is non-trivial to define the envelope constraints for time-varying systems over a finite horizon; and 2) it is challenging to develop appropriate methodology to analyze the transient dynamics of the filtering error due to the time-varying nature. Obviously, it is of both theoretical significance and practical importance to design an envelope-constrained filter for the time-varying systems, especially for ones with network-enhanced complexities.

On the other hand, in the past few decades, the $\mathcal{H}_\infty$ index has been proven to be an effective criterion to evaluate the disturbance attenuation and rejection behaviors of complex dynamic systems. As such, the $\mathcal{H}_\infty$ control and state estimation problems have received a great deal of research interest, see [14,17,39,41,46,56,64] and the references therein. From a technical point of view, there are generally two approaches to designing the $\mathcal{H}_\infty$ estimator or controller, namely, the linear matrix inequality (LMI) approach [28,46] and the Riccati differential/difference equation (RDE) approach [42,56]. Recently, to handle the time-varying systems, the so-called differential/difference linear matrix inequality (DLMI) and recursive linear matrix inequality (RLMI) methods have been employed in [42,108,109] to effectively solve the finite-horizon control and state estimation problems in a recursive manner. Very recently, the concept of the traditional $\mathcal{H}_\infty$-performance has been applied to evaluate the behaviors of emerging complex dynamic systems such as complex networks, sensor networks and multi-agent systems. Accordingly, several novel $\mathcal{H}_\infty$-like research problems (e.g., the $\mathcal{H}_\infty$ consensus and the distributed $\mathcal{H}_\infty$ state estimation) have appeared that have quickly gained attention from a variety of research communities [29]. It can be concluded that $\mathcal{H}_\infty$ control and state estimation remain as ongoing research issues, especially for the systems with network-enhanced complexities.

Distributed Filtering

Sensor networks have recently received increasing research interests due to their extensive applications in a variety of areas such as information collection, environmental monitoring, industrial automation and intelligent buildings. In particular, distributed filtering over sensor networks has been an ongoing research issue that attracts special attention from researchers in the area. Different from traditional filtering techniques based on single or centralized structured/located sensors, the information available for the filter design on an individual node of the sensor network is not only from its own measurement but also from its neighboring sensors' measurements according to the given topology. As such, the objective of filtering based on a sensor network can be achieved in a distributed yet collaborative way. Such a problem is usually referred to as the distributed filtering problem. It is worth mentioning that one of the main challenges in designing distributed filters lies in how to cope with the complicated couplings between one sensor and its neighboring sensors by reflecting such couplings in the filter structure specification.

In the past decade, distributed filtering issues through sensor networks have gained considerable research attention and a wealth of literature has appeared on this topic, see e.g., [19, 23, 69, 89, 99–101, 110, 118, 136] and the references therein. So far, four arguably representative distributed estimation strategies can be summarized as follows: 1) the distributed Kalman filtering algorithm [23, 89, 101] with dynamic consensus protocols; 2) the optimal distributed estimation algorithm [19, 118, 136] for minimizing the estimation mean-square error (MSE) or the estimation error variance; 3) the distributed estimation algorithm [69, 109] to guarantee the convergence or $\mathcal{H}_\infty$ performance of the estimation error dynamics; and 4) the maximum-likelihood approach [99, 100] for achieving the best possible variance for a given bandwidth constraint.

In the context of sensor networks, the network-enhanced complexities become more severe due primarily to the network size, communication constraints, limited battery storage, strong coupling and spatial deployment, which would unavoidably deteriorate the overall network performance. In view of the specific features of sensor networks, network-induced complexities may occur in a probabilistic way. In addition, the most reported results have been concerned with the distributed estimation algorithms for linear and/or deterministic systems over an infinite horizon, and therefore there is a great need to examine how the network-enhanced complexities impact the performance of the distributed estimation for a sensor network over a finite horizon. On the other hand, the sensor nodes in wireless communication usually own the limited computation capacity and energy store. From the viewpoint of engineering, in order to reduce the computation burden and to prolong the lifetime, event-triggered scheduling seems to be more suitable for distributed filtering over sensor networks and, accordingly, the necessity of designing the schemes of distributed filtering with event-triggering communication protocol arises naturally in situations where randomly occurring network-enhanced complexities are also inevitable.

Consensus and Consensus Control

For multi-agent systems, one of the major research lines is consensus control, whose objective is to design a consensus algorithm (or protocol) using local neighboring information such that the states of a team of agents reach some common features. Here, the common features are dependent on the states of all agents and examples of the features include positions, phases, velocities, attitudes, and so on. Up to now, a number of consensus analysis issues have been extensively investigated for first- or second-order multi-agent systems in the framework of fixed topology or switching topology subject to stochastic noises, communication delays, or limited communication data rate, see [48, 54, 67, 68, 81, 88, 90, 98, 121, 140, 147] and the references therein. Generally, the agents are usually distributed spatially to form a large-scale networked system and have the ability of computation and wireless communications. As such, the possible network-enhanced complexities in multi-agent systems are typical due to the unreliable signal transmission through wireless communication networks. It is necessary to examine the impact from the given topology and the network-enhanced complexities on the consensus performance.

It is worth mentioning that, as discussed in [2], the behavior of the individual entities of animal swarms (i.e., the agents) is influenced by a large number of factors, most of which are time-varying. For instance, the evolution of the outside temperature in time may be characterized as periodic functions (with periods one day or one year) and a residual stochastic component. In fact, all real-world multi-agent systems should literally possess time-varying parameters. Unfortunately, a literature search reveals that only scattered results have been reported on the consensus control problem for time-varying systems due mainly to the conceptual challenges. For example, when we consider the finite-horizon $\mathcal{H}_\infty$ consensus problem for discrete time-varying uncertain multi-agent systems with missing measurements, the following two questions emerge naturally: 1) how do we define an appropriate $\mathcal{H}_\infty$ consensus performance index in a finite horizon? and 2) how do we understand the impact from both parameter uncertainties and missing measurements on the existence of the consensus protocol?

In addition, in order to reduce the communication burden, an event-triggered mechanism has been taken into consideration in consensus control issues. It should be noted that the challenges in designing event-triggered controllers for multi-agent systems stem mainly from proper definition of the consensus performance, the mathematical difficulty in consensus analysis and the effects from network topology and external disturbances. So far, most research focuses have been on the *steady-state* behaviors of the consensus, i.e., the consensus performance when time tends to infinity. Another important system behavior, namely, *transient* consensus performance at specific times, has been largely overlooked despite its significance in reflecting the consensus dynamics. On the other hand, rather than requiring the objectives of the system performance to be met accurately, it is quite common for multi-agent systems design to have their individual performance objective being described in terms of the desired probability of attaining that objective. In this sense, the concept of *consensus in probability* appears to be of engineering significance since 1) it is almost impossible to assure that certain performances are achieved with probability 1 because of uncontrolled external forces; and 2) in some cases, the commonly used *mean-square* measure might be crude to quantify the consensus performance and it would be satisfactory if certain performances are achieved with an acceptable probability [29].

Security and Security Control

Network security is of utmost importance in modern society and has therefore received ever-increasing attention in recent years. In order to guarantee security, one of the effective strategies for handling network attacks is to discard or retransmit the destroyed data when they are detected or identified by utilizing some hardware or software tools. Such a strategy would not be suitable for real-world industrial control systems. The main reasons can be summarized as follows: 1) the attacks are usually stealthy and cannot be easily detected; and 2) the retransmission method inevitably increases the network burden and also reduces the real-time performance of the addressed systems. Therefore, an issue of crucial importance is how to design a control system whose performance is insensitive to the effects from the malicious attacks. On the other hand, since nonlinearities are ubiquitous in practice, when the nonlinearities and randomly occurring deception attacks come together for NCSs, the security control issue has become quite intractable due primarily to lack of appropriate methodology. For instance, although significant progress has been made for the input-to-state stability (ISS) theory of continuous-time stochastic nonlinear systems [53,76,129,130], the results for their discrete-time counterparts have been very few.

When it comes to the distributed control/filtering issues subject to cyber-attacks, only a limited number of results have been available in the literature, see e.g., [83, 133, 134, 152]. For instance, under binary hypotheses with quantized sensor observations, the optimal attacking distributions have been estimated in [83] to minimize the detection error exponent and the fraction of Byzantine sensors (i.e., compromised sensors for adversaries). Recently, in [152], the identification and categorization issues of attacked sensors have been discussed by utilizing the joint estimation of the statistical description of the attacks and the estimated parameter. Note that most existing results have been concerned with static target plants despite the fact that dynamic target plants are more often encountered in engineering practice, which is due probably to the difficulties in analyzing the dynamics in collaborative nature and spatial structure when designing distributed controller/filters.

1.4   Outline

This book is concerned with performance analysis and synthesis for several classes of discrete-time stochastic systems with various network-enhanced complexities. According to the source of complexities and the addressed system characteristic, the content can be divided into two parts conceptually. In the first part, a series of finite-horizon state estimation and control issues are investigated for discrete time-varying stochastic systems with network-enhanced complexities due to limited network bandwidth. A novel analysis and design framework, namely, backward recursive Riccati difference equations (RDEs) is developed in Chapter 2, and then utilized to deal with the $\mathcal{H}_\infty$ consensus control and $\mathcal{H}_\infty$ distributed filtering issues in Chapter 3 and Chapter 4, respectively. Furthermore, the dissipative control and $\mathcal{H}_\infty$ filtering issues are, respectively, addressed in Chapter 5 and Chapter 6 for state-saturated time-varying systems by virtue of recursive linear matrix inequalities (RLMIs); and the envelope-constrained $\mathcal{H}_\infty$ filtering problem is discussed in Chapter 7 in the framework of set-membership estimation.

In the second part, the impact on the control or estimation performance over an infinite-horizon is deeply discussed from complexities induced by both event-triggered communication protocols and cyber-attacks for discrete-time stochastic systems. First, the distributed recursive filter is designed in Chapter 8 for discrete time-delayed stochastic systems subject to deception attack effects. Then, Chapter 9 concerns distributed $\mathcal{H}_\infty$ state estimation in mean square are systematically investigated, and Chapter 10 is concerned with the event-triggered consensus control in probability. When the communication protocols combined with cyber-attacks becomes a focus of research, Chapter 11 develops a new security control scheme, where a new concept of a mean-square security domain is put forward to quantify the security degree. Such a scheme is successfully extended to the consensus control issue in Chapter 12 with the security perspective. The organization structure of this book is shown in Fig. 1.1 and the outline of the book is as follows:

- In Chapter 1, the research background, motivations and research problems are firstly introduced from the following aspects, i.e., several classes of discrete-time stochastic systems, network-enhanced complexities, and system performance and design purposes, then the outline of the book is listed.

- In Chapter 2, the $\mathcal{H}_\infty$ control problem is investigated for a class of discrete time-varying nonlinear systems with both randomly occurring nonlinearities (RONs) and fading measurements over a finite horizon. The system measurements are transmitted through fading channels described by a modified stochastic Rice fading model. The model transformation technique is first employed to simplify the addressed problem, and then the stochastic analysis in combination with the completing squares method are carried out to obtain *necessary and sufficient* conditions of an auxiliary index which is closely related to the desired $\mathcal{H}_\infty$ performance over a finite horizon. Moreover, the time-varying controller parameters are characterized via solving coupled backward recursive Riccati difference equations (RDEs).

- Chapter 3 deals with the $\mathcal{H}_\infty$ consensus control problem for a class of discrete time-varying multi-agent systems with both missing measurements and parameter uncertainties. A directed graph is used to represent the communication topology of the multi-agent network, and a binary switching sequence satisfying a conditional probability distribution is employed to describe the missing measurements. According to the given topology, the measurement output available for the controller is not only from the individual agent but also from its neighboring agents. By using the completing squares method and stochastic analysis techniques, some sufficient conditions are derived such that the $\mathcal{H}_\infty$ consensus is ensured, and then the time-varying controller parameters are obtained by solving coupled backward recursive RDEs.

- Chapter 4 addresses the distributed $\mathcal{H}_\infty$ state estimation problem for a class of discrete time-varying nonlinear systems with both stochastic parameters and stochastic nonlinearities. The system measurements are collected through sensor networks with sensors distributed according to a given topology. Through available measurements from not only the individual sensor but also its neighboring sensors, a necessary and sufficient

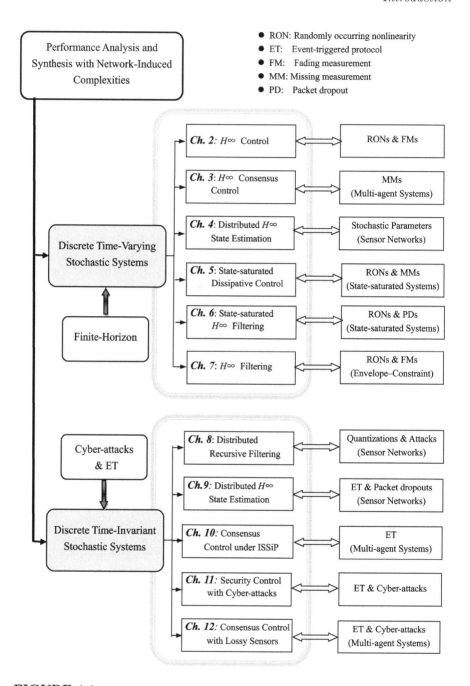

FIGURE 1.1
The architecture of the book.

condition is established to achieve the $\mathcal{H}_\infty$ performance constraint, and a novel estimator design scheme is proposed via a certain $\mathcal{H}_2$-type criterion combined with the established necessary and sufficient condition. The desired estimator parameters can be obtained by solving coupled backward recursive RDEs.

- Chapter 5 is concerned with the dissipative control problem for a class of discrete time-varying systems with simultaneous presence of state saturations, RONs, as well as multiple missing measurements. By introducing a free matrix with its infinity norm less than or equal to 1, the system state is bounded by a convex hull so that some sufficient conditions can be obtained in the form of recursive nonlinear matrix inequalities. Furthermore, a novel controller design algorithm is then developed to deal with the recursive nonlinear matrix inequalities. Finally, the obtained results are extended to the case when the state saturations are partial.

- Chapter 6 investigates the $\mathcal{H}_\infty$ filtering problem for a class of discrete time-varying systems with simultaneous presence of state saturations, RONs, as well as successive packet dropouts. Two mutually independent sequences of random variables obeying two given Bernoulli distributions are employed to describe the random occurrence of the nonlinearities and packet dropouts. By introducing a free matrix with its infinity norm less than or equal to 1, some sufficient conditions are obtained and then the filter parameter gains are characterized by the solution of a certain set of recursive nonlinear matrix inequalities. Furthermore, the obtained results are extended to the case when state saturations are partial.

- In Chapter 7, the envelope-constrained $\mathcal{H}_\infty$ filtering problem is investigated for a class of discrete time-varying stochastic systems over a finite horizon. The system under consideration involves fading measurements, RONs and mixed (multiplicative and additive) noises. A novel envelope-constrained performance criterion is proposed to better quantify the transient dynamics of the filtering error process over the finite horizon. By utilizing the ellipsoid description of the estimation errors, sufficient conditions are established in the form of recursive matrix inequalities (RMIs) reflecting the given envelope information. Furthermore, the filter parameters are characterized by means of the solvability of RMIs.

- Chapter 8 discusses the distributed recursive filtering problem for a class of discrete time-delayed stochastic systems subject to both uniform quantization and deception attack effects on the measurement outputs. Attention is focused on the design of a distributed recursive filter such that, in the simultaneous presence of time delays, deception attacks and uniform quantization effects, an upper bound for the filtering error covariance is guaranteed and subsequently minimized by properly designing the filter parameters via a gradient-based method at each sampling instant.

Furthermore, by utilizing mathematical induction, a sufficient condition is established to ensure the asymptotic boundedness of the sequence of the error covariance.

- In Chapter 9, the distributed $\mathcal{H}_\infty$ state estimation problem is discussed for a class of discrete-time stochastic nonlinear systems with packet dropouts over sensor networks. Considering the sparsity of neighboring information due to the adopted event-triggered protocol, a novel distributed state estimator with two filtering parameters is designed to adequately utilize the available innovations from not only itself (without an event-triggered mechanism) but also its neighboring sensors (with an event-triggered mechanism). By utilizing the property of the Kronecker product combined with stochastic analysis approaches, sufficient conditions are proposed under which the desired parameters are dependent on the solution of a set of matrix inequalities.

- Chapter 10 is concerned with the event-triggered consensus control problem for discrete-time stochastic multi-agent systems with state-dependent noise. A novel definition of *consensus in probability* is proposed to better describe the dynamics of the consensus process. An event-triggered mechanism is adopted to attempt to reduce the communication burden, where the control input on each agent is updated only when certain triggering condition is violated. First of all, a theoretical framework is established for analyzing the so-called *input-to-state stability in probability* (ISSiP) for general discrete-time nonlinear stochastic systems. Within such a theoretical framework, some sufficient conditions are derived under which the consensus in probability is reached. Furthermore, both the controller parameter and the triggering threshold are obtained in terms of the solution of certain matrix inequalities.

- In Chapter 11, the event-based security control problem is investigated for a class of discrete-time stochastic systems with multiplicative noises subject to both randomly occurring Denial-of-Service (DoS) attacks and randomly occurring deception attacks. A novel attack model is proposed to reflect the randomly occurring behaviors of DoS attacks as well as deception attacks within a unified framework via two sets of Bernoulli distributed white sequences with known conditional probabilities. A new concept of mean-square security domain is put forward to quantify the security degree. By using stochastic analysis techniques, some sufficient conditions are established to guarantee the desired security requirement and the control gain is obtained by solving some linear matrix inequalities with nonlinear constraints.

- Chapter 12 is concerned with the observer-based event-triggering consensus control problem for a class of discrete-time multi-agent systems with lossy sensors and cyber-attacks. A novel distributed observer is proposed to estimate the relative full states, and the estimated states are then used in the feedback protocol in order to achieve the overall consensus. An event-triggered mechanism with a state-independent threshold is adopted to update the control input signals so as to reduce unnecessary data communications. The success ratio of the launched attacks is taken into account to reflect the probabilistic failures of the attacks passing through the protection devices subject to limited resources and network fluctuations. By making use of eigenvalues and eigenvectors of the Laplacian matrix, the closed-loop system is transformed into an easy-to-analyze setting and then a sufficient condition is derived to guarantee the desired consensus. Furthermore, the controller gain is obtained in terms of the solution to a certain matrix inequality which is independent of the number of agents.

2

Finite-Horizon $\mathcal{H}_\infty$ Control with Randomly Occurring Nonlinearities and Fading Measurements

In practical systems, nonlinearities are ubiquitous and their strength or occurring types could be subject to random abrupt changes due possibly to some abrupt phenomena such as random failures and repairs of the components and changes in the interconnections of subsystems, to name just a few. Therefore, the $\mathcal{H}_\infty$ control problem for dynamic systems with randomly occurring nonlinearities has attracted considerable research attention in the past two decades. On the other hand, almost all plants are subject to time-varying changes or fluctuations stemming from time-varying temperature, operating point periodic shifting, as well as complex environments. For this kind of plant, a performance index over a finite horizon should be defined to describe better transient performance in comparison with the specified steady-state case for traditional time-invariant plants. Additionally, it is of great significance to analyze the energy cost of controlled states and nominal controller inputs.

This chapter deals with the $\mathcal{H}_\infty$ control problem for a class of discrete time-varying nonlinear systems with both randomly occurring nonlinearities and fading measurements over a finite horizon. The system measurements are transmitted through fading channels described by a modified stochastic Rice fading model. The purpose of the discussed issue is to design a set of time-varying controllers such that, in the presence of channel fading and randomly occurring nonlinearities, the $\mathcal{H}_\infty$ performance is guaranteed over a given finite-horizon. The model transformation technique is first employed to simplify the addressed problem, and then the stochastic analysis in combination with the completing squares method are carried out to obtain a *necessary and sufficient* condition under which the finite-horizon $\mathcal{H}_\infty$ control performance is ensured. Moreover, by resorting to a novel nominal energy cost, the time-varying controller parameters are characterized via solving coupled backward recursive Riccati difference equations (RDEs).

2.1 Modeling and Problem Formulation

Let us investigate the following discrete time-varying stochastic systems defined on $k \in [0, N]$:

$$\begin{cases} x_{k+1} = A_k x_k + \alpha_k h_k(x_k) + B_k u_k + D_k w_k, \\ y_k = C_k x_k + E_k v_k, \\ z_k = L_k x_k \end{cases} \qquad (2.1)$$

where $x_k \in \mathbb{R}^{n_x}$, $y_k \in \mathbb{R}^{n_y}$, and $z_k \in \mathbb{R}^{n_z}$ stand for the state vector that cannot be observed directly, the measurement output, and the controlled output, respectively. w_k and $v_k \in \mathcal{L}_{[0,N]}$ are two unknown external disturbances and belong to $\mathcal{L}_{[0,N]}$. A_k, B_k, C_k, D_k, E_k and L_k are known time-varying real-valued matrices with compatible dimensions. α_k is a stochastic variable obeying the well-known Bernoulli distribution and takes values of 0 or 1 with the probabilities:

$$\text{Prob} \{\alpha_k = 0\} = 1 - \bar{\alpha}, \quad \text{Prob} \{\alpha_k = 1\} = \bar{\alpha}.$$

The nonlinear vector-valued function $h_k : \mathbb{R}^{n_x} \to \mathbb{R}^{n_x}$ is assumed to be continuously differentiable in x and satisfies the following sector-bound condition

$$(h_k(x) - \Phi_k x)^T (h_k(x) - \Psi_k x) \le 0 \qquad (2.2)$$

where Φ_k and Ψ_k are known matrices with compatible dimensions with $\Phi_k > \Psi_k$ for all k.

It follows easily from (2.2) that

$$\left[h_k(x) - \left(\frac{\Phi_k + \Psi_k}{2} + \frac{\Phi_k - \Psi_k}{2} \right)x \right]^T \left[h_k(x) - \left(\frac{\Phi_k + \Psi_k}{2} - \frac{\Phi_k - \Psi_k}{2} \right)x \right] \le 0.$$

Then, denoting

$$\Delta(h_k) := h_k(x) - \frac{\Phi_k + \Psi_k}{2} x, \quad \mathcal{N}_k := \frac{\Phi_k - \Psi_k}{2},$$

one has

$$\Delta^T(h_k) \Delta(h_k) \le x^T \mathcal{N}_k^T \mathcal{N}_k x,$$

immediately. Therefore, there exists at least a function $\Theta_{k,x}$ satisfying $\Delta(h_k) = \Theta_{k,x} x$ and $\Theta_{k,x}^T \Theta_{k,x} \le \mathcal{N}_k^T \mathcal{N}_k$. Furthermore, the sector-bound condition can be transformed into the sector-bound uncertainties described by

$$h_k(x) = \frac{\Phi_k + \Psi_k}{2} x + F_{k,x} \mathcal{N}_k x_k \qquad (2.3)$$

with $F_{k,x} := \Theta_{k,x} \mathcal{N}_k^{-1}$ satisfying $F_{k,x}^T F_{k,x} \le I$.

In consideration of the fading channels, in this chapter, the measurement signal received by the controller is described by

$$\tilde{y}_k = \sum_{i=0}^{\ell_k} \vartheta_k^i y_{k-i} + M\xi_k \tag{2.4}$$

with $\ell_k = \min\{\ell, k\}$, where ϑ_k^i ($i = 0, 1, \ldots, \ell_k$) are the channel coefficients which are mutually independent and have probability density functions on $[0, 1]$ with mathematical expectations $\bar{\vartheta}_i$ and variances $\tilde{\vartheta}_i$. $\xi_k \in l_2([0, +\infty); \mathbb{R}^{n_y})$ is an external disturbance. M is a known real-valued matrix with appropriate dimensions. For simplicity, we will set $\{y_k\}_{k \in [-\ell, -1]} = 0$, i.e., $\{x_k\}_{k \in [-\ell, -1]} = 0$ and $\{v_k\}_{k \in [-\ell, -1]} = 0$.

Remark 2.1 *By separating the received signals into three levels of spatial variation, the available fading models can be divided into three classes: fast fading, slow fading (or shadowing) and path loss. It is worth pointing out that, as a mixed model of fast fading and slow fading, the fading channels with erasure and delay can be described by an ℓth-order Rice fading model or AR model (2.4), which have been widely exploited in the area of signal processing and remote control. It should be mentioned that, in the traditional ℓth-order Rice fading model or AR model, the channel coefficients are usually assumed to be independent and identically distributed Gaussian random variables. However, in practice, the channel coefficients in the fading channel typically take values over the interval $[0, 1]$. Therefore, in this chapter, we make an assumption that the channel coefficients in (2.4) have a probability density function on $[0, 1]$.*

For the given receiver model (2.4), we consider the following output feedback controller for the discrete time-varying nonlinear system (2.1):

$$u_k = K_k \tilde{y}_k = \sum_{i=0}^{\ell} \vartheta_k^i K_k y_{k-i} + K_k M\xi_k. \tag{2.5}$$

Setting

$$\bar{x}_k = [\begin{array}{cccccc} x_k^T & x_{k-1}^T & x_{k-2}^T & \cdots & x_{k-\ell}^T \end{array}]^T,$$
$$\eta_k = [\begin{array}{ccccc} w_k & v_k & \cdots & v_{k-\ell} & \xi_k \end{array}]^T,$$

we obtain an augmented system from (2.1) and (2.5) as follows:

$$\begin{cases} \bar{x}_{k+1} = \left(\mathscr{A}_k + \mathscr{B}_k K_k \bar{\vartheta} \mathscr{C}_k + (\alpha_k - \bar{\alpha})\tilde{\mathscr{A}}_k + \alpha_k \tilde{\Theta}_k\right) \bar{x}_k \\ \qquad\quad + \mathscr{B}_k K_k \vartheta_k \mathscr{C}_k \bar{x}_k + (\mathscr{B}_k K_k \bar{v} \mathscr{E}_k + \mathscr{D}_k)\eta_k + \mathscr{B}_k K_k v_k \mathscr{E}_k \eta_k, \\ z_k = \mathcal{L}_k \bar{x}_k \end{cases} \tag{2.6}$$

where

$$\mathscr{A}_k = \begin{bmatrix} A_k + \bar{\alpha}\mathscr{M}_k & 0 & 0 & \cdots & 0 & 0 \\ I & 0 & 0 & \cdots & 0 & 0 \\ 0 & I & 0 & \cdots & 0 & 0 \\ \vdots & \vdots & \vdots & \ddots & \vdots & \vdots \\ 0 & 0 & 0 & \cdots & I & 0 \end{bmatrix},$$

$$\tilde{\mathscr{A}}_k = \operatorname{diag}\{\mathscr{M}_k, 0, 0, \ldots, 0\}, \quad \mathscr{B}_k = [\; B_k^T \;\; 0 \;\; 0 \;\; \cdots \;\; 0 \;]^T,$$

$$\mathscr{C}_k = \operatorname{diag}\{C_k, C_{k-1}, \ldots, C_{k-\ell}\}, \quad \tilde{\mathscr{D}}_k = [\mathscr{D}_k \;\; [\varepsilon_k^{-1} I \; 0 \; \cdots \; 0]^T],$$

$$\mathscr{E}_k = \operatorname{diag}\{0, E_k, E_{k-1}, \ldots, E_{k-\ell}, M\}, \quad \tilde{\Theta}_k = \operatorname{diag}\{\Theta_{k,x}, 0, 0, \ldots, 0\},$$

$$\mathcal{L}_k = [\; L_k \;\; 0 \;\; 0 \;\; \cdots \;\; 0 \;], \quad \bar{\vartheta} = [\; \bar{\vartheta}_0 I \;\; \bar{\vartheta}_1 I \;\; \cdots \;\; \bar{\vartheta}_\ell I \;],$$

$$\mathscr{M}_k = (\Phi_k + \Psi_k)/2, \quad \bar{\upsilon} = [\; 0 \;\; \bar{\vartheta} \;\; I \;], \quad \upsilon_k = [\; 0 \;\; \vartheta_k \;\; 0 \;],$$

$$\vartheta_k = [\; (\vartheta_k^0 - \bar{\vartheta}_0)I \;\; (\vartheta_k^1 - \bar{\vartheta}_1)I \;\; \cdots \;\; (\vartheta_k^\ell - \bar{\vartheta}_\ell)I \;].$$

Our aim in this chapter is to design a finite-horizon output feedback controller of the form (2.5) such that, for the given disturbance attenuation level $\gamma > 0$, the positive definite matrix W, and the initial state x_0, the controlled output z_k satisfies the following $\mathcal{H}_\infty$ performance constraint :

$$\mathbb{E}\{\|z_k\|_{[0,N]}^2\} < \gamma^2 \|\eta_k\|_{[0,N]}^2 + \gamma^2 \mathbb{E}\left\{x_0^T W x_0\right\} \tag{2.7}$$

where $\|x_k\|_{[0,N]}^2 := \sum_{k=0}^N \|x_k\|^2$ for any vector sequences x_k.

2.2 $\mathcal{H}_\infty$ Performance Analysis

In this section, let us investigate the $\mathcal{H}_\infty$ performance analysis of system (2.1) with fading channels (2.4). The following four lemmas will be used in deriving our main results, where Lemma 2.2 can be obtained along the similar line of the proof of Theorem 2 in [55].

Lemma 2.1 *[95] Let $\mathcal{U}$, $\mathcal{V}$ and $\mathcal{W}$ be known nonzero matrices with appropriate dimensions. The solution X to $\min_X \|\mathcal{U}X\mathcal{W} - \mathcal{V}\|_F$ is $\mathcal{U}^\dagger \mathcal{V} \mathcal{W}^\dagger$.*

To cope with the parameter uncertainties in (2.6), a convenient way is to regard them as one of the sources of the disturbances. Therefore, what we need to do is to reject the influence from all the disturbances of the controlled output according to the prescribed $\mathcal{H}_\infty$ requirement. For this purpose, we rewrite (2.6) as follows:

$$\begin{cases} \bar{x}_{k+1} = \left(\mathscr{A}_k + \mathscr{B}_k K_k \bar{\vartheta}\mathscr{C}_k\right)\bar{x}_k + (\alpha_k - \bar{\alpha})\tilde{\mathscr{A}}_k \bar{x}_k \\ \qquad + \mathscr{B}_k K_k \vartheta_k \mathscr{C}_k \bar{x}_k + (\mathscr{B}_k K_k \bar{\theta}\tilde{\mathscr{E}}_k + \tilde{\mathscr{D}}_k)\tilde{\eta}_k + \mathscr{B}_k K_k \theta_k \tilde{\mathscr{E}}_k \tilde{\eta}_k, \\ z_k = \mathcal{L}_k \bar{x}_k \end{cases} \tag{2.8}$$

where

$$\tilde{\eta}_k = [\eta_k^T \quad (\varepsilon_k \alpha_k F_{k,x} \mathcal{N}_k x_k)^T]^T, \quad \tilde{\mathscr{D}}_k = [\mathscr{D}_k \quad \varepsilon_k^{-1} I],$$
$$\tilde{\mathscr{E}}_k = \text{diag}\{\mathscr{E}_k, 0\}, \quad \bar{\theta} = [\bar{\upsilon} \quad 0], \quad \theta_k = [\upsilon_k \quad 0].$$

Here, ε_k is a positive function representing the scaling of the perturbation, which is introduced to provide more flexibility in the controller design (see [55] for more details). Furthermore we introduce the following *auxiliary index*:

$$\mathbb{E}\left\{\|z_k\|^2_{[0,N]}\right\} < \mathbb{E}\left\{\gamma^2 \|\tilde{\eta}_k\|^2_{[0,N]} - \bar{\alpha}\gamma^2 \|\varepsilon_k \aleph_k \bar{x}_k\|^2_{[0,N]}\right\} + \gamma^2 \mathbb{E}\left\{\bar{x}_0^T W \bar{x}_0\right\} \quad (2.9)$$

with $\aleph_k := [\ \mathcal{N}_k \quad 0 \quad \dots \quad 0\ ]$. The following lemma reveals the relationship between (2.7) and (2.9).

Lemma 2.2 *For the performance indices defined in (2.7) and (2.9), (2.7) is satisfied when (2.9) holds.*

Lemma 2.3 *Considering the corresponding solution $\bar{x}_k$ of system (2.8) with initial value $\bar{x}_0$ subject to the external disturbances $\tilde{\eta}_k$, one has*

$$\mathcal{J}_1(\bar{x}_0, \tilde{\eta}_k) = \mathbb{E}\left\{\|z_k\|^2_{[0,N]} - \gamma^2 \|\tilde{\eta}_k\|^2_{[0,N]} + \bar{\alpha}\gamma^2 \|\varepsilon_k \aleph_k \bar{x}_k\|^2_{[0,N]}\right\}$$
$$= \sum_{k=0}^N \mathbb{E}\left\{\begin{bmatrix} \bar{x}_k \\ \tilde{\eta}_k \end{bmatrix}^T \begin{bmatrix} \mathcal{R}^{11}_{k+1} - \mathcal{P}_k & \mathcal{R}^{12}_{k+1} \\ * & -\mathcal{R}^{22}_{k+1} \end{bmatrix} \begin{bmatrix} \bar{x}_k \\ \tilde{\eta}_k \end{bmatrix}\right\} \quad (2.10)$$
$$+ \mathbb{E}\left\{\bar{x}_0^T \mathcal{P}_0 \bar{x}_0 - \bar{x}_{N+1}^T \mathcal{P}_{N+1} \bar{x}_{N+1}\right\}.$$

Furthermore, if $|\mathcal{R}^{22}_k| \neq 0$ for all $k \in [0,N]$, by selecting $\tilde{\eta}_k = (\mathcal{R}^{22}_{k+1})^{-1}(\mathcal{R}^{12}_{k+1})^T \bar{x}_k$ and denoting $\bar{u}_k = K_k \bar{\vartheta} \mathscr{C}_k \bar{x}_k$, one has

$$\mathcal{J}_2(\bar{u}_k; \tilde{\eta}_k) = \mathbb{E}\{\|z_k\|^2_{[0,N]} + \|\bar{u}_k\|^2_{[0,N]}\}$$
$$= \mathbb{E}\sum_{k=0}^N \left\{\begin{bmatrix} \bar{x}_k \\ \bar{u}_k \end{bmatrix}^T \begin{bmatrix} \Delta_{k+1} + \mathcal{L}_k^T \mathcal{L}_k - \mathcal{Q}_k & \mathcal{S}^1_{k+1} \\ * & \mathcal{S}^2_{k+1} \end{bmatrix} \begin{bmatrix} \bar{x}_k \\ \bar{u}_k \end{bmatrix}\right\} \quad (2.11)$$
$$+ \mathbb{E}\left\{\bar{x}_0^T \mathcal{Q}_0 \bar{x}_0 - \bar{x}_{N+1}^T \mathcal{Q}_{N+1} \bar{x}_{N+1}\right\}$$

where $\{\mathcal{P}_k\}_{0 \leq k \leq N+1}$ and $\{\mathcal{Q}_k\}_{0 \leq k \leq N+1}$ are two families of matrices with partitioning

$$\mathcal{P}_k = \begin{pmatrix} \mathcal{P}^{11}_k & \mathcal{P}^{12}_k & \dots & \mathcal{P}^{1,\ell+1}_k \\ \mathcal{P}^{21}_k & \mathcal{P}^{22}_k & \dots & \mathcal{P}^{2,\ell+1}_k \\ \dots & \dots & \dots & \dots \\ \mathcal{P}^{\ell+1,1}_k & \mathcal{P}^{\ell+1,2}_k & \dots & \mathcal{P}^{\ell+1,\ell+1}_k \end{pmatrix},$$

$$\mathcal{Q}_k = \begin{pmatrix} \mathcal{Q}^{11}_k & \mathcal{Q}^{12}_k & \dots & \mathcal{Q}^{1,\ell+1}_k \\ \mathcal{Q}^{21}_k & \mathcal{Q}^{22}_k & \dots & \mathcal{Q}^{2,\ell+1}_k \\ \dots & \dots & \dots & \dots \\ \mathcal{Q}^{\ell+1,1}_k & \mathcal{Q}^{\ell+1,2}_k & \dots & \mathcal{Q}^{\ell+1,\ell+1}_k \end{pmatrix},$$

and

$$\Pi_{k+1}^1 = diag\{\tilde{\vartheta}_0, \tilde{\vartheta}_1, \ldots, \tilde{\vartheta}_\ell\} \otimes (K_k^T B_k^T \mathcal{P}_{k+1}^{11} B_k K_k),$$

$$\Pi_{k+1}^2 = [0 \quad diag\{\tilde{\vartheta}_0, \tilde{\vartheta}_1, \ldots, \tilde{\vartheta}_\ell\} \quad 0 \quad 0] \otimes (K_k^T B_k^T \mathcal{P}_{k+1}^{11} B_k K_k),$$

$$\Pi_{k+1}^3 = diag\{0, \tilde{\vartheta}_0, \tilde{\vartheta}_1, \ldots, \tilde{\vartheta}_\ell, 0, 0\} \otimes (K_k^T B_k^T \mathcal{P}_{k+1}^{11} B_k K_k),$$

$$\Xi_{k+1}^1 = diag\{\tilde{\vartheta}_0, \tilde{\vartheta}_1, \ldots, \tilde{\vartheta}_\ell\} \otimes (K_k^T B_k^T \mathcal{Q}_{k+1}^{11} B_k K_k),$$

$$\Xi_{k+1}^2 = [0 \quad diag\{\tilde{\vartheta}_0, \tilde{\vartheta}_1, \ldots, \tilde{\vartheta}_\ell\} \quad 0 \quad 0] \otimes (K_k^T B_k^T \mathcal{Q}_{k+1}^{11} B_k K_k),$$

$$\Xi_{k+1}^3 = diag\{0, \tilde{\vartheta}_0, \tilde{\vartheta}_1, \ldots, \tilde{\vartheta}_\ell, 0, 0\} \otimes (K_k^T B_k^T \mathcal{Q}_{k+1}^{11} B_k K_k),$$

$$\mathcal{R}_{k+1}^{11} = (\mathscr{A}_k + \mathscr{B}_k K_k \bar{\vartheta} \mathscr{C}_k)^T \mathcal{P}_{k+1} (\mathscr{A}_k + \mathscr{B}_k K_k \bar{\vartheta} \mathscr{C}_k)$$
$$\quad + \tilde{\alpha} \tilde{\mathscr{A}}_k^T \mathcal{P}_{k+1} \tilde{\mathscr{A}}_k + \mathscr{C}_k^T \Pi_{k+1}^1 \mathscr{C}_k + \bar{\alpha} \gamma^2 \varepsilon_k^2 \aleph_k^T \aleph_k + \mathcal{L}_k^T \mathcal{L}_k,$$

$$\mathcal{R}_{k+1}^{12} = (\mathscr{A}_k + \mathscr{B}_k K_k \bar{\vartheta} \mathscr{C}_k)^T \mathcal{P}_{k+1} (\mathscr{B}_k K_k \bar{\theta} \tilde{\mathscr{E}}_k + \tilde{\mathscr{D}}_k) + \mathscr{C}_k^T \Pi_{k+1}^2 \tilde{\mathscr{E}}_k,$$

$$\mathcal{R}_{k+1}^{22} = \gamma^2 I - (\mathscr{B}_k K_k \bar{\theta} \tilde{\mathscr{E}}_k + \tilde{\mathscr{D}}_k)^T \mathcal{P}_{k+1} (\mathscr{B}_k K_k \bar{\theta} \tilde{\mathscr{E}}_k + \tilde{\mathscr{D}}_k) - \tilde{\mathscr{E}}_k^T \Pi_{k+1}^3 \tilde{\mathscr{E}}_k,$$

$$\Lambda_{k+1} = (\mathscr{B}_k K_k \bar{\theta} \tilde{\mathscr{E}}_k + \tilde{\mathscr{D}}_k)(\mathcal{R}_{k+1}^{22})^{-1}(\mathcal{R}_{k+1}^{12})^T, \quad \tilde{\alpha} = \bar{\alpha}(1 - \bar{\alpha}),$$

$$\Delta_{k+1} = (\mathscr{A}_k + \Lambda_{k+1})^T \mathcal{Q}_{k+1}(\mathscr{A}_k + \Lambda_{k+1}) + \tilde{\alpha} \tilde{\mathscr{A}}_k^T \mathcal{Q}_{k+1} \tilde{\mathscr{A}}_k$$
$$\quad + \mathscr{C}_k^T \Xi_{k+1}^1 \mathscr{C}_k + 2\mathscr{C}_k^T \Xi_{k+1}^2 \tilde{\mathscr{E}}_k (\mathcal{R}_{k+1}^{22})^{-1}(\mathcal{R}_{k+1}^{12})^T$$
$$\quad + \mathcal{R}_{k+1}^{12}(\mathcal{R}_{k+1}^{22})^{-1T} \tilde{\mathscr{E}}_k^T \Xi_{k+1}^3 \tilde{\mathscr{E}}_k (\mathcal{R}_{k+1}^{22})^{-1}(\mathcal{R}_{k+1}^{12})^T,$$

$$\mathcal{S}_{k+1}^1 = (\mathscr{A}_k + \Lambda_k)^T \mathcal{Q}_{k+1} \mathscr{B}_k, \quad \mathcal{S}_{k+1}^2 = \mathscr{B}_k^T \mathcal{Q}_{k+1} \mathscr{B}_k + I.$$

Proof *Along the trajectory of system (2.8), one has*

$$\mathbb{E}\left\{\bar{x}_{k+1}^T \mathcal{P}_{k+1} \bar{x}_{k+1} - \bar{x}_k^T \mathcal{P}_k \bar{x}_k\right\}$$

$$= \mathbb{E}\left\{\left(\bar{x}_k^T(\mathscr{A}_k + \mathscr{B}_k K_k \bar{\vartheta} \mathscr{C}_k)^T + \bar{x}_k^T(\alpha_k - \bar{\alpha})\tilde{\mathscr{A}}_k^T + \bar{x}_k^T \mathscr{C}_k^T \vartheta_k^T K_k^T \mathscr{B}_k^T\right.\right.$$
$$\quad + \tilde{\eta}_k^T \tilde{\mathscr{E}}_k^T \theta_k^T K_k^T \mathscr{B}_k^T + \tilde{\eta}_k^T(\mathscr{B}_k K_k \bar{\theta} \tilde{\mathscr{E}}_k + \tilde{\mathscr{D}}_k)^T\Big) \mathcal{P}_{k+1}$$
$$\quad \times \Big((\mathscr{A}_k + \mathscr{B}_k K_k \bar{\vartheta} \mathscr{C}_k) \bar{x}_k + (\alpha_k - \bar{\alpha})\tilde{\mathscr{A}}_k \bar{x}_k + \mathscr{B}_k K_k \vartheta_k \mathscr{C}_k \bar{x}_k$$
$$\quad + (\mathscr{B}_k K_k \bar{\theta} \tilde{\mathscr{E}}_k + \tilde{\mathscr{D}}_k)\tilde{\eta}_k + \mathscr{B}_k K_k \theta_k \tilde{\mathscr{E}}_k \tilde{\eta}_k\Big) - \bar{x}_k^T \mathcal{P}_k \bar{x}_k\Big\}$$

$$= \mathbb{E}\left\{\bar{x}_k^T(\mathscr{A}_k + \mathscr{B}_k K_k \bar{\vartheta} \mathscr{C}_k)^T \mathcal{P}_{k+1}(\mathscr{A}_k + \mathscr{B}_k K_k \bar{\vartheta} \mathscr{C}_k) \bar{x}_k - \bar{x}_k^T \mathcal{P}_k \bar{x}_k\right.$$
$$\quad + 2\bar{x}_k^T(\mathscr{A}_k + \mathscr{B}_k K_k \bar{\vartheta} \mathscr{C}_k)^T \mathcal{P}_{k+1}(\mathscr{B}_k K_k \bar{\theta} \tilde{\mathscr{E}}_k + \tilde{\mathscr{D}}_k)\tilde{\eta}_k$$
$$\quad + \bar{x}_k^T \tilde{\alpha} \tilde{\mathscr{A}}_k^T \mathcal{P}_{k+1} \tilde{\mathscr{A}}_k \bar{x}_k + \bar{x}_k^T \mathscr{C}_k^T \vartheta_k^T K_k^T \mathscr{B}_k^T \mathcal{P}_{k+1} \mathscr{B}_k K_k \vartheta_k \mathscr{C}_k \bar{x}_k$$
$$\quad + 2\bar{x}_k^T \mathscr{C}_k^T \vartheta_k^T K_k^T \mathscr{B}_k^T \mathcal{P}_{k+1} \mathscr{B}_k K_k \theta_k \tilde{\mathscr{E}}_k \tilde{\eta}_k$$
$$\quad + \tilde{\eta}_k^T \tilde{\mathscr{E}}_k^T \theta_k^T K_k^T \mathscr{B}_k^T \mathcal{P}_{k+1} \mathscr{B}_k K_k \theta_k \tilde{\mathscr{E}}_k \tilde{\eta}_k$$
$$\quad \left. + \tilde{\eta}_k^T(\mathscr{B}_k K_k \bar{\theta} \tilde{\mathscr{E}}_k + \tilde{\mathscr{D}}_k)^T \mathcal{P}_{k+1}(\mathscr{B}_k K_k \bar{\theta} \tilde{\mathscr{E}}_k + \tilde{\mathscr{D}}_k)\tilde{\eta}_k\right\}.$$

$$(2.12)$$

Taking (2.12) into consideration, it can be derived that

$$\mathbb{E}\left\{\|z_k\|^2_{[0,N]}\right\}$$

$$= \sum_{k=0}^{N}\mathbb{E}\left\{\bar{x}_k^T\mathcal{L}_k^T\mathcal{L}_k\bar{x}_k\right\} + \mathbb{E}\left\{\bar{x}_0^T\mathcal{P}_0\bar{x}_0 - \bar{x}_{N+1}^T\mathcal{P}_{N+1}\bar{x}_{N+1}\right\}$$

$$+ \sum_{k=0}^{N}\mathbb{E}\left\{\bar{x}_k^T\left[(\mathscr{A}_k + \mathscr{B}_kK_k\bar{\vartheta}\mathscr{C}_k)^T\mathcal{P}_{k+1}(\mathscr{A}_k + \mathscr{B}_kK_k\bar{\vartheta}\mathscr{C}_k)\right.\right.$$

$$+ \tilde{\alpha}\tilde{\mathscr{A}}_k^T\mathcal{P}_{k+1}\tilde{\mathscr{A}}_k + \mathscr{C}_k^T\Pi_{k+1}^1\mathscr{C}_k - \mathcal{P}_k\Big]\bar{x}_k$$

$$+ 2\bar{x}_k^T\left[(\mathscr{A}_k + \mathscr{B}_kK_k\bar{\vartheta}\mathscr{C}_k)^T\mathcal{P}_{k+1}(\mathscr{B}_kK_k\bar{\theta}\tilde{\mathscr{E}}_k + \tilde{\mathscr{D}}_k) + \mathscr{C}_k^T\Pi_{k+1}^2\tilde{\mathscr{E}}_k\right]\tilde{\eta}_k$$

$$+ \tilde{\eta}_k^T\left[(\mathscr{B}_kK_k\bar{\theta}\tilde{\mathscr{E}}_k + \tilde{\mathscr{D}}_k)^T\mathcal{P}_{k+1}(\mathscr{B}_kK_k\bar{\theta}\tilde{\mathscr{E}}_k + \tilde{\mathscr{D}}_k) + \tilde{\mathscr{E}}_k^T\Pi_{k+1}^3\tilde{\mathscr{E}}_k\right]\tilde{\eta}_k\bigg\}$$

$$= \mathbb{E}\left\{\bar{x}_0^T\mathcal{P}_0\bar{x}_0 - \bar{x}_{N+1}^T\mathcal{P}_{N+1}\bar{x}_{N+1}\right\}$$

$$+ \mathbb{E}\left\{\gamma^2\|\tilde{\eta}_k\|^2_{[0,N]} - \bar{\alpha}\gamma^2\|\varepsilon_k\aleph_k\bar{x}_k\|^2_{[0,N]}\right\}$$

$$+ \sum_{k=0}^{N}\mathbb{E}\left\{\begin{bmatrix}\bar{x}_k\\\tilde{\eta}_k\end{bmatrix}^T\begin{bmatrix}\mathcal{R}_{k+1}^{11} - \mathcal{P}_k & \mathcal{R}_{k+1}^{12}\\ * & -\mathcal{R}_{k+1}^{22}\end{bmatrix}\begin{bmatrix}\bar{x}_k\\\tilde{\eta}_k\end{bmatrix}\right\}. \quad (2.13)$$

Similarly, noticing that $\bar{u}_k = K_k\bar{\vartheta}\mathscr{C}_k\bar{x}_k$, *one has*

$$\mathbb{E}\left\{\bar{x}_{k+1}^T\mathcal{Q}_{k+1}\bar{x}_{k+1} - \bar{x}_k^T\mathcal{Q}_k\bar{x}_k\right\}$$

$$= \mathbb{E}\left\{\bar{x}_k^T\left(\mathscr{A}_k^T\mathcal{Q}_{k+1}\mathscr{A}_k + \tilde{\alpha}\tilde{\mathscr{A}}_k^T\mathcal{Q}_{k+1}\tilde{\mathscr{A}}_k + \mathscr{C}_k^T\Xi_{k+1}^1\mathscr{C}_k - \mathcal{Q}_k\right)\bar{x}_k\right.$$

$$+ 2\bar{x}_k^T\mathscr{A}_k^T\mathcal{Q}_{k+1}\mathscr{B}_k\bar{u}_k + \bar{u}_k^T\mathscr{B}_k^T\mathcal{Q}_{k+1}\mathscr{B}_k\bar{u}_k$$

$$+ 2\bar{x}_k^T\left[\mathscr{A}_k^T\mathcal{Q}_{k+1}(\mathscr{B}_kK_k\bar{\theta}\tilde{\mathscr{E}}_k + \tilde{\mathscr{D}}_k) + \mathscr{C}_k^T\Xi_{k+1}^2\tilde{\mathscr{E}}_k\right]\tilde{\eta}_k \quad (2.14)$$

$$+ 2\bar{u}_k^T\mathscr{B}_k^T\mathcal{Q}_{k+1}(\mathscr{B}_kK_k\bar{\theta}\tilde{\mathscr{E}}_k + \tilde{\mathscr{D}}_k)\tilde{\eta}_k$$

$$+ \tilde{\eta}_k^T\left[(\mathscr{B}_kK_k\bar{\theta}\tilde{\mathscr{E}}_k + \tilde{\mathscr{D}}_k)^T\mathcal{Q}_{k+1}(\mathscr{B}_kK_k\bar{\theta}\tilde{\mathscr{E}}_k + \tilde{\mathscr{D}}_k) + \tilde{\mathscr{E}}_k^T\Xi_{k+1}^3\tilde{\mathscr{E}}_k\right]\tilde{\eta}_k\bigg\}.$$

Moreover, under $|\mathcal{R}_{k+1}^{22}| \neq 0$ *for all* $k \in [0,N]$, *by selecting* $\tilde{\eta}_k = (\mathcal{R}_{k+1}^{22})^{-1}(\mathcal{R}_{k+1}^{12})^T\bar{x}_k$, *it is easy to obtain that*

$$\mathbb{E}\left\{\|z_k\|^2_{[0,N]}\right\}$$

$$= \sum_{k=0}^{N}\mathbb{E}\{\|z_k\|^2 + \|\bar{u}_k\|^2 - \|\bar{u}_k\|^2\} + \mathbb{E}\left\{\bar{x}_0^T\mathcal{Q}_0\bar{x}_0 - \bar{x}_{N+1}^T\mathcal{Q}_{N+1}\bar{x}_{N+1}\right\}$$

$$+ \sum_{k=0}^{N}\mathbb{E}\left\{\bar{x}_k^T\Delta_{k+1}\bar{x}_k + 2\bar{x}_k^T(\mathscr{A}_k + \Lambda_{k+1})\mathcal{Q}_{k+1}\mathscr{B}_k\bar{u}_k\right.$$

$$+ \bar{u}_k^T \mathscr{B}_k^T \mathcal{Q}_{k+1} \mathscr{B}_k \bar{u}_k \Big\}$$

$$= \mathbb{E}\Big\{ \bar{x}_0^T \mathcal{Q}_0 \bar{x}_0 - \bar{x}_{N+1}^T \mathcal{Q}_{N+1} \bar{x}_{N+1} \Big\} - \sum_{k=0}^{N} \mathbb{E}\Big\{ \|\bar{u}_k\|^2 \Big\}$$

$$+ \sum_{k=0}^{N} \mathbb{E}\Big\{ \bar{x}_k^T (\Delta_{k+1} + \mathcal{L}_k^T \mathcal{L}_k) \bar{x}_k + 2 \bar{x}_k^T (\mathscr{A}_k + \Lambda_{k+1})^T \mathcal{Q}_{k+1} \mathscr{B}_k \bar{u}_k$$

$$+ \bar{u}_k^T (\mathscr{B}_k^T \mathcal{Q}_{k+1} \mathscr{B}_k + I) \bar{u}_k \Big\}$$

$$= \mathbb{E}\Big\{ \bar{x}_0^T \mathcal{Q}_0 \bar{x}_0 - \bar{x}_{N+1}^T \mathcal{Q}_{N+1} \bar{x}_{N+1} \Big\} - \sum_{k=0}^{N} \mathbb{E}\Big\{ \|\bar{u}_k\|^2 \Big\}$$

$$+ \sum_{k=0}^{N} \mathbb{E}\left\{ \begin{bmatrix} \bar{x}_k \\ \bar{u}_k \end{bmatrix}^T \begin{bmatrix} \Delta_{k+1} + \mathcal{L}_k^T \mathcal{L}_k - \mathcal{Q}_k & (\mathscr{A}_k + \Lambda_{k+1})^T \mathcal{Q}_{k+1} \mathscr{B}_k \\ * & \mathscr{B}_k^T \mathcal{Q}_{k+1} \mathscr{B}_k + I \end{bmatrix} \begin{bmatrix} \bar{x}_k \\ \bar{u}_k \end{bmatrix} \right\}. \tag{2.15}$$

Obviously, equalities (2.10) and (2.11) are guaranteed by (2.13) and (2.15), respectively. Therefore, the proof is complete.

Lemma 2.4 *Given the disturbance attenuation level $\gamma > 0$ and the positive definite matrix W, for the augmented system (2.8) with any nonzero $\{\tilde{\eta}_k\}_{0 \leq k \leq N} \in \mathcal{L}_{[0,N]}$, the following two statements are equivalent:*

(i) *The auxiliary index (2.9) is satisfied.*

(ii) *There exists a set of real-valued matrices $\{K_k\}_{0 \leq k \leq N}$, positive scalars $\{\varepsilon_k\}_{0 \leq k \leq N}$ and matrices $\{\mathcal{P}_k\}_{0 \leq k \leq N+1}$ (with the final condition $\mathcal{P}_{N+1} = 0$) such that the following backward recursive Riccati difference equations (RDE):*

$$\mathcal{R}_{k+1}^{11} + \mathcal{R}_{k+1}^{12} (\mathcal{R}_{k+1}^{22})^{-1} (\mathcal{R}_{k+1}^{12})^T = \mathcal{P}_k \tag{2.16}$$

with

$$\mathcal{R}_{k+1}^{22} > 0 \quad and \quad \mathcal{P}_0 < \gamma^2 W \tag{2.17}$$

where the corresponding matrix parameters are defined in Lemma 2.3.

Proof *(ii) $\Rightarrow$ (i). For non-negative definite matrices $\{\mathcal{P}_k\}_{0 \leq k \leq N+1}$ satisfying the recursive RDE (2.16), it can be obtained from Lemma 2.3 that*

$$\mathbb{E}\left\{||z_k||^2_{[0,N]}\right\} - \mathbb{E}\left\{\gamma^2||\tilde{\eta}_k||^2_{[0,N]} - \bar{\alpha}\gamma^2||\varepsilon_k\aleph_k\bar{x}_k||^2_{[0,N]}\right\}$$

$$= \mathbb{E}\left\{\bar{x}_0^T\mathcal{P}_0\bar{x}_0 - \bar{x}_{N+1}^T\mathcal{P}_{N+1}\bar{x}_{N+1}\right\}$$

$$+ \sum_{k=0}^{N}\mathbb{E}\left\{\bar{x}_k^T(\mathcal{R}_{k+1}^{11} - \mathcal{P}_k)\bar{x}_k + 2\bar{x}_k^T\mathcal{R}_{k+1}^{12}\tilde{\eta}_k - \tilde{\eta}_k^T\mathcal{R}_{k+1}^{22}\tilde{\eta}_k\right\} \qquad (2.18)$$

$$= \sum_{k=0}^{N}\mathbb{E}\left\{\bar{x}_k^T\left(\mathcal{R}_{k+1}^{11} - \mathcal{P}_k + \mathcal{R}_{k+1}^{12}(\mathcal{R}_{k+1}^{22})^{-1}(\mathcal{R}_{k+1}^{12})^T\right)\bar{x}_k\right.$$

$$\left. - (\tilde{\eta}_k - \tilde{\eta}_k^*)^T\mathcal{R}_{k+1}^{22}(\tilde{\eta}_k - \tilde{\eta}_k^*)\right\} + \mathbb{E}\left\{\bar{x}_0^T\mathcal{P}_0\bar{x}_0 - \bar{x}_{N+1}^T\mathcal{P}_{N+1}\bar{x}_{N+1}\right\}$$

where

$$\tilde{\eta}_k^* = (\mathcal{R}_{k+1}^{22})^{-1}(\mathcal{R}_{k+1}^{12})^T\bar{x}_k. \qquad (2.19)$$

Since $\mathcal{R}_k^{22} > 0$ and $\mathcal{P}_0 < \gamma^2W$, for any nonzero $\{\tilde{\eta}_k\}_{0\le k\le N} \in \mathcal{L}_{[0,N]}$, it can be derived from the final condition $\mathcal{P}_{N+1} = 0$ that

$$\mathbb{E}\left\{||z_k||^2_{[0,N]}\right\} - \mathbb{E}\left\{\gamma^2||\tilde{\eta}_k||^2_{[0,N]} - \bar{\alpha}\gamma^2||\varepsilon_k\aleph_k\bar{x}_k||^2_{[0,N]}\right\} - \gamma^2\mathbb{E}\left\{\bar{x}_0^TW\bar{x}_0\right\}$$

$$< \mathbb{E}\left\{||z_k||^2_{[0,N]}\right\} - \mathbb{E}\left\{\gamma^2||\tilde{\eta}_k||^2_{[0,N]} - \bar{\alpha}\gamma^2||\varepsilon_k\aleph_k\bar{x}_k||^2_{[0,N]}\right\} - \mathbb{E}\left\{\bar{x}_0^T\mathcal{P}_0\bar{x}_0\right\}$$

$$= -\mathbb{E}\left\{\sum_{k=0}^{N}(\tilde{\eta}_k - \tilde{\eta}_k^*)^T\mathcal{R}_{k+1}^{22}(\tilde{\eta}_k - \tilde{\eta}_k^*)\right\}$$

$$< 0. \qquad (2.20)$$

(i) $\Rightarrow$ (ii). Noting $\mathcal{R}_{N-1}^{22} = \gamma^2I > 0$, we can solve $\mathcal{P}_{N-1}$ from the recursion (2.16). By the same procedure, the recursion (2.16) can be solved backward if and only if $\mathcal{R}_k^{22} > 0$ for all $k \in [0, N)$. If (2.16) fails to proceed for some $k = k_0 \in [0, N)$, then $\mathcal{R}_{k_0}^{22}$ has at least one zero or negative eigenvalue. Therefore, the proof of necessity reduces to proving the following proposition:

$$\mathcal{J}(\bar{x}_0, \tilde{\eta}) < 0 \Longrightarrow \lambda_i(\mathcal{R}_k^{22}) > 0, \quad \forall k \in [0, N), \quad i = 1, 2, \cdots, n_x. \qquad (2.21)$$

The rest of the proof is carried out by contradiction. That is, assuming that at least one eigenvalue of $\mathcal{R}_k^{22}$ is either equal to 0 or negative at some time point $k = k_0 \in [0, N)$, then $\mathcal{J}(\bar{x}_0, \tilde{\eta}) < 0$ is not true. For simplicity, we denote such an eigenvalue of $\mathcal{R}_k^{22}$ at time k_0 as λ_{k_0}. In the following, we shall use such a non-positive λ_{k_0} to reveal that there exist certain $(\bar{x}_0, \tilde{\eta}) \ne 0$ such that $\mathcal{J}(\bar{x}_0, \tilde{\eta}) \ge 0$. First, we can choose $\bar{x}_0 = 0$ and

$$\tilde{\eta}_k = \begin{cases} \psi_{k_0}, & k = k_0, \\ \tilde{\eta}_k^*, & k_0 < k \le N, \\ 0, & 0 \le k < k_0 \end{cases} \qquad (2.22)$$

where ψ_{k_0} is the eigenvector of $\mathcal{R}^{22}_{k_0+1}$ with respect to λ_{k_0}. Since $\bar{x}_0 = 0$ and $\tilde{\eta}_k = 0$ when $0 \leq k < k_0$, we can see from (2.18) that

$$\sum_{k=0}^{k_0-1} \mathbb{E}\left\{ ||z_k||^2 - \gamma^2||\tilde{\eta}_k||^2 + \bar{\alpha}\gamma^2||\varepsilon_k \aleph_k \bar{x}_k||^2 \right\} = 0. \tag{2.23}$$

Furthermore, for this given sequence $(\bar{x}_0, \tilde{\eta})$, one has from (2.18) that

$$\mathcal{J}(\bar{x}_0, \tilde{\eta}) = \sum_{k=0,k\neq k_0}^{N} \mathbb{E}\left\{ ||z_k||^2 - \gamma^2||\tilde{\eta}_k||^2 + \bar{\alpha}\gamma^2||\varepsilon_k \aleph_k \bar{x}_k||^2 \right\}$$
$$+ \mathbb{E}\left\{ ||z_{k_0}||^2 - \gamma^2||\tilde{\eta}_{k_0}||^2 + \bar{\alpha}\gamma^2||\varepsilon_{k_0} \aleph_{k_0} \bar{x}_{k_0}||^2 \right\} \tag{2.24}$$

with

$$\sum_{k=k_0+1}^{N} \mathbb{E}\left\{ ||z_k||^2 - \gamma^2||\tilde{\eta}_k||^2 + \bar{\alpha}\gamma^2||\varepsilon_k \aleph_k \bar{x}_k||^2 \right\}$$
$$= -\mathbb{E}\left\{ \sum_{k=k_0+1}^{N} (\tilde{\eta}_k - \tilde{\eta}_k^*)^T \mathcal{R}_k^{22} (\tilde{\eta}_k - \tilde{\eta}_k^*) \right\} \tag{2.25}$$
$$+ \mathbb{E}\left\{ \bar{x}_{k_0+1}^T \mathcal{P}_{k_0+1} \bar{x}_{k_0+1} - \bar{x}_{N+1}^T \mathcal{P}_{N+1} \bar{x}_{N+1} \right\}$$

and

$$\mathbb{E}\left\{ ||z_{k_0}||^2 - \gamma^2||\tilde{\eta}_{k_0}||^2 + \bar{\alpha}\gamma^2||\varepsilon_{k_0} \aleph_{k_0} \bar{x}_{k_0}||^2 \right\}$$
$$= \mathbb{E}\left\{ \bar{x}_{k_0}^T \mathcal{P}_{k_0} \bar{x}_{k_0} - \bar{x}_{k_0+1}^T \mathcal{P}_{k_0+1} \bar{x}_{k_0+1} \right\}$$
$$+ \mathbb{E}\left\{ \bar{x}_{k_0}^T \mathcal{R}_{k_0}^{11} \bar{x}_{k_0} + 2\bar{x}_{k_0}^T \mathcal{R}_{k_0}^{12} \tilde{\eta}_{k_0} - \tilde{\eta}_{k_0}^T \mathcal{R}_{k_0}^{22} \tilde{\eta}_{k_0} \right\} \tag{2.26}$$
$$= -\mathbb{E}\left\{ \tilde{\eta}_{k_0}^T \tilde{\eta}_{k_0}^T \mathcal{R}_{k_0}^{22} \tilde{\eta}_{k_0} \tilde{\eta}_{k_0} \right\} - \mathbb{E}\left\{ \bar{x}_{k_0+1}^T \mathcal{P}_{k_0+1} \bar{x}_{k_0+1} \right\}.$$

Finally, substituting (2.23), (2.25), and (2.26) into (2.24) yields

$$\mathcal{J}(\bar{x}_0, \tilde{\eta}) = -\mathbb{E}\left\{ \tilde{\eta}_{k_0}^T \mathcal{R}_{k_0}^{22} \tilde{\eta}_{k_0} \right\}$$
$$= -\mathbb{E}\left\{ \psi_{k_0}^T \mathcal{R}_{k_0}^{22} \psi_{k_0} \right\} = -\lambda_{k_0} \mathbb{E}\left\{ ||\psi_{k_0}||^2 \right\} \geq 0.$$

This contradicts the condition $\mathcal{J}(\bar{x}_0, \tilde{\eta}) < 0$. Therefore, all the eigenvalues of $\mathcal{R}_k^{22}$ should be positive, i.e., $\mathcal{R}_k^{22} > 0$. The proof is complete.

It should be pointed out that Lemma 2.4 provides a necessary and sufficient condition of the auxiliary index (2.9). Such a condition serves as a key to solve the addressed stochastic disturbance attenuation problems. So far, we have analyzed the $\mathcal{H}_\infty$ performance in terms of the solvability to a backward RDE in Lemma 2.4. In the next stage, we shall proceed to tackle the design problem of the controller (2.5) such that the closed-loop system (2.6) satisfies the $\mathcal{H}_\infty$ performance requirement (2.7).

Theorem 2.1 *For the given disturbance attenuation level $\gamma > 0$ and positive definite matrix W, the closed-loop system (2.6) satisfies the $\mathcal{H}_\infty$ performance constraint (2.7) for any nonzero disturbance sequence $\{\eta_k\}_{0 \leq k \leq N} \in \mathcal{L}_{[0,N]}$ if there exist a set of solutions $\{(\varepsilon_k, \mathcal{P}_k, \mathcal{Q}_k, K_k)\}_{0 \leq k \leq N}$ with $\varepsilon_k > 0$ satisfying the following recursive RDEs:*

$$
\begin{cases}
\mathcal{R}_{k+1}^{11} + \mathcal{R}_{k+1}^{12}(\mathcal{R}_{k+1}^{22})^{-1}(\mathcal{R}_{k+1}^{12})^T = \mathcal{P}_k, & (2.27a) \\
\Delta_{k+1} + \mathcal{L}_k^T \mathcal{L}_k - \mathcal{S}_{k+1}^1(\mathcal{S}_{k+1}^2)^{-1}(\mathcal{S}_{k+1}^1)^T = \mathcal{Q}_k & (2.27b)
\end{cases}
$$

subject to

$$
\begin{cases}
\mathcal{P}_{N+1} = \mathcal{Q}_{N+1} = 0, & (2.28a) \\
\mathcal{S}_{k+1}^2 > 0, & (2.28b) \\
\mathcal{R}_{k+1}^{22} > 0, & (2.28c) \\
\mathcal{P}_0 < \gamma^2 W, & (2.28d) \\
K_k^* = \arg\min_{K_k} \left\| K_k \bar{\vartheta}\mathscr{C}_k + (\mathcal{S}_{k+1}^2)^{-1}(\mathcal{S}_{k+1}^1)^T \right\|_F & (2.28e)
\end{cases}
$$

where corresponding matrix parameters are defined in Lemma 2.3.

    Proof *Firstly, if there exists $\{\mathcal{P}_k\}_{0 \leq k \leq N}$ satisfying (2.27a) and (2.28a), it can be easily seen from Lemma 2.4 that the system (2.8) satisfies the pre-specified $\mathcal{H}_\infty$ performance (2.9) and, according to Lemma 2.2, the pre-specified $\mathcal{H}_\infty$ performance (2.7) is satisfied for the closed-loop system (2.6). In this case, the worst-case disturbance can be expressed as $\tilde{\eta}_k^* = (\mathcal{R}_{k+1}^{22})^{-1}(\mathcal{R}_{k+1}^{12})^T \bar{x}_k$.*

    Next, by employing the worst-case disturbance, we aim to provide a design scheme of the controller parameter K_k. For this purpose, by using the completing squares method, it follows from Lemma 2.3 that

$$
\begin{aligned}
&\mathcal{J}_2(\bar{u}_k; \tilde{\eta}_k^*) \\
&= \mathbb{E}\left\{ \bar{x}_0^T \mathcal{Q}_0 \bar{x}_0 - \bar{x}_{N+1}^T \mathcal{Q}_{N+1} \bar{x}_{N+1} \right\} \\
&\quad + \sum_{k=0}^{N} \mathbb{E}\left\{ \bar{x}_k^T(\Delta_{k+1} + \mathcal{L}_k^T \mathcal{L}_k - \mathcal{Q}_k)\bar{x}_k + 2\bar{x}_k^T \mathcal{S}_{k+1}^1 \bar{u}_k + \bar{u}_k^T \mathcal{S}_{k+1}^2 \bar{u}_k \right\} \\
&= \mathbb{E}\left\{ \bar{x}_0^T \mathcal{Q}_0 \bar{x}_0 - \bar{x}_{N+1}^T \mathcal{Q}_{N+1} \bar{x}_{N+1} \right\} + \sum_{k=0}^{N} \mathbb{E}\left\{ \bar{x}_k^T \left(\Delta_{k+1} + \mathcal{L}_k^T \mathcal{L}_k \right. \right. \\
&\quad \left. \left. -\mathcal{Q}_k - \mathcal{S}_{k+1}^1(\mathcal{S}_{k+1}^2)^{-1}(\mathcal{S}_{k+1}^1)^T \right)\bar{x}_k + (\bar{u}_k - \bar{u}_k^*)^T \mathcal{S}_{k+1}^2(\bar{u}_k - \bar{u}_k^*) \right\} \\
&\leq \mathbb{E}\left\{ \bar{x}_0^T \mathcal{Q}_0 \bar{x}_0 - \bar{x}_{N+1}^T \mathcal{Q}_{N+1} \bar{x}_{N+1} \right\} + \sum_{k=0}^{N} \mathbb{E}\left\{ \bar{x}_k^T \left(\Delta_{k+1} + \mathcal{L}_k^T \mathcal{L}_k \right. \right. \\
&\quad \left. - \mathcal{S}_{k+1}^1(\mathcal{S}_{k+1}^2)^{-1}(\mathcal{S}_{k+1}^1)^T - \mathcal{Q}_k \right)\bar{x}_k \\
&\quad \left. + \left\| K_k \bar{\vartheta}\mathscr{C}_k + (\mathcal{S}_{k+1}^2)^{-1}\mathcal{S}_{k+1}^{1T} \right\|_F^2 \left\| \mathcal{S}_{k+1}^2 \right\|_F \|\bar{x}_k\|^2 \right\}
\end{aligned}
\tag{2.29}
$$

where $\bar{u}_k^* = -(\mathcal{S}_{k+1}^2)^{-1}(\mathcal{S}_{k+1}^1)^T \bar{x}_k$. Furthermore, the controller parameter K_k can be selected to satisfy (2.27b) and (2.28b) simultaneously, which ends the proof.

2.3 $\mathcal{H}_\infty$ Controller Design

Clearly, it is generally difficult to solve the optimization problem (2.28b). For convenience in application, the expression of the parameter K_k can be acquired by using the Moore-Penrose pseudoinverse in this section.

Theorem 2.2 *For the given disturbance attenuation level $\gamma > 0$ and positive definite matrix W, the closed-loop system (2.6) satisfies the $\mathcal{H}_\infty$ performance constraint (2.7) for any nonzero disturbance sequence $\{\eta_k\}_{0 \le k \le N} \in \mathcal{L}_{[0,N]}$, if there exists a set of solutions $\{(\varepsilon_k, \lambda_k, \delta_k, \mathcal{P}_k, \mathcal{Q}_k, K_k)\}_{0 \le k \le N}$ (with $\lambda_k > 0$, $\delta_k > 0$ and $\varepsilon_k > 0$) satisfying the following recursive RDEs:*

$$\begin{cases} \mathcal{R}_{k+1}^{11} + \bar{\mathcal{R}}_{k+1}^{12}(\bar{\mathcal{R}}_{k+1}^{22})^{-1}(\bar{\mathcal{R}}_{k+1}^{12})^T = \mathcal{P}_k, & (2.30a) \\ \bar{\Delta}_{k+1} + \mathcal{L}_k^T \mathcal{L}_k - \bar{\mathcal{S}}_{k+1}^1(\mathcal{S}_{k+1}^2)^{-1}(\bar{\mathcal{S}}_{k+1}^1)^T = \mathcal{Q}_k & (2.30b) \end{cases}$$

subject to

$$\begin{cases} \mathcal{P}_{N+1} = \mathcal{Q}_{N+1} = 0, & (2.31a) \\ \bar{\mathcal{R}}_{k+1}^{22} > 0, & (2.31b) \\ \mathcal{P}_0 < \gamma^2 W, & (2.31c) \\ \mathcal{S}_{k+1}^2 > 0, & (2.31d) \\ K_k^* = \Upsilon_{k+1}^\dagger \Gamma_{k+1}(\bar{\vartheta}\mathscr{C}_k)^\dagger, & (2.31e) \\ \mathscr{W}_k \le \delta_k I & (2.31f) \end{cases}$$

where

$$\mathscr{G}_k = \left[\ [D_k \ \lambda_k^{-1} B_k \ \varepsilon_k^{-1} I]^T \quad 0 \quad 0 \quad \cdots \quad 0 \right]^T,$$

$$\bar{\mathcal{R}}_{k+1}^{12} = (\mathscr{A}_k + \mathscr{B}_k K_k \bar{\vartheta}\mathscr{C}_k)^T \mathcal{P}_{k+1} \mathscr{G}_k, \quad \bar{\mathcal{R}}_{k+1}^{22} = \gamma^2 I - \mathscr{G}_k^T \mathcal{P}_{k+1}\mathscr{G}_k,$$

$$\bar{\Lambda}_{k+1} = \mathscr{G}_k(\bar{\mathcal{R}}_{k+1}^{22})^{-1}(\bar{\mathcal{R}}_{k+1}^{12})^T, \quad \bar{\mathcal{S}}_{k+1}^1 = (\mathscr{A}_k + \bar{\Lambda}_{k+1})^T \mathcal{Q}_{k+1}\mathscr{B}_k,$$

$$\bar{\Delta}_{k+1} = (\mathscr{A}_k + \bar{\Lambda}_{k+1})^T \mathcal{Q}_{k+1}(\mathscr{A}_k + \bar{\Lambda}_{k+1}) + \tilde{\alpha}\tilde{\mathscr{A}}_k^T \mathcal{Q}_{k+1}\tilde{\mathscr{A}}_k + \mathscr{C}_k^T \Xi_{k+1}^1 \mathscr{C}_k,$$

$$\Gamma_{k+1} = -(\mathcal{S}_{k+1}^2)^{-1}\mathscr{B}_k^T \mathcal{Q}_{k+1}(I + \mathscr{G}_k(\bar{\mathcal{R}}_{k+1}^{22})^{-1}\mathscr{G}_k^T \mathcal{P}_{k+1})\mathscr{A}_k,$$

$$\Upsilon_{k+1} = I + (\mathcal{S}_{k+1}^2)^{-1}\mathscr{B}_k^T \mathcal{Q}_{k+1}\mathscr{G}_k(\bar{\mathcal{R}}_{k+1}^{22})^{-1}\mathscr{G}_k^T \mathcal{P}_{k+1}\mathscr{B}_k,$$

$$\mathscr{W}_k = \gamma^2 \lambda_k \mathscr{E}_k^T \big(\bar{v}^T K_k^{*T} K_k^* \bar{v} + diag\{0, \tilde{\vartheta}_0, \ldots, \tilde{\vartheta}_\ell, 0\} \otimes (K_k^{*T} K_k^*)\big)\mathscr{E}_k,$$

and the other corresponding matrix parameters are defined as in Lemma 2.3.

Proof *Denote $\tilde{v}_k = \lambda_k\left(\sum_{i=0}^{\ell} \vartheta_k^i K_k E_{k-i} v_{k-i} + K_k \xi_k\right)$ where $\lambda_k > 0$ is introduced to offer more flexibility in the controller design. Next, selecting $\zeta_k = [w_k \quad \tilde{v}_k \quad (\varepsilon_k \alpha_k F_k \mathcal{N}_k x_k)^T]^T$, we rewrite (2.8) as follows:*

$$\begin{cases} \bar{x}_{k+1} = \big(\mathscr{A}_k + \mathscr{B}_k K_k \bar{\vartheta}\mathscr{C}_k\big)\bar{x}_k \\ \qquad\qquad + (\alpha_k - \bar{\alpha})\tilde{\mathscr{A}}_k \bar{x}_k + \mathscr{B}_k K_k \vartheta_k \mathscr{C}_k \bar{x}_k + \mathscr{G}_k \zeta_k, & (2.32) \\ z_k = \mathcal{L}_k \bar{x}_k. \end{cases}$$

It can be easily seen that the nonzero disturbance sequence $\{\zeta_k\}_{0 \leq k \leq N}$ belongs to $\mathcal{L}_{[0,N]}$. On the other hand, it follows from Lemma 2.1 that (2.31c) is the solution of the optimization problem

$$\min_{K_k} \left\| \Upsilon_{k+1} K_k \bar{\vartheta} \mathscr{C}_k - \Gamma_{k+1} \right\|_F,$$

which can be rewritten as

$$\min_{K_k} \left\| K_k \bar{\vartheta} \mathscr{C}_k + (\mathcal{S}^2_{k+1})^{-1} (\bar{\mathcal{S}}^1_{k+1})^T \right\|_F. \tag{2.33}$$

According to Theorem 2.1, if there exists a set of solutions satisfying the recursive RDEs (2.30a) and (2.30b) with (2.31a)-(2.31d), one has

$$\mathbb{E}\left\{ \|z_k\|^2_{[0,N]} \right\} < \mathbb{E}\left\{ \gamma^2 \|\zeta_k\|^2_{[0,N]} - \delta_k \|\mathcal{U}\zeta_k\|^2_{[0,N]} \right. \\ \left. - \bar{\alpha}\gamma^2 \|\varepsilon_k \aleph_k \bar{x}_k\|^2_{[0,N]} \right\} + \gamma^2 \mathbb{E}\left\{ \bar{x}_0^T W \bar{x}_0 \right\}. \tag{2.34}$$

Furthermore, in light of (2.31d), the above inequality yields to

$$\mathbb{E}\left\{ \|z_k\|^2_{[0,N]} \right\} < \mathbb{E}\left\{ \gamma^2 \|\tilde{\eta}_k\|^2_{[0,N]} + \gamma^2 \|\tilde{v}_k\|^2_{[0,N]} \right. \\ \left. - \delta_k \|w_k\|^2_{[0,N]} - \bar{\alpha}\gamma^2 \|\varepsilon_k \aleph_k \bar{x}_k\|^2_{[0,N]} \right\} + \gamma^2 \mathbb{E}\left\{ \bar{x}_0^T W \bar{x}_0 \right\} \tag{2.35}$$

$$\leq \mathbb{E}\left\{ \gamma^2 \|\tilde{\eta}_k\|^2_{[0,N]} - \bar{\alpha}\gamma^2 \|\varepsilon_k \aleph_k \bar{x}_k\|^2_{[0,N]} \right\} + \gamma^2 \mathbb{E}\left\{ \bar{x}_0^T W \bar{x}_0 \right\},$$

which implies that the closed-loop system (2.6) achieves the $\mathcal{H}_\infty$ performance constraint (2.7). The proof is complete.

Remark 2.2 *In this chapter, the model transformation is conducted several times to simplify the addressed problem by reducing the complexity of the system analysis and synthesis. In particular, 1) the sector-bounded condition (2.2) has been transformed into the sector-bound uncertainties (2.3), 2) the systems (2.6) have been reformulated as (2.8), and further changed into (2.32) by introducing a new disturbance, and 3) the performance index (2.8) has been used to replace (2.7) in the $\mathcal{H}_\infty$ control problem.*

Corollary 2.1 *Let C_k be of column full rank for all $k \in [0, N]$. Assume that the disturbance attenuation level $\gamma > 0$ and positive definite matrix $W > 0$ are given. If there exist a set of solutions $\{(\varepsilon_k, \lambda_k, \mathcal{P}_k, \mathcal{Q}_k, K_k)\}_{0 \leq k \leq N}$ (with $\varepsilon_k > 0$ and $\lambda_k > 0$) to the coupled backward recursive RDEs (2.30a) and (2.30b) subject to (2.31a)-(2.31c) and $|\Upsilon_{k+1}| \neq 0$, then*
(i) the worst-case disturbance ζ_k^* and the controller gain K_k^* are given by

$$\zeta_k^* = (\mathcal{R}^{22}_{k+1})^{-1} (\mathcal{R}^{12}_{k+1})^T \bar{x}_k, \quad K_k^* = \Upsilon_{k+1}^{-1} \Gamma_{k+1} (\bar{\vartheta}\mathscr{C}_k)^\dagger + Y_k - Y_k \bar{\vartheta}\mathscr{C}_k (\bar{\vartheta}\mathscr{C}_k)^\dagger$$

where Y_k is any vector with appropriate dimension;
(ii) the closed-loop system (2.32) satisfies the auxiliary index (2.9) for any nonzero disturbance sequence $\{\zeta_k\}_{0 \leq k \leq N} \in \mathcal{L}_{[0,N]}$.

Noting from Theorem 2.2 that the controller gain matrices are calculated by solving the proposed coupled RDEs, we summarize the $\mathcal{H}_\infty$ controller design algorithm as follows.

Controller design algorithm

Step 1. For given disturbance attenuation level $\gamma > 0$ and positive definite matrix W, set $k = N$. Go to the next step.

Step 2. Calculate $\bar{\mathcal{R}}^{22}_{k+1}$ and $\mathcal{S}^2_{k+1}$. If $\bar{\mathcal{R}}^{22}_{k+1} > 0$ and $\mathcal{S}^2_{k+1} > 0$, then go to the next step, else go to *Step 6*.

Step 3. Calculate Υ_{k+1} and Γ_{k+1}, and then obtain the controller gain matrix K_k by (2.31c), go to the next stop.

Step 4. Solve the equation (2.30a) and (2.30b) to get $\mathcal{P}_k$ and $\mathcal{Q}_k$, respectively.

Step 5. If $k \neq 0$, set $k = k - 1$ and go to *Step 2*, else go to the next step.

Step 6. If either $\bar{\mathcal{R}}^{22}_{k+1} > 0$ for some $k \neq 0$ or $\mathcal{P}_0 < \gamma^2 W$ does not hold, this algorithm is infeasible. Stop.

Remark 2.3 *In this chapter, we examine how the channel fading and randomly occurring nonlinearities influence the $\mathcal{H}_\infty$ control performance over a finite-horizon $[0, N]$. It is worth mentioning that the conditions in Lemma 2.4 and Theorem 2.1 are obtained mainly by the "completing the square" technique, which results in little conservatism. Compared to the existing literature, our results have the following three distinguishing features: 1) the system under investigation is in the discrete time-varying form; 2) the technology of model transformation is employed to reduce the complexity of system analysis; and 3) this chapter represents one of the first attempts to address both channel fading and RONs for the $\mathcal{H}_\infty$ control problems by using backward recursive RDEs. Furthermore, in Theorem 2.2, all the system parameters, the probability for channel coefficients, as well as RONs are reflected in the backward recursive RDEs. Also, an efficient algorithm is provided to solve the addressed controller design problem. Due to the recursive nature of the algorithm, the proposed design procedure is suitable for online application without the need of increasing the problem size.*

2.4 Simulation Examples

In this section, we present a simulation example in order to 1) illustrate the effectiveness of the proposed controller design scheme for networked discrete time-varying systems with RONs and fading channels, and 2) show the relationship between the disturbance attenuation level γ and the channel number ℓ.

Consider system (2.1) with the fading measurement (2.4) with

$$A_k = \begin{bmatrix} 0.55 + \sin(k-1) & -0.40 \\ -0.45 + e^{-4k} & 0.72 \end{bmatrix}, \quad B_k = \begin{bmatrix} 0.60 \\ -0.75 \end{bmatrix},$$

$$C_k = \begin{bmatrix} -0.62 & 0.70 \end{bmatrix}, \quad D_k = \begin{bmatrix} -0.02 & 0.015 \end{bmatrix}^T, \quad E_k = 0.01, \quad L_k = \begin{bmatrix} 0.20 & 0.20 \end{bmatrix}.$$

Let the nonlinear vector-valued function $f_k(x_k)$ be

$$h_k(x_k) = \begin{cases} \begin{bmatrix} -0.55x_k^1 + 0.30x_k^2 + \tanh(0.25x_k^1) \\ 0.58x_k^2 - \tanh(0.24x_k^2) \end{bmatrix}, & 0 \le k < 15, \\[2ex] \begin{bmatrix} 0.40x_k^1 - \tanh(0.3x_k^1) \\ 0.48x_k^2 \end{bmatrix}, & 15 \le k \le 41 \end{cases}$$

where x_k^i ($i = 1, 2$) denotes the i-th element of the system state x_k. The probability of RONs is taken as $\bar{a} = 0.10$. The order of the fading model is $\ell = 2$ and channel coefficients ϑ_k^0, ϑ_k^1 and ϑ_k^2 obey the Gaussian distributions $\mathcal{N}(0.9, 0.1^2)$, $\mathcal{N}(0.2, 0.5^2)$ and $\mathcal{N}(0.2, 0.5^2)$, respectively. Meanwhile, it is easy to see that the constraint (2.2) can be met with

$$\Phi_k = \begin{cases} \begin{bmatrix} -0.30 & 0.30 \\ 0 & 0.58 \end{bmatrix}, & 0 \le k < 15, \\[2ex] \begin{bmatrix} 0.10 & 0 \\ 0 & 0.48 \end{bmatrix}, & 15 \le k < 41, \end{cases}$$

$$\Psi_k = \begin{cases} \begin{bmatrix} -0.55 & 0.30 \\ 0 & 0.34 \end{bmatrix}, & 0 \le k < 15, \\[2ex] \begin{bmatrix} 0.40 & 0 \\ 0 & 0.48 \end{bmatrix}, & 15 \le k < 41. \end{cases}$$

In this example, the $\mathcal{H}_\infty$ performance level γ, positive definite matrix W, and time-horizon N are taken as 0.98, diag{0.50, 0.50}, and 40, respectively. Using the given algorithm and MATLAB® software, the set of solutions to the recursive RDEs in Theorem 2.2 are obtained and the controller gain matrices are shown in Table 2.1, where ε_k and λ_k are selected as $\varepsilon_k = 1.251$ and $\lambda_k = 0.51$, respectively. In the simulation, the exogenous disturbance inputs are selected as

$$w_k = 5\sin(2k), \quad v_k = 0.85\cos(0.6k), \quad \xi_k = 0.52\cos(0.25k).$$

The simulation results are shown in Fig. 2.1 and Fig. 2.2, where Fig. 2.1 plots the output trajectories of the open-loop and closed-loop system, and Fig. 2.2 depicts the measurement outputs and the received signals by controller, respectively. The simulation results have confirmed that the designed controller performs very well.

It would be interesting to see the relationship between the disturbance attenuation level γ and the probability $\bar{a}$. For the same parameters λ_k, ε_k and δ_k, the permitted minimum γ is shown in Table 2.2. It is easy to find that the disturbance attenuation performance deteriorates with increased $\bar{a}$.

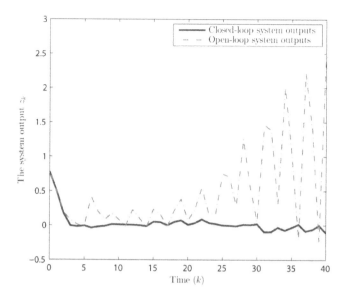

FIGURE 2.1
The system output.

TABLE 2.1
The distributed state estimator gain matrices.

k	0	1	2	3	4	5	6	$\cdots$
K_k	0.0436	0.3267	0.2987	0.1956	0.1454	-0.1669	0.1350	$\cdots$

TABLE 2.2
The permitted minimum γ.

$\bar{\alpha}$	0.05	0.06	0.07	0.08	0.09	0.10	0.11	0.12	0.13	0.14
γ	0.90	0.91	0.92	0.93	0.94	0.95	0.97	0.98	0.99	1.00

2.5 Summary

In this chapter, we have made an attempt to investigate the finite-horizon $\mathcal{H}_\infty$ control problem for a class of discrete time-varying nonlinear systems.

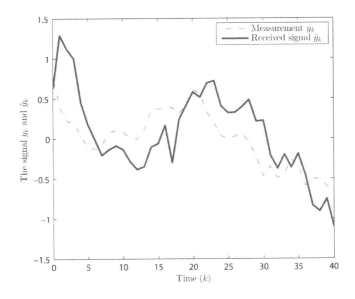

FIGURE 2.2
The measurement signal and the received signal.

To reflect more realistic communication situations, the fading channels described by modified stochastic Rice fading models and randomly occurring nonlinearities governed by Bernoulli distributed random variables have been simultaneously taken into account. Then, the nonlinearities satisfying the sector-bounded condition have been transformed into sector-bound uncertainties, and the augmented systems have also been simplified to reduce the complexity in system analysis. By employing the completing squares method and the stochastic analysis techniques, some sufficient conditions have been provided to ensure that the closed-loop system satisfies the $\mathcal{H}_\infty$ performance constraint. Furthermore, the desired controller gains can be obtained by solving two coupled backward recursive RDEs.

3

Finite-Horizon $\mathcal{H}_\infty$ Consensus Control for Multi-Agent Systems with Missing Measurements

---

Multi-agent systems are made up of multiple autonomous entities with distributed information, computational ability and possibly divergent interests. This kind of system may be cooperative (i.e., mobile robots in a warehouse), competitive (i.e., in electronic commerce or in settings of resource), or task allocated. The past decade has seen a surge of research interest in the consensus problems for multi-agent systems due to their extensive applications in various engineering systems such as cooperative control of unmanned air vehicles (UAVs) or unmanned underwater vehicles (UUVs) [98], formation control for multi-robot systems [1, 37, 140], collective behavior of flocks or swarms [88, 90, 144], distributed sensor networks [63, 109, 149], attitude alignment of clusters of satellites, and synchronization of complex networks [111, 117]. In Chapter 2, a novel design scheme with the form of backward RDEs was developed to deal with the $\mathcal{H}_\infty$ control issue of discrete time-varying stochastic systems. Obviously, it is very interesting and significant to extend such an approach to the $\mathcal{H}_\infty$ consensus of multi-agent systems subject to various undesirable phenomena.

This chapter deals with the $\mathcal{H}_\infty$ consensus control problem for a class of discrete time-varying multi-agent systems with both missing measurements and parameter uncertainties. A directed graph is used to represent the communication topology of the multi-agent network, and a binary switching sequence satisfying a conditional probability distribution is employed to describe the missing measurements. The purpose of the addressed problem is to design a time-varying controller such that, for all probabilistic missing observations and admissible parameter uncertainties, the $\mathcal{H}_\infty$ consensus performance is guaranteed over a given finite horizon for closed-loop networked multi-agent systems. According to the given topology, the measurement output available for the controller is not only from the individual agent but also from its neighboring agents. By using the completing squares method and stochastic analysis techniques, a *necessary and sufficient* condition is derived for an auxiliary index which is closely related the $\mathcal{H}_\infty$ consensus performance, and then the time-varying controller parameters are designed by solving coupled backward recursive Riccati difference equations (RDEs).

3.1   Modeling and Problem Formulation

In this chapter, it is assumed that the multi-agent system has N agents which communicate with each other according to a fixed network topology represented by a directed graph $\mathcal{G} = (\mathcal{V}, \mathcal{E}, \mathcal{H})$ of order N with the set of agents $\mathcal{V} = \{1, 2, \cdots, N\}$, the set of edges $\mathcal{E} \in \mathcal{V} \times \mathcal{V}$, and the weighted adjacency matrix $\mathcal{H} = [h_{ij}]$ with nonnegative adjacency element h_{ij}. An edge of $\mathcal{G}$ is denoted by the ordered pair (i, j). The adjacency elements associated with the edges of the graph are positive, i.e., $h_{ij} > 0 \iff (i, j) \in \mathcal{E}$, which means that agent i can obtain information from agent j. The neighborhood of agent i is denoted by $\mathcal{N}_i = \{j \in \mathcal{V} : (j, i) \in \mathcal{E}\}$. An element of $\mathcal{N}_i$ is called a neighbor of agent i. The ith agent is called a source, if it has no neighbors but is a neighbor of another agent in $\mathcal{V}$. An agent is called an isolated agent, if it has no neighbor and it is not a neighbor of any other agents.

Consider a multi-agent system consisting of N agents, in which the dynamics of agent i is described by the following discrete time-varying stochastic system defined on $k \in [0, M]$:

$$\begin{cases} x_i(k+1) = (A(k) + \Delta A_i(k))x_i(k) + B(k)u_i(k) + D(k)w_i(k), \\ y_i(k) = \alpha_i(k)C(k)x_i(k) + E(k)\nu_i(k), \\ z_i(k) = L(k)x_i(k) \end{cases} \tag{3.1}$$

where $x_i(k) \in \mathbb{R}^n$, $y_i(k) \in \mathbb{R}^q$, $u_i(k) \in \mathbb{R}^p$ and $z_i(k) \in \mathbb{R}^r$ are the state, measurement output, control input, and controlled output of agent i, respectively. $w_i(k) \in \ell_2([0, M]; \mathbb{R}^s)$ and $\nu_i(k) \in \ell_2([0, M]; \mathbb{R}^m)$ are the external disturbances. The stochastic variables $\alpha_i(k)$ $(i \in \mathcal{V})$ are mutually independent Bernoulli distributed white sequences taking values of 0 and 1 with the following probabilities:

$$\text{Prob}\{\alpha_i(k) = 1\} = \bar{\alpha} \quad \text{and} \quad \text{Prob}\{\alpha_i(k) = 0\} = 1 - \bar{\alpha}$$

where $\bar{\alpha} \in [0, 1]$ is a known constant. $A(k)$, $B(k)$, $C(k)$, $D(k)$ and $E(k)$ are known time-varying matrices with compatible dimensions. The matrices $\Delta A_i(k)$ $(i \in \mathcal{V})$ represent time-varying norm-bounded parameter uncertainties that satisfy

$$\Delta A_i(k) = M_i(k)F_i(k)G(k) \tag{3.2}$$

where $M_i(k)$ and $G(k)$ are known time-varying matrices of appropriate dimensions, and $F_i(k)$ is an unknown matrix function satisfying

$$F_i^T(k)F_i(k) \leq I, \quad \forall k \in [0, M]. \tag{3.3}$$

The parameter uncertainties $\Delta A_i(k)$ $(i \in \mathcal{V})$ are said to be admissible if both (3.2) and (3.3) hold.

In this chapter, self-edges (i, i) are not allowed, i.e., $(i, i) \notin \mathscr{E}$ for any $i \in \mathcal{V}$. The in-degree of agent i is defined as $\deg_{in}(i) = \sum_{j \in \mathcal{N}_i} h_{ij}$, and the weighted adjacency matrix $\mathscr{H} = [h_{ij}]$ can be described as

$$\mathscr{H} = [h_{ij}] = \begin{cases} h_{ij} > 0, & j \in \mathcal{N}_i, \\ h_{ij} = 0, & j \notin \mathcal{N}_i. \end{cases}$$

Therefore, the control protocol of the following form is adopted:

$$u_i(k) = K(k) \sum_{j \in \mathcal{N}_i} h_{ij}\Big(y_j(k) - y_i(k)\Big) \tag{3.4}$$

where $K(k) \in \mathbb{R}^{n \times q}$ $(0 \le k \le M)$ are feedback gain matrices to be determined. For purposes of simplicity, we introduce the following notations:

$$X_k = [\; x_1^T(k) \quad x_2^T(k) \quad \cdots \quad x_N^T(k) \;]^T,$$
$$Z_k = [\; z_1^T(k) \quad z_2^T(k) \quad \cdots \quad z_N^T(k) \;]^T,$$
$$\eta_k = [\; w_1^T(k) \quad \cdots \quad w_N^T(k) \quad \nu_1^T(k) \quad \cdots \quad \nu_N^T(k) \;]^T,$$
$$\bar{u}_i(k) = \bar{\alpha} K(k) \mathcal{C}_k^{(i)} X_k, \quad \mathcal{A}_k = I_N \otimes A(k),$$
$$\mathcal{B}_k = I_N \otimes B(k), \quad \mathcal{C}_k = (\mathscr{H} - \bar{\mathscr{H}}) \otimes C(k),$$
$$\mathcal{D}_k = [I_N \otimes D(k) \quad \mathcal{B}_k \mathcal{K}_k \mathcal{E}_k], \quad \mathcal{E}_k = (\mathscr{H} - \bar{\mathscr{H}}) \otimes E(k),$$
$$\mathcal{G}_k = I_N \otimes G(k), \quad \mathcal{K}_k = I_N \otimes K(k), \quad \mathcal{L}_k = I_N \otimes L(k),$$
$$\bar{\mathscr{H}} = \text{diag}\{\deg_{in}(1), \deg_{in}(2), \cdots, \deg_{in}(N)\},$$
$$\mathcal{F}_k = \text{diag}\{F_1(k), F_2(k), \cdots, F_N(k)\}, \quad \Xi_k = [\Xi_k^{(ij)}]_{N \times N},$$
$$\mathcal{M}_k = \text{diag}\{M_1(k), M_2(k), \cdots, M_N(k)\},$$
$$\bar{U}_k \triangleq [\; \bar{u}_1^T(k) \quad \bar{u}_2^T(k) \quad \cdots \quad \bar{u}_N^T(k) \;]^T = \bar{\alpha} \mathcal{K}_k \mathcal{C}_k X_k,$$
$$\Xi_k^{(ij)} = \begin{cases} -(\alpha_i(k) - \bar{\alpha})\deg_{in}(i) C(k), & i = j, \\ (\alpha_j(k) - \bar{\alpha}) h_{ij} C(k), & i \ne j. \end{cases}$$

Then, by using the Kronecker product, the closed-loop system resulting from (3.1)-(3.4) can be written as

$$\begin{cases} X_{k+1} = (\mathcal{A}_k + \mathcal{M}_k \mathcal{F}_k \mathcal{G}_k + \mathcal{B}_k \mathcal{K}_k \Xi_k) X_k + \mathcal{B}_k \bar{U}_k + \mathcal{D}_k \eta_k, \\ Z_k = \mathcal{L}_k X_k. \end{cases} \tag{3.5}$$

Remark 3.1 *It should be pointed out that the multi-agent model under consideration in this chapter is very general, which includes the first- or second-order systems as its special cases. For example, the second- and high-order multi-agent systems (see, respectively, [72] and [120]) can be deduced from our model (3.1) with*

$$A(k) = \begin{bmatrix} 1 & t \\ 0 & 1 \end{bmatrix}, \quad A(k) = \begin{bmatrix} 1 & t & 0 \\ 0 & 1 & t \\ 0 & 0 & 1 \end{bmatrix}$$

where t is a sampling period. The concept of "missing measurements" refers to any observation with a nonzero probability that it consists of noise alone. Such a phenomenon does occur in practice when there are certain failures in the measurement, intermittent sensor failures, accidental loss of some collected data, some of the data jammed or coming from a high-noise environment, or high maneuverability of the tracked target. For example, in each of our agents, there might be a case that the individual sensors (for the particular agent itself) experience missing probabilistic measurement due to randomly occurring sensor failures or noise interferences, and our research aims to provide for such a case. Furthermore, there appear to be two main differences between missing measurements and fading channels. First, in terms of the occurrence mechanism, missing measurements usually stem from probabilistic sensor failure or limited bandwidths, while fading channels result mainly from multipath propagation and shadowing from obstacles affecting the wave propagation. Second, in terms of information integrity, missing measurements imply a complete loss of the valid information, and fading channels only degrade the signal quality.

Before proceeding further, we introduce the following definition.

Definition 3.1 *Given the disturbance attenuation level $\gamma > 0$ and the positive definite matrix $W = W^T > 0$, the multi-agent system (3.1) is said to satisfy the $\mathcal{H}_\infty$ consensus performance constraint over a finite horizon $[0, M]$ if the following inequality holds*

$$\mathbb{E}\left\{\sum_{i=1}^{N} \|z_i(k) - z^*(k)\|_{[0,M]}^2\right\}$$
$$< \gamma^2 \sum_{i=1}^{N}\left\{\|w_i(k)\|_{[0,M]}^2 + \|\nu_i(k)\|_{[0,M]}^2\right\} + \gamma^2 \mathbb{E}\left\{\sum_{i=1}^{N} x_i^T(0)Wx_i(0)\right\} \quad (3.6)$$

or, equivalently

$$\mathbb{E}\left\{\|Z_k - \mathcal{I}Z_k\|_{[0,M]}^2\right\} < \gamma^2 \|\eta_k\|_{[0,M]}^2 + \gamma^2 \mathbb{E}\left\{X_0^T(I_N \otimes W)X_0\right\}, \quad (3.7)$$

where $z^*(k) = \frac{1}{N}\sum_{i=1}^{N} z_i(k)$ and $\mathcal{I} = (\frac{1}{N}\mathbf{1}_N\mathbf{1}_N^T) \otimes I_n$.

Remark 3.2 *In the past few years, the consensus problems of multi-agent systems have been well studied over the infinite time horizon, where consensus errors between the outputs of all agents and the group decision value $1/N\sum_{i=0}^{N} z_i(k)$ (see [90] for more details) are required to asymptotically approach zero. However, for the inherently time-varying multi-agent systems addressed in this chapter, we are more interested in the transient consensus behavior over a specified time interval. In other words, we like to examine the transient behavior over a finite horizon rather than the steady-state property over an infinite horizon. For this purpose, we define the notion of* *finite-horizon $\mathcal{H}_\infty$ consensus with a disturbance attenuation level γ so as to characterize the consensus performance requirement over a finite horizon. It is easy to see that, if the constraint (3.6) or (3.7) is met, then the overall consensus error is guaranteed to be bounded. Notice that a similar definition of the bounded $\mathcal{H}_\infty$-synchronization for discrete time-varying stochastic complex networks can be found in [111].*

In this chapter, we aim to design the parameters $K(k)$ $(0 \leq k \leq M)$ for the time-varying controller such that, for all probabilistic missing observations and admissible parameter uncertainties, the $\mathcal{H}_\infty$ consensus performance constraint (3.6) or (3.7) is satisfied over a given finite horizon $[0, M]$ for closed-loop networked multi-agent systems.

3.2 Consensus Performance Analysis

Due to the time-varying nature of the addressed multi-agent system, the performance objectives of the consensus control are actually the boundedness and the disturbance rejection attenuation level of the controlled system over the given finite horizon. In this sense, the parameter uncertainties in (3.5) can be considered as one of the sources of disturbances, and what we need to do is to reject the influence from the disturbances on the controlled output through the prescribed $\mathcal{H}_\infty$ requirement. For this purpose, we rewrite (3.5) as follows:

$$\begin{cases} X_{k+1} = (\mathcal{A}_k + \mathcal{B}_k \mathcal{K}_k \Xi_k) X_k + \mathcal{B}_k \bar{U}_k + \tilde{\mathcal{D}}_k \tilde{\eta}_k, \\ Z_k = \mathcal{L}_k X_k \end{cases} \tag{3.8}$$

where $\tilde{\mathcal{D}}_k = [\mathcal{D}_k \quad \varepsilon_k^{-1} \mathcal{M}_k]$ and $\tilde{\eta}_k = [\eta_k^T \quad (\varepsilon_k \mathcal{F}_k \mathcal{G}_k X_k)^T]^T$. ε_k is a positive function which represents a scaling of the perturbation, and is introduced to provide more flexibility in the controller design (see [55] for more details).

In view of system (3.8), an auxiliary index is defined as follows:

$$\begin{aligned} &\mathbb{E}\left\{ \|Z_k - \mathcal{I} Z_k\|_{[0,M]}^2 \right\} \\ &< \mathbb{E}\left\{ \gamma^2 \|\tilde{\eta}_k\|_{[0,M]}^2 - \gamma^2 \|\varepsilon_k \mathcal{G}_k X_k\|_{[0,M]}^2 \right\} + \gamma^2 \mathbb{E}\left\{ X_0^T (I_N \otimes W) X_0 \right\}. \end{aligned} \tag{3.9}$$

The partitioning matrix

$$\mathcal{P}_k = \begin{pmatrix} P_{11}^k & P_{12}^k & \cdots & P_{1,N}^k \\ P_{21}^k & P_{22}^k & \cdots & P_{2,N}^k \\ \vdots & \vdots & \vdots & \vdots \\ P_{N,1}^k & P_{N,2}^k & \cdots & P_{N,N}^k \end{pmatrix}$$

and the following lemmas will be utilized in the subsequent developments.

Lemma 3.1 *Consider the performance indices defined in (3.7) and (3.9), respectively. If (3.9) is satisfied, then (3.7) is satisfied, too.*

Proof *Denote*

$$
\begin{cases}
\mathcal{J}_1 = \mathbb{E}\left\{\|Z_k - \mathcal{I}Z_k\|^2_{[0,M]}\right\} - \gamma^2\|\eta_k\|^2_{[0,M]} - \gamma^2\mathbb{E}\left\{X_0^T(I_N \otimes W)X_0\right\}, \\
\tilde{\mathcal{J}}_1 = \mathbb{E}\left\{\|Z_k - \mathcal{I}Z_k\|^2_{[0,M]}\right\} - \mathbb{E}\left\{\gamma^2\|\tilde{\eta}_k\|^2_{[0,M]}\right. \\
\qquad \left. - \gamma^2\|\varepsilon_k\mathcal{G}_kX_k\|^2_{[0,M]}\right\} - \gamma^2\mathbb{E}\left\{X_0^T(I_N \otimes W)X_0\right\}.
\end{cases}
\tag{3.10}
$$

It is easily seen that

$$
\begin{aligned}
\mathcal{J}_1 - \tilde{\mathcal{J}}_1 &= -\gamma^2\|\eta_k\|^2_{[0,M]} + \mathbb{E}\left\{\gamma^2\|\tilde{\eta}_k\|^2_{[0,M]} - \gamma^2\|\varepsilon_k\mathcal{G}_kX_k\|^2_{[0,M]}\right\} \\
&= \gamma^2\mathbb{E}\left\{\|\varepsilon_k\mathcal{F}_k\mathcal{G}_kX_k\|^2_{[0,M]} - \|\varepsilon_k\mathcal{G}_kX_k\|^2_{[0,M]}\right\} \\
&= -\gamma^2\mathbb{E}\left\{\|\varepsilon_k[I - \mathcal{F}_k^T\mathcal{F}_k]^{1/2}\mathcal{G}_kX_k\|^2_{[0,M]}\right\} \le 0.
\end{aligned}
$$

Therefore, if $\tilde{\mathcal{J}}_1 < 0$, then $\mathcal{J}_1 < 0$ is true and the proof is complete.

Lemma 3.2 *Let the disturbance attenuation level $\gamma > 0$ and the positive definite matrix W be given. For any nonzero $\{\tilde{\eta}_k\}_{0 \le k \le M} \in \ell_2$, the augmented system (3.8) satisfies the constraint (3.9) if and only if there exist a set of real-valued matrices $\{\mathcal{K}_k\}_{0 \le k \le M}$, a set of positive scales $\{\varepsilon_k\}_{0 \le k \le M}$, and a family of non-negative definite matrices $\{\mathcal{P}_k\}_{0 \le k \le M+1}$ with the final condition $\mathcal{P}_{M+1} = 0$ to the following backward recursive RDE:*

$$
\begin{aligned}
\mathcal{P}_k &= \mathcal{A}_k^T\mathcal{P}_{k+1}\mathcal{A}_k + \bar{\alpha}\mathcal{A}_k^T\mathcal{P}_{k+1}\mathcal{B}_k\mathcal{K}_k\mathcal{C}_k \\
&\quad + \bar{\alpha}\mathcal{C}_k^T\mathcal{K}_k^T\mathcal{B}_k^T\mathcal{P}_{k+1}\mathcal{A}_k + \Pi_k^P + \gamma^2\varepsilon_k^2\mathcal{G}_k^T\mathcal{G}_k \\
&\quad + \bar{\alpha}^2\mathcal{C}_k^T\mathcal{K}_k^T\mathcal{B}_k^T\mathcal{P}_{k+1}\mathcal{B}_k\mathcal{K}_k\mathcal{C}_k + (\mathcal{L}_k - \mathcal{I}\mathcal{L}_k)^T(\mathcal{L}_k - \mathcal{I}\mathcal{L}_k) \\
&\quad + (\mathcal{A}_k + \bar{\alpha}\mathcal{B}_k\mathcal{K}_k\mathcal{C}_k)^T\mathcal{P}_{k+1}\tilde{\mathcal{D}}_k\Phi_k^{-1}\tilde{\mathcal{D}}_k^T\mathcal{P}_{k+1}(\mathcal{A}_k + \bar{\alpha}\mathcal{B}_k\mathcal{K}_k\mathcal{C}_k)
\end{aligned}
\tag{3.11}
$$

subject to

$$
\begin{cases}
\Phi_k = \gamma^2 I - \tilde{\mathcal{D}}_k^T\mathcal{P}_{k+1}\tilde{\mathcal{D}}_k > 0, \\
\mathcal{P}_0 < \gamma^2(I_N \otimes W)
\end{cases}
\tag{3.12}
$$

where $\Pi_k^P \triangleq diag\{\Pi_k^{P(1)}, \Pi_k^{P(2)}, \cdots, \Pi_k^{P(N)}\}$ with $\tilde{\alpha}^* = \bar{\alpha}(1 - \bar{\alpha})$ and

$$
\begin{aligned}
\Pi_k^{P(s)} &= \tilde{\alpha}^*(deg_{in}(s))^2 C^T(k)K^T(k)B^T(k)P_{ss}^{k+1}B(k)K(k)C(k) \\
&\quad - \sum_{i=1, \; i\ne s}^{N} \tilde{\alpha}^* h_{is} deg_{in}(s) C^T(k)K^T(k)B^T(k)(P_{is}^{k+1} + P_{si}^{k+1}) \\
&\quad \times B(k)K(k)C(k) + \sum_{i,j=1, i,j\ne s}^{N} \tilde{\alpha}^* h_{is}h_{js}C^T(k)K^T(k)B^T(k) \\
&\quad \times P_{ij}^{k+1}B(k)K(k)C(k), \; s = 1, \cdots, N.
\end{aligned}
$$

Proof Sufficiency: *For non-negative definite matrices* $\{\mathcal{P}_k\}_{0 \leq k \leq M+1}$ *satisfying the recursive RDE (3.11), it can be obtained from (3.8) that*

$$
\mathbb{E}\left\{ X_{k+1}^T \mathcal{P}_{k+1} X_{k+1} - X_k^T \mathcal{P}_k X_k \right\}
$$

$$
= \mathbb{E}\left\{ \left(X_k^T (\mathcal{A}_k + \mathcal{B}_k \mathcal{K}_k \Xi_k)^T + \bar{U}_k^T \mathcal{B}_k^T + \tilde{\eta}_k^T \tilde{\mathcal{D}}_k^T \right) \mathcal{P}_{k+1} \right.
$$

$$
\left. \times \left((\mathcal{A}_k + \mathcal{B}_k \mathcal{K}_k \Xi_k) X_k + \mathcal{B}_k \bar{U}_k + \tilde{\mathcal{D}}_k \tilde{\eta}_k \right) - X_k^T \mathcal{P}_k X_k \right\}
$$

$$
= \mathbb{E}\left\{ X_k^T \mathcal{A}_k^T \mathcal{P}_{k+1} \mathcal{A}_k X_k + X_k^T \Xi_k^T \mathcal{K}_k^T \mathcal{B}_k^T \mathcal{P}_{k+1} \mathcal{B}_k \mathcal{K}_k \Xi_k X_k \right.
$$

$$
+ 2 X_k^T \mathcal{A}_k^T \mathcal{P}_{k+1} \mathcal{B}_k \bar{U}_k + 2 X_k^T \mathcal{A}_k^T \mathcal{P}_{k+1} \tilde{\mathcal{D}}_k \tilde{\eta}_k + \bar{U}_k^T \mathcal{B}_k^T \mathcal{P}_{k+1} \mathcal{B}_k \bar{U}_k \quad (3.13)
$$

$$
\left. + 2 \bar{U}_k^T \mathcal{B}_k^T \mathcal{P}_{k+1} \tilde{\mathcal{D}}_k \tilde{\eta}_k + \tilde{\eta}_k^T \tilde{\mathcal{D}}_k^T \mathcal{P}_{k+1} \tilde{\mathcal{D}}_k \tilde{\eta}_k - X_k^T \mathcal{P}_k X_k \right\}
$$

$$
= \mathbb{E}\left\{ X_k^T \left(\mathcal{A}_k^T \mathcal{P}_{k+1} \mathcal{A}_k + 2 \bar{\alpha} \mathcal{A}_k^T \mathcal{P}_{k+1} \mathcal{B}_k \mathcal{K}_k \mathcal{C}_k \right.\right.
$$

$$
\left. + \bar{\alpha}^2 \mathcal{C}_k^T \mathcal{K}_k^T \mathcal{B}_k^T \mathcal{P}_{k+1} \mathcal{B}_k \mathcal{K}_k \mathcal{C}_k + \Pi_k^P - \mathcal{P}_k \right) X_k
$$

$$
\left. + 2 X_k^T (\mathcal{A}_k^T + \bar{\alpha} \mathcal{C}_k^T \mathcal{K}_k^T \mathcal{B}_k^T) \mathcal{P}_{k+1} \tilde{\mathcal{D}}_k \tilde{\eta}_k + \tilde{\eta}_k^T \tilde{\mathcal{D}}_k^T \mathcal{P}_{k+1} \tilde{\mathcal{D}}_k \tilde{\eta}_k \right\}.
$$

Let us now deal with the $\mathcal{H}_\infty$ *performance of the system (3.8) for any nonzero disturbance sequence* $\{\tilde{\eta}_k\}_{0 \leq k \leq M} \in \ell_2$. *Firstly, it is easy to see from (3.13) that*

$$
J_1 := \mathbb{E}\left\{ \|Z_k - \mathcal{I} Z_k\|_{[0,M]}^2 - \gamma^2 \|\tilde{\eta}_k\|_{[0,M]}^2 + \gamma^2 \|\varepsilon_k \mathcal{G}_k X_k\|_{[0,M]}^2 \right\}
$$

$$
= \mathbb{E}\left\{ X_0^T \mathcal{P}_0 X_0 - X_{M+1}^T \mathcal{P}_{M+1} X_{M+1} \right\} + \sum_{k=0}^{M} \mathbb{E}\left\{ \|Z_k - \mathcal{I} Z_k\|^2 \right.
$$

$$
- \gamma^2 \|\tilde{\eta}_k\|^2 + \gamma^2 \|\varepsilon_k \mathcal{G}_k X_k\|^2 + X_k^T \left(\mathcal{A}_k^T \mathcal{P}_{k+1} \mathcal{A}_k - \mathcal{P}_k \right.
$$

$$
+ 2 \bar{\alpha} \mathcal{A}_k^T \mathcal{P}_{k+1} \mathcal{B}_k \mathcal{K}_k \mathcal{C}_k + \bar{\alpha}^2 \mathcal{C}_k^T \mathcal{K}_k^T \mathcal{B}_k^T \mathcal{P}_{k+1} \mathcal{B}_k \mathcal{K}_k \mathcal{C}_k + \Pi_k^P \Big) X_k
$$

$$
\left. + 2 X_k^T (\mathcal{A}_k^T + \bar{\alpha} \mathcal{C}_k^T \mathcal{K}_k^T \mathcal{B}_k^T) \mathcal{P}_{k+1} \tilde{\mathcal{D}}_k \tilde{\eta}_k + \tilde{\eta}_k^T \tilde{\mathcal{D}}_k^T \mathcal{P}_{k+1} \tilde{\mathcal{D}}_k \tilde{\eta}_k \right\} \quad (3.14)
$$

$$
= \mathbb{E}\left\{ X_0^T \mathcal{P}_0 X_0 - X_{M+1}^T \mathcal{P}_{M+1} X_{M+1} \right\} + \sum_{k=0}^{M} \mathbb{E}\left\{ X_k^T \left(\mathcal{A}_k^T \mathcal{P}_{k+1} \mathcal{A}_k \right.\right.
$$

$$
+ 2 \bar{\alpha} \mathcal{A}_k^T \mathcal{P}_{k+1} \mathcal{B}_k \mathcal{K}_k \mathcal{C}_k + \bar{\alpha}^2 \mathcal{C}_k^T \mathcal{K}_k^T \mathcal{B}_k^T \mathcal{P}_{k+1} \mathcal{B}_k \mathcal{K}_k \mathcal{C}_k + \Pi_k^P
$$

$$
+ (\mathcal{L}_k - \mathcal{I} \mathcal{L}_k)^T (\mathcal{L}_k - \mathcal{I} \mathcal{L}_k) + \gamma^2 \varepsilon_k^2 \mathcal{G}_k^T \mathcal{G}_k - \mathcal{P}_k \Big) X_k
$$

$$
+ 2 X_k^T (\mathcal{A}_k^T + \bar{\alpha} \mathcal{C}_k^T \mathcal{K}_k^T \mathcal{B}_k^T) \mathcal{P}_{k+1} \tilde{\mathcal{D}}_k \tilde{\eta}_k
$$

$$
\left. - \tilde{\eta}_k^T (\gamma^2 I - \tilde{\mathcal{D}}_k^T \mathcal{P}_{k+1} \tilde{\mathcal{D}}_k) \tilde{\eta}_k \right\}.
$$

Then, by applying the completing squares method, we have

$$
\begin{aligned}
J_1 = \sum_{k=0}^{M} \mathbb{E}\Big\{ & X_k^T \Big(\mathcal{A}_k^T \mathcal{P}_{k+1} \mathcal{A}_k + 2\bar{\alpha} \mathcal{A}_k^T \mathcal{P}_{k+1} \mathcal{B}_k \mathcal{K}_k \mathcal{C}_k \\
& + \gamma^2 \varepsilon_k^2 \mathcal{G}_k^T \mathcal{G}_k + \bar{\alpha}^2 \mathcal{C}_k^T \mathcal{K}_k^T \mathcal{B}_k^T \mathcal{P}_{k+1} \mathcal{B}_k \mathcal{K}_k \mathcal{C}_k + \Pi_k^P \\
& + (\mathcal{L}_k - \mathcal{I}\mathcal{L}_k)^T (\mathcal{L}_k - \mathcal{I}\mathcal{L}_k) - \mathcal{P}_k \\
& + (\mathcal{A}_k + \bar{\alpha} \mathcal{B}_k \mathcal{K}_k \mathcal{C}_k)^T \mathcal{P}_{k+1} \tilde{\mathcal{D}}_k \Phi_k^{-1} \tilde{\mathcal{D}}_k^T \mathcal{P}_{k+1} (\mathcal{A}_k + \bar{\alpha} \mathcal{B}_k \mathcal{K}_k \mathcal{C}_k) \Big) X_k \\
& - (\tilde{\eta}_k - \tilde{\eta}_k^*)^T \Phi_k (\tilde{\eta}_k - \tilde{\eta}_k^*) \Big\} \\
& + \mathbb{E}\big\{ X_0^T \mathcal{P}_0 X_0 - X_{M+1}^T \mathcal{P}_{M+1} X_{M+1} \big\}
\end{aligned}
\tag{3.15}
$$

where

$$
\tilde{\eta}_k^* = \Phi_k^{-1} \tilde{\mathcal{D}}_k^T \mathcal{P}_{k+1} (\mathcal{A}_k + \bar{\alpha} \mathcal{B}_k \mathcal{K}_k \mathcal{C}_k) X_k.
\tag{3.16}
$$

Since $\Phi_k > 0$ and $\mathcal{P}_0 < \gamma^2 (I_N \otimes W)$, it follows from the final condition $\mathcal{P}_{M+1} = 0$ that

$$
\begin{aligned}
\tilde{\mathcal{J}}_1 &= \mathbb{E}\Big\{ \|Z_k - \mathcal{I}Z_k\|_{[0,M]}^2 - \gamma^2 \|\tilde{\eta}_k\|_{[0,M]}^2 + \gamma^2 \|\varepsilon_k \mathcal{G}_k X_k\|_{[0,M]}^2 \Big\} \\
&\quad - \gamma^2 \mathbb{E}\big\{ X_0^T (I_N \otimes W) X_0 \big\} \\
&\leq \mathbb{E}\Big\{ \|Z_k - \mathcal{I}Z_k\|_{[0,M]}^2 - \gamma^2 \|\tilde{\eta}_k\|_{[0,M]}^2 + \gamma^2 \|\varepsilon_k \mathcal{G}_k X_k\|_{[0,M]}^2 \Big\} \\
&\quad - \mathbb{E}\big\{ X_0^T \mathcal{P}_0 X_0 \big\} \\
&= J_1 - \mathbb{E}\big\{ X_0^T \mathcal{P}_0 X_0 \big\} \leq -\mathbb{E}\Big\{ \sum_{k=0}^{M} (\tilde{\eta}_k - \tilde{\eta}_k^*)^T \Phi_k (\tilde{\eta}_k - \tilde{\eta}_k^*) \Big\} \leq 0.
\end{aligned}
\tag{3.17}
$$

It is not difficult to verify that, if $X_0 \neq 0$, then $\tilde{\mathcal{J}}_1 < 0$ due to $\mathcal{P}_0 < \gamma^2 (I_N \otimes W)$. On the other hand, if $X_0 = 0$, substituting $\tilde{\eta}_k^* = \Phi_k^{-1} \tilde{\mathcal{D}}_k^T \mathcal{P}_{k+1} (\mathcal{A}_k + \bar{\alpha} \mathcal{B}_k \mathcal{K}_k \mathcal{C}_k) X_k$ into the system dynamics (3.8) leads to $X_k \equiv 0$ for $\forall k \in [0, M]$, and therefore $\tilde{\eta}_k^* \equiv 0$ for $\forall k \in [0, M]$. It can now be concluded that $\tilde{\mathcal{J}}_1 = 0$ if and only if $\tilde{\eta}_k = \tilde{\eta}_k^* = 0$ for $\forall k \in [0, M]$, which contradicts the assumption of $\tilde{\eta} \neq 0$. It follows that there exists at least one $\tilde{\eta}_k$ that cannot be equal to $\tilde{\eta}_k^*$ in (3.8) or, in other words, $\tilde{\mathcal{J}}_1 < 0$ for any nonzero $\tilde{\eta} \in \ell_2$. We now reach the conclusion that (3.17) holds as a strict inequality, which means that the pre-specified auxiliary index (3.9) is satisfied.

Necessity. *The proof follows directly from that of Lemma 2.3 in Chapter 2 and is therefore omitted.*

To evaluate the impact on the solvability of the RDE (3.11) from different $\bar{\alpha}$ and γ, let us denote $\mathcal{U}_k \triangleq \bar{\alpha} \mathcal{B}_k \mathcal{K}_k \mathcal{C}_k$. Therefore, (3.11) can be rewritten as

$$
\begin{aligned}
\mathcal{P}_k = {} & (\mathcal{A}_k + \mathcal{U}_k)^T \mathcal{P}_{k+1} \Big(\mathcal{P}_{k+1}^{-1} + \tilde{\mathcal{D}}_k \Phi_k^{-1} \tilde{\mathcal{D}}_k^T \Big) \mathcal{P}_{k+1} (\mathcal{A}_k + \mathcal{U}_k) \\
& + \frac{1 - \bar{\alpha}}{\bar{\alpha}} \Pi_k^P + \gamma^2 \varepsilon_k^2 \mathcal{G}_k^T \mathcal{G}_k + (\mathcal{L}_k - \mathcal{I}\mathcal{L}_k)^T (\mathcal{L}_k - \mathcal{I}\mathcal{L}_k).
\end{aligned}
$$

We have two observations here. 1) With increased γ, the constraint (3.12) is more easily satisfied, which enhances the feasibility of the RDR (3.11). Obviously, in engineering practice, we like to minimize the disturbance attenuation level γ, and this constitutes a tradeoff between the performance and the feasibility. 2) With decreased $\bar{\alpha}$, the missing probability becomes larger, and the norm of the term $\frac{1-\bar{\alpha}}{\bar{\alpha}}\Pi_k^P$ increases. Accordingly, the norm of the possible solution $\mathcal{P}_k$ increases and such an increase will accumulate with the recursion leading to possible divergence of the algorithm. This gives another tradeoff between the missing probability and the performance.

So far, we have analyzed the $\mathcal{H}_\infty$ performance in terms of the solvability to a backward Riccati equation in Lemma 3.2. In the next stage, we shall proceed to tackle the design problem of the consensus controller (3.4) such that the closed-loop system (3.5) satisfies the $\mathcal{H}_\infty$ performance requirement (3.6) or (3.7).

Theorem 3.1 *Let the disturbance attenuation level $\gamma > 0$ and the positive definite matrix W be given. There exist parameters $\{K(k)\}_{0 \le k \le M}$ for the control protocol (3.4) such that the closed-loop system (3.5) satisfies the $\mathcal{H}_\infty$ performance constraint (3.6) or (3.7) for any nonzero disturbance sequence $\{\eta_k\}_{0 \le k \le M} \in \ell_2$, if the recursive RDE (3.11) has a solution $(\varepsilon_k, \mathcal{P}_k, \mathcal{K}_k)$ $(0 \le k \le M)$ satisfying (3.12) and the following recursive RDE*

$$
\begin{cases}
\mathcal{Q}_k = \Delta_k^T \mathcal{Q}_{k+1} \Delta_k + \Pi_k^Q \\
\qquad + (\mathcal{L}_k - \mathcal{IL}_k)^T (\mathcal{L}_k - \mathcal{IL}_k) - \Delta_k^T \mathcal{Q}_{k+1} \mathcal{B}_k \Psi_k^{-1} \mathcal{B}_k^T \mathcal{Q}_{k+1}^T \Delta_k, \quad (3.18) \\
\mathcal{Q}_{M+1} = 0
\end{cases}
$$

has a solution $(\mathcal{Q}_k, \mathcal{K}_k)$ $(0 \le k \le M)$ satisfying

$$
\begin{cases}
\Psi_k = \mathcal{B}_k^T \mathcal{Q}_{k+1} \mathcal{B}_k + I > 0, \\
\mathcal{K}_k = \arg\min\limits_{\mathcal{K}_k} norm\left(\bar{\alpha}\mathcal{K}_k \mathcal{C}_k + (\mathcal{B}_k^T \mathcal{Q}_{k+1} \mathcal{B}_k + I)^{-1} \mathcal{B}_k^T \mathcal{Q}_{k+1}^T \Delta_k\right)
\end{cases} \quad (3.19)
$$

where "arg" is short for "argument", which means the value of $\mathcal{K}_k$ minimizing the norm, and Π_k^Q and Δ_k are defined by

$$
\Pi_k^Q := diag\left\{\Pi_k^{Q(1)}, \Pi_k^{Q(2)}, \cdots, \Pi_k^{Q(N)}\right\},
$$

$$
\Delta_k = \mathcal{A}_k + \tilde{\mathcal{D}}_k \Phi_k^{-1} \tilde{\mathcal{D}}_k^T \mathcal{P}_{k+1}(\mathcal{A}_k + \bar{\alpha}\mathcal{B}_k \mathcal{K}_k \mathcal{C}_k)
$$

with

$$
\Pi_k^{Q(s)} = \tilde{\alpha}^* (deg_{in}(s))^2 C^T(k) K^T(k) B^T(k) Q_{ss}^{k+1} B(k) K(k) C(k)
$$

$$
- \sum_{i=1, \, i\neq s}^N \tilde{\alpha}^* h_{is} deg_{in}(s) C^T(k) K^T(k) B^T(k)(Q_{is}^{k+1} + Q_{si}^{k+1})
$$

$$
\times B(k) K(k) C(k) + \sum_{i,j=1, i, j\neq s}^N \tilde{\alpha}^* h_{is} h_{js} C^T(k) K^T(k) B^T(k)
$$

$$
\times Q_{ij}^{k+1} B(k) K(k) C(k), \quad s = 1, \cdots, N.
$$

Here, Q_{ij}^{k+1} $(i,j = 1,2,\cdots,N)$ are block elements of the matrix $\mathcal{Q}_{k+1}$ (same as the form $\mathcal{P}_{k+1}$).

Proof *First, it follows easily from Lemma 3.2 that the system (3.8) satisfies the pre-specified auxiliary index (3.9) when there exists a solution $\mathcal{P}_k$ to (3.11) such that $\Phi_k > 0$ and $\mathcal{P}_0 < \gamma^2(I_N \otimes W)$. Meanwhile, according to (3.16), the worst-case disturbance can be expressed as $\tilde{\eta}_k^* = \Phi_k^{-1}\tilde{\mathcal{D}}_k^T\mathcal{P}_{k+1}(\mathcal{A}_k + \bar{\alpha}\mathcal{B}_k\mathcal{K}_k\mathcal{C}_k)X_k$, and we know from Lemma 3.1 that the pre-specified $\mathcal{H}_\infty$ performance (3.6) or (3.7) is satisfied.*

In what follows, we shall propose an approach for determining the control protocol parameter $K(k)$ under the situation of worst-case disturbance. Firstly, a cost functional is defined as

$$\mathcal{J}_2 = \mathbb{E}\left\{ \sum_{i=1}^N \|z_i(k) - z^*(k)\|_{[0,M]}^2 + \sum_{i=1}^N \|\bar{u}_i(k)\|_{[0,M]}^2 \right\} \tag{3.20}$$

or, equivalently

$$\mathcal{J}_2 = \mathbb{E}\left\{ \|Z_k - \mathcal{I}Z_k\|_{[0,M]}^2 + \|\bar{U}_k\|_{[0,M]}^2 \right\}. \tag{3.21}$$

For non-negative definite matrices $\{\mathcal{Q}_k\}_{0 \le k \le M+1}$ satisfying the iterative RDE (3.18), it follows from (3.8) that

$$\mathbb{E}\left\{ X_{k+1}^T \mathcal{Q}_{k+1} X_{k+1} - X_k^T \mathcal{Q}_k X_k \right\}$$

$$= \mathbb{E}\Big\{ X_k^T \left(\mathcal{A}_k + \mathcal{B}_k\mathcal{K}_k\Xi_k + \tilde{\mathcal{D}}_k\Phi_k^{-1}\tilde{\mathcal{D}}_k^T\mathcal{P}_{k+1}(\mathcal{A}_k + \bar{\alpha}\mathcal{B}_k\mathcal{K}_k\mathcal{C}_k) \right)^T$$

$$\times \mathcal{Q}_{k+1}\left(\mathcal{A}_k + \mathcal{B}_k\mathcal{K}_k\Xi_k + \tilde{\mathcal{D}}_k\Phi_k^{-1}\tilde{\mathcal{D}}_k^T\mathcal{P}_{k+1}(\mathcal{A}_k + \bar{\alpha}\mathcal{B}_k\mathcal{K}_k\mathcal{C}_k) \right) X_k$$

$$+ 2X_k^T \left(\mathcal{A}_k + \tilde{\mathcal{D}}_k\Phi_k^{-1}\tilde{\mathcal{D}}_k^T\mathcal{P}_{k+1}(\mathcal{A}_k + \bar{\alpha}\mathcal{B}_k\mathcal{K}_k\mathcal{C}_k) \right)^T \mathcal{Q}_{k+1}\mathcal{B}_k\bar{U}_k$$

$$- X_k^T \mathcal{Q}_k X_k + \bar{U}_k^T \mathcal{B}_k^T \mathcal{Q}_{k+1}\mathcal{B}_k\bar{U}_k \Big\}$$

$$= \mathbb{E}\Big\{ X_k^T \Delta_k^T \mathcal{Q}_{k+1}\Delta_k X_k + X_k^T(\Pi_k^Q - \mathcal{Q}_k)X_k$$

$$+ 2X_k^T \Delta_k^T \mathcal{Q}_{k+1}\mathcal{B}_k\bar{U}_k + \bar{U}_k^T \mathcal{B}_k^T \mathcal{Q}_{k+1}\mathcal{B}_k\bar{U}_k \Big\}. \tag{3.22}$$

Similar to the proof of Lemma 3.2, one has

$$\mathcal{J}_2 = \sum_{k=0}^M \mathbb{E}\Big\{ \|Z_k - \mathcal{I}Z_k\|^2 + \|\bar{U}_k\|^2 + X_k^T \Delta_k^T \mathcal{Q}_{k+1}\Delta_k X_k$$

$$+ X_k^T(\Pi_k^Q - \mathcal{Q}_k)X_k + 2X_k^T \Delta_k^T \mathcal{Q}_{k+1}\mathcal{B}_k\bar{U}_k + \bar{U}_k^T \mathcal{B}_k^T \mathcal{Q}_{k+1}\mathcal{B}_k\bar{U}_k \Big\}$$

$$+ \mathbb{E}\left\{ X_0^T \mathcal{Q}_0 X_0 - X_{M+1}^T \mathcal{Q}_{M+1} X_{M+1} \right\}$$

$$= \sum_{k=0}^M \mathbb{E}\Big\{ X_k^T \left(\Delta_k^T \mathcal{Q}_{k+1}\Delta_k + \Pi_k^Q - \mathcal{Q}_k + (\mathcal{L}_k - \mathcal{I}\mathcal{L}_k)^T(\mathcal{L}_k - \mathcal{I}\mathcal{L}_k) \right) X_k$$

$$+ 2X_k^T \Delta_k^T \mathcal{Q}_{k+1} \mathcal{B}_k \bar{U}_k + \bar{U}_k^T (\mathcal{B}_k^T \mathcal{Q}_{k+1} \mathcal{B}_k + I) \bar{U}_k \Big\}$$

$$+ \mathbb{E} \left\{ X_0^T \mathcal{Q}_0 X_0 - X_{M+1}^T \mathcal{Q}_{M+1} X_{M+1} \right\}. \tag{3.23}$$

By the completing squares method again, it follows that

$$
\begin{aligned}
\mathcal{J}_2 = \sum_{k=0}^{M} \mathbb{E} \Big\{ & X_k^T \Big(\Delta_k^T \mathcal{Q}_{k+1} \Delta_k + \Pi_k^Q - \mathcal{Q}_k + (\mathcal{L}_k - \mathcal{IL}_k)^T (\mathcal{L}_k - \mathcal{IL}_k) \\
& - \Delta_k^T \mathcal{Q}_{k+1} \mathcal{B}_k (\mathcal{B}_k^T \mathcal{Q}_{k+1} \mathcal{B}_k + I)^{-1} \mathcal{B}_k^T \mathcal{Q}_{k+1}^T \Delta_k \Big) X_k \\
& + (\bar{U}_k - \bar{U}_k^*)^T (\mathcal{B}_k^T \mathcal{Q}_{k+1} \mathcal{B}_k + I)(\bar{U}_k - \bar{U}_k^*) \Big\}
\end{aligned}
\tag{3.24}
$$

$$+ \mathbb{E} \left\{ X_0^T \mathcal{Q}_0 X_0 - X_{M+1}^T \mathcal{Q}_{M+1} X_{M+1} \right\}$$

where $\bar{U}_k^* = -(\mathcal{B}_k^T \mathcal{Q}_{k+1} \mathcal{B}_k + I)^{-1} \mathcal{B}_k^T \mathcal{Q}_{k+1}^T \Delta_k X_k$.

Under the final condition $\mathcal{Q}_{M+1} = 0$, *the best choice of* $\mathcal{K}_k$ *suppressing the cost of* $\mathcal{J}_2$ *is the* $\mathcal{K}_k$ *that satisfies (3.18) and (3.19), and this ends the proof.*

Remark 3.3 *Note that the parameter uncertainties in system (3.1) are considered as an additional perturbation* $\tilde{\eta}_k$, *and the newly introduced parameter* ε_k *provides more flexibility for the consensus controller design. As can be seen from (3.10),* $\tilde{\mathcal{J}}_1$ *is an upper bound of* $\mathcal{J}_1$, *i.e.,* $\tilde{\mathcal{J}}_1 = \max \mathcal{J}_1$. *Therefore, the performance index (3.9) can be used to replace (3.7) in the* $\mathcal{H}_\infty$ *consensus control problem. Furthermore, it can be observed that the functional* $\tilde{\mathcal{J}}_1$ *is associated with the* $\mathcal{H}_\infty$-*constraint criterion, whereas the functional* $\mathcal{J}_2$ *is related to the energies of the control input as well as the consensus error output of the system (3.8) in the mean-square sense (or* $\mathcal{H}_2$-*type criterion). Because of the fixed topology of multi-agent systems and the existence of missing measurements, it is extremely difficult to design a globally optimal control protocol via minimizing the cost* $\mathcal{J}_2$ *in the mean-square sense. A seemingly more realistic way is to seek a sub-optimal controller in terms of (3.19) in order to realize the* $\mathcal{H}_\infty$-*constraint criterion.*

3.3 $\mathcal{H}_\infty$ Controller Design

Notice that the optimization problem in (3.19) is not easily solved because the optimal cost is actually a cubic polynomial about $\mathcal{K}_k$. In the following, we proceed to deal with this problem. Denoting $\tilde{\nu}_i(k) \triangleq \lambda_k^{-1} K(k) E(k) \nu_i(k)$ $(i = 1, 2, \cdots, N)$, one has $\tilde{\nu}_i(k) \in \ell_2([0, M]; \mathbb{R}^p)$ due to $\nu_i(k) \in \ell_2([0, M]; \mathbb{R}^m)$, where $\lambda_k > 0$ is introduced to offer more flexibility. Next, rewrite (3.8) as follows:

$$
\begin{cases}
X_{k+1} = (\mathcal{A}_k + \mathcal{B}_k \mathcal{K}_k \Xi_k) X_k + \mathcal{B}_k \bar{U}_k + \mathcal{R}_k \tilde{\xi}_k, \\
Z_k = \mathcal{L}_k X_k
\end{cases}
\tag{3.25}
$$

where

$$\xi_k = [\ w_1^T(k) \quad w_2^T(k) \quad \cdots \quad w_N^T(k) \quad \tilde{\nu}_1^T(k) \quad \tilde{\nu}_2^T(k) \quad \cdots \quad \tilde{\nu}_N^T(k)\ ]^T,$$

$$\tilde{\xi}_k = [\xi_k^T \quad (\varepsilon_k \mathcal{F}_k \mathcal{G}_k X_k)^T]^T, \quad \bar{\mathcal{D}}_k = [I_N \otimes D(k) \quad \lambda_k \mathcal{B}_k \bar{\mathcal{E}}_k],$$

$$\bar{\mathcal{E}}_k^{(i)} = [\ h_{i1}I \quad \cdots \quad h_{i,i-1}I \quad -\deg_{in}(i)I \quad h_{i,i+1}I \quad \cdots \quad h_{iN}I\ ],$$

$$\bar{\mathcal{E}}_k = \left[\ \bar{\mathcal{E}}_k^{(1)T} \quad \bar{\mathcal{E}}_k^{(2)T} \quad \cdots \quad \bar{\mathcal{E}}_k^{(N)T}\ \right]^T, \quad \mathcal{R}_k = [\bar{\mathcal{D}}_k \quad \varepsilon_k^{-1} \mathcal{M}_k].$$

It is easy to see that, if the system (3.25) satisfies the constraint (3.9), so does the system (3.8). According to Theorem 3.1, we have the following result.

Theorem 3.2 *Let the disturbance attenuation level $\gamma > 0$ and the positive definite matrix W be given. There exist parameters $\{K(k)\}_{0 \le k \le M}$ for the control protocol (3.4) such that the closed-loop system (3.5) satisfies the $\mathcal{H}_\infty$ performance constraint (3.6) (or (3.7)) for any nonzero disturbance sequence $\{\eta_k\}_{0 \le k \le M} \in \ell_2$, if the following coupled backward recursive RDEs have solutions $(\varepsilon_k, \lambda_k, \mathcal{P}_k, \mathcal{K}_k)$ and $(\mathcal{Q}_k, \mathcal{K}_k)$ $(0 \le k \le M)$:*

$$\begin{cases} \mathcal{P}_k = \mathcal{A}_k^T \mathcal{P}_{k+1} \mathcal{A}_k + \bar{\alpha} \mathcal{A}_k^T \mathcal{P}_{k+1} \mathcal{B}_k \mathcal{K}_k \mathcal{C}_k + \bar{\alpha} \mathcal{C}_k^T \mathcal{K}_k^T \mathcal{B}_k^T \mathcal{P}_{k+1} \mathcal{A}_k \\ \qquad + \bar{\alpha}^2 \mathcal{C}_k^T \mathcal{K}_k^T \mathcal{B}_k^T \mathcal{P}_{k+1} \mathcal{B}_k \mathcal{K}_k \mathcal{C}_k + \gamma^2 \varepsilon_k^2 \mathcal{G}_k^T \mathcal{G}_k \\ \qquad + (\mathcal{L}_k - \mathcal{IL}_k)^T (\mathcal{L}_k - \mathcal{IL}_k) + \Pi_k^P + (\mathcal{A}_k \\ \qquad + \bar{\alpha} \mathcal{B}_k \mathcal{K}_k \mathcal{C}_k)^T \mathcal{P}_{k+1} \mathcal{R}_k \bar{\Phi}_k^{-1} \mathcal{R}_k^T \mathcal{P}_{k+1} (\mathcal{A}_k + \bar{\alpha} \mathcal{B}_k \mathcal{K}_k \mathcal{C}_k), \\ \bar{\Phi}_k = \gamma^2 I - \mathcal{R}_k^T \mathcal{P}_{k+1} \mathcal{R}_k > 0, \\ \mathcal{P}_k \ge 0, \quad \mathcal{P}_{M+1} = 0, \quad \mathcal{P}_0 < \gamma^2 (I_N \otimes W), \end{cases} \tag{3.26}$$

$$\begin{cases} \mathcal{Q}_k = \bar{\Delta}_k^T \mathcal{Q}_{k+1} \bar{\Delta}_k + \Pi_k^Q + (\mathcal{L}_k - \mathcal{IL}_k)^T (\mathcal{L}_k - \mathcal{IL}_k) \\ \qquad - \bar{\Delta}_k^T \mathcal{Q}_{k+1} \mathcal{B}_k \Psi_k^{-1} \mathcal{B}_k^T \mathcal{Q}_{k+1}^T \bar{\Delta}_k, \\ \Psi_k = \mathcal{B}_k^T \mathcal{Q}_{k+1} \mathcal{B}_k + I > 0, \quad \mathcal{Q}_{M+1} = 0 \end{cases} \tag{3.27}$$

where

$$\begin{cases} K(k) = [\ \mathcal{S}_1^k \quad \mathcal{S}_2^k \quad \cdots \quad \mathcal{S}_N^k\ ][\ \mathcal{C}_k^{(1)} \quad \mathcal{C}_k^{(2)} \quad \cdots \quad \mathcal{C}_k^{(N)}\ ]^\dagger, \\ \mathscr{S}_k = -\mathscr{A}_k^\dagger \mathscr{B}_k \triangleq [\ \mathcal{S}_1^{kT} \quad \mathcal{S}_2^{kT} \quad \cdots \quad \mathcal{S}_N^{kT}\ ]^T, \\ \mathscr{A}_k = \bar{\alpha}(I + \Psi_k^{-1} \mathcal{B}_k^T \mathcal{Q}_{k+1}^T \mathcal{R}_k \bar{\Phi}_k^{-1} \mathcal{R}_k^T \mathcal{P}_{k+1} \mathcal{B}_k), \\ \mathscr{B}_k = \Psi_k^{-1} \mathcal{B}_k^T \mathcal{Q}_{k+1}^T (I + \mathcal{R}_k \bar{\Phi}_k^{-1} \mathcal{R}_k^T \mathcal{P}_{k+1}) \mathcal{A}_k, \\ \bar{\Delta}_k = \mathcal{A}_k + \mathcal{R}_k \bar{\Phi}_k^{-1} \mathcal{R}_k^T \mathcal{P}_{k+1} (\mathcal{A}_k + \bar{\alpha} \mathcal{B}_k \mathcal{K}_k \mathcal{C}_k). \end{cases} \tag{3.28}$$

Here, Q_{ij}^{k+1} and $\mathcal{S}_i^k$ $(i, j = 1, 2, \cdots, N)$ are block elements of the matrix $\mathcal{Q}_{k+1}$ and $\mathscr{S}_k$, respectively, where $\mathcal{Q}_{k+1}$ has the same structure as the form $\mathcal{P}_{k+1}$.

Proof *According to Theorem 3.1, the closed-loop system (3.5) satisfies* the $\mathcal{H}_\infty$ *performance constraint (3.6) (or (3.7)) for any nonzero disturbance sequence* $\{\eta_k\}_{0 \le k \le M} \in \ell_2$, *if*

$$\mathcal{K}_k = \arg \min_{\mathcal{K}_k} norm\left(\bar{a}\mathcal{K}_k \mathcal{C}_k + (\mathcal{B}_k^T \mathcal{Q}_{k+1} \mathcal{B}_k + I)^{-1} \mathcal{B}_k^T \mathcal{Q}_{k+1}^T \bar{\Delta}_k\right). \quad (3.29)$$

Moreover, one has

$$\mathcal{K}_k = \arg \min_{\mathcal{K}_k} norm\left(\mathscr{A}_k \mathcal{K}_k \mathcal{C}_k + \mathscr{B}_k\right) \quad (3.30)$$

and

$$K(k) = \arg \min_{K(k)} norm\left(\bar{a}K(k)[\ \mathcal{C}_k^{(1)} \ \cdots \ \mathcal{C}_k^{(N)} \ ] + [\ \mathcal{S}_1^k \ \cdots \ \mathcal{S}_N^k \ ]\right).$$

$$(3.31)$$

According to the Moore-Penrose pseudoinverse, the protocol parameter $K(k)$ can be easily obtained.

Remark 3.4 *The least-squares problem for* $\min \|\mathscr{A}_k X \mathscr{B}_k - \mathscr{C}_k\|$ *of the matrix equation* $\mathscr{A}_k X \mathscr{B}_k = \mathscr{C}_k$ *has been studied in depth and the corresponding results have been widely applied in engineering. Generally, the least-square solution in a minimal-norm sense can be easily found by using the Moore-Penrose pseudoinverse or iterative algorithms. In this chapter, due to the special structure of* $\mathcal{K}_k$, *we have to modify the algorithm by using the partitioning matrix method such that the parameter expression can be acquired.*

Noting that the controller gain matrices are involved in the proposed coupled RDEs, by means of Theorem 3.2, we can summarize the consensus protocol design algorithm as follows.

Design algorithm of distributed controller

Step 1. Set $k = M$. Then $\mathcal{P}_{M+1} = \mathcal{Q}_{M+1} = 0$ are available.

Step 2. Calculate $\bar{\Phi}_k$ and Ψ_k via (3.26) and (3.27), and then obtain the controller gain matrix $K(k)$ by (3.28) and the matrix's generalized inverse.

Step 3. Calculate $\Theta_k^{(1)}$, $\Theta_k^{(2)}$, $\Xi_k^{(1)}$ and $\Xi_k^{(2)}$ in (3.21) and Lemma 3.1, respectively.

Step 4. Solve the first equations of (3.26) and (3.27) to get $\mathcal{P}_k$ and $\mathcal{Q}_k$, respectively.

Step 5. If $\mathcal{P}_k \ge 0$, then $K(k)$ are suitable estimator parameters, and go to the next step, else go to *Step 7*.

Step 6. If $k \ne 0$, set $k = k - 1$ and go to *Step 2*, else go to the next step.

Step 7. If either $\mathcal{P}_k \ge 0$ (for some $k \ne 0$) or $\mathcal{P}_0 < \gamma^2 W$ does not hold, then this algorithm is infeasible. Stop.

Remark 3.5 *In real-world systems, the length M of the time-horizon is typically determined by the transient performance requirement of the systems under consideration. Generally speaking, the horizon length should reflect the tradeoff between the transient and steady-state performance constraints where a larger horizon places more emphasis on the steady-state performance. On the other hand, the topological structure plays a crucial role in the dynamic behavior of multi-agents, where the spanning-tree serves as a frequently used structure. Traditionally, for linear time-invariant multi-agent systems, the second eigenvalue of the Laplacian matrix (reflecting the topological structure) is usually employed to analyze the consensus performance. Unfortunately, such an approach is no longer valid for the consensus problem for the systems considered in this chapter exhibiting a time-varying nature, parameter uncertainties, and probabilistic missing measurement. Accordingly, a novel RDE approach is proposed to overcome the emerging difficulties. In the RDE (3.11), the topological information is explicitly included in matrices $\mathcal{C}_k$, $\mathcal{E}_k$ and Π_k^P. Compared to the single-loop case (i.e., one agent), our algorithm development exhibits the following distinguishing characteristics: 1) the Kronecker product is employed for facilitating the derivation in a more compact way; 2) the introduced parameters ε_k and λ_k, which provide extra design freedom, are utilized to deal with parameter uncertainties and handle $K(k)E(k)\nu_i(k)$ $(i = 1, 2, \cdots, N)$; and 3) the Moore-Penrose pseudoinverse is used to reflect the topology structure in order to obtain the corresponding controller gain in Theorem 3.2.*

Remark 3.6 *In this chapter, we examine how the missing measurements and parameter uncertainties influence the performance of the consensus control for multi-agent systems with a given topology. In Theorem 3.1, all the system parameters, the occurrence probability for missing measurements, as well as the norm bound for parameter uncertainties are reflected in the backward recursive RDEs. Also, an efficient algorithm CPD is proposed to solve the addressed consensus controller design problem. Due to the recursive nature of the algorithm, the proposed design procedure is suitable for online application without the need of increasing the problem size. Compared to existing literature, the model considered in this chapter is more general and covers both the missing measurements and parameter uncertainties. The main technical contributions lie in that 1) the proposed consensus control approach is capable of handling linearization error, missing measurements, and other non-Gaussian disturbances which do not have exact noise statistics; and 2) a new "stochastic"-version of the bounded real lemma is developed specifically for handling the $\mathcal{H}_\infty$ consensus control of the multi-agent systems.*

3.4 Simulation Examples

In this section, we present a simulation example to illustrate the effectiveness of the proposed consensus protocol design scheme for multi-agent systems with simultaneous presence of missing measurements, parameter uncertainties, as well as external disturbances.

Consider the system (3.1) with

$$A(k) = \begin{bmatrix} 1.02 + 0.15\cos(k) & 0.30 & -0.40 \\ 0.20 & -0.68 & 0.21 \\ 0.05 & 0.34 & 0.55 \end{bmatrix},$$

$$B(k) = \begin{bmatrix} 0.85 \\ -0.50 \\ 0.70 \end{bmatrix}, \quad D(k) = \begin{bmatrix} 0.04 \\ -0.02 \\ 0.02 \end{bmatrix},$$

$$C(k) = \begin{bmatrix} 0.60 & 0.40 & 0 \\ 0 & 0.15 & 0.50 \end{bmatrix}, \quad E(k) = \begin{bmatrix} 0.04 \\ -0.04 \end{bmatrix},$$

$$L(k) = \begin{bmatrix} 0.80 & -0.60 & 0.80 \end{bmatrix}, \quad G(k) = \begin{bmatrix} 0.02 & 0.05 & -0.08 \end{bmatrix},$$

$$M_i(k) = \begin{cases} \begin{bmatrix} -0.10 & 0.04 & 0.08 \\ 0.08 & -0.02 & 0.04 \end{bmatrix}, & i = 1, 3, 5, \\ & i = 2, 4, 6. \end{cases}$$

Suppose that there are six agents with two directed communication graphs $\mathscr{G}$ shown in Figs. 3.1-3.2, where the topology in Fig. 3.2 corresponds to an uncooperative leader, and the associated adjacency matrices are

$$\mathscr{H}_1 = \begin{bmatrix} 0 & 0 & 0 & 1 & 0 & 1 \\ 1 & 0 & 0 & 0 & 0 & 0 \\ 0 & 0 & 0 & 0 & 1 & 0 \\ 0 & 0 & 1 & 0 & 0 & 0 \\ 0 & 1 & 0 & 1 & 0 & 0 \\ 0 & 1 & 1 & 0 & 0 & 0 \end{bmatrix}, \quad \mathscr{H}_2 = \begin{bmatrix} 0 & 0 & 0 & 0 & 0 & 0 \\ 0 & 0 & 1 & 1 & 0 & 0 \\ 1 & 0 & 0 & 0 & 0 & 1 \\ 1 & 0 & 0 & 0 & 0 & 0 \\ 0 & 1 & 0 & 0 & 0 & 0 \\ 1 & 0 & 0 & 0 & 1 & 0 \end{bmatrix}.$$

In this example, the $\mathcal{H}_\infty$ consensus performance level γ, missing measurement's probability $\bar{\alpha}$, positive definite matrix W and time-horizon N are taken as 1.98, 0.99, $\mathrm{diag}_{18}\{82\}$ and 45, respectively. Using the developed computational algorithm and MATLAB® (with YALMIP 3.0), we can check the feasibility of the coupled recursive RDEs and obtain the desired controller parameters as shown in Table 3.1, and ε_k and λ_k are set as

$$\varepsilon_k = \begin{cases} 1.22, & 0 \le k < 30, \\ 1.25, & 30 \le k \le 45, \end{cases} \quad \lambda_k = \begin{cases} 0.012, & 0 \le k < 30, \\ 0.022, & 30 \le k \le 45. \end{cases}$$

In the simulation, the exogenous disturbance inputs and unknown matrix functions are selected as

$$w_1(k) = w_3(k) = -1.2\exp(-0.2k)\cos(k),$$

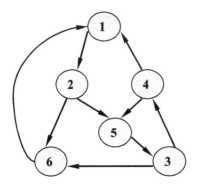

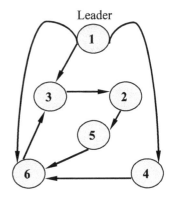

FIGURE 3.1
The directed communication graph without leader.

FIGURE 3.2
The directed communication graph with leader.

$$
\begin{aligned}
w_2(k) &= w_4(k) = 1.5\sin(0.8k)/(k+1),\\
w_5(k) &= w_6(k) = 1.2\cos(-k)/(4k+1),\\
\nu_1(k) &= \nu_3(k) = 1.5\exp(-0.4k)\sin(k),\\
\nu_2(k) &= \nu_4(k) = 1.6\sin(-0.4k)/(3k+1),\\
\nu_5(k) &= \nu_6(k) = -1.7\cos(0.8k)/(k+1),\\
F_1(k) &= F_6(k) = 0.95\cos(k),\\
F_2(k) &= F_5(k) = 0.9\sin(2k),\\
F_3(k) &= F_4(k) = \cos(3k)\sin(-2k).
\end{aligned}
$$

The initial values $x_i(0)$ $(i = 1,2,\cdots,6)$ are generated by MATLAB software and obey uniform distribution over $[-4.0,\ 4.0]$. Simulation results are shown in Figs. 3.3-3.10, where Fig. 3.3 and Fig. 3.8 plot the state trajectories of $x_i(k)$ $(i = 1,2,\cdots,6)$, and Fig. 3.9 and Fig. 3.10 depict the consensus errors $z_i(k) - z^*(k)$. In the figures, x_i^j $(j = 1,2,3)$ denotes the j-th element of the system state x_i. Note that the consensus issue for multi-agent systems with the topology $\mathscr{H}_2$ is essentially a distributed tracking control problem, and our proposed consensus protocol design scheme is also effective for such a tracking problem. The simulation results have confirmed that the designed $\mathcal{H}_\infty$ consensus controllers perform very well.

3.5 Summary

In this chapter, we have dealt with the $\mathcal{H}_\infty$ consensus control problem for a class of discrete time-varying multi-agent systems with missing measurements

TABLE 3.1
The distributed state estimator gain matrices.

k	0	1	2	3
$K^T(k)$ ($\mathscr{H}_1$)	$\begin{bmatrix} 0.3313 \\ -0.0241 \end{bmatrix}$	$\begin{bmatrix} 0.3019 \\ -0.0001 \end{bmatrix}$	$\begin{bmatrix} 0.2720 \\ 0.0104 \end{bmatrix}$	$\begin{bmatrix} 0.2663 \\ -0.0029 \end{bmatrix}$
$K^T(k)$ ($\mathscr{H}_2$)	$\begin{bmatrix} 0.3326 \\ -0.0225 \end{bmatrix}$	$\begin{bmatrix} 0.2931 \\ -0.0010 \end{bmatrix}$	$\begin{bmatrix} 0.2640 \\ 0.0098 \end{bmatrix}$	$\begin{bmatrix} 0.2590 \\ -0.0044 \end{bmatrix}$

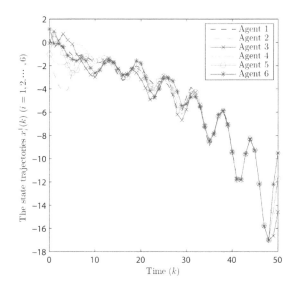

FIGURE 3.3
The trajectories of the first state with $\mathscr{H}_1$.

and parameter uncertainties. The missing measurements have been modeled by a set of Bernoulli distributed white sequence with a known conditional probability. Also, parameter uncertainties satisfying norm-bounded conditions have been taken into account. By employing the completing squares method and the stochastic analysis technique, a necessary and sufficient condition has been established to ensure the dynamics of consensus errors $z_i(k) - z^*(k)$ to satisfy the auxiliary performance index. Furthermore, the controller gains have been explicitly characterized by means of the solutions to two coupled backward recursive RDEs. Finally, an illustrative example has been provided that highlights the usefulness of the developed consensus control approach.

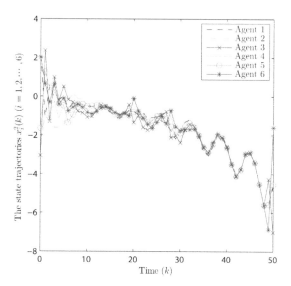

FIGURE 3.4
The trajectories of the second state with $\mathscr{H}_1$.

Further research topics include the extension of the main results to the mobile robot localization problem with missing measurements and also to the $\mathcal{H}_\infty$ control/filtering problem for network control systems with missing measurements or fading channels, and so on.

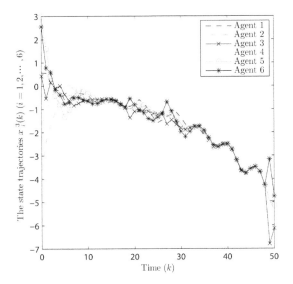

FIGURE 3.5
The trajectories of the third state with $\mathscr{H}_1$.

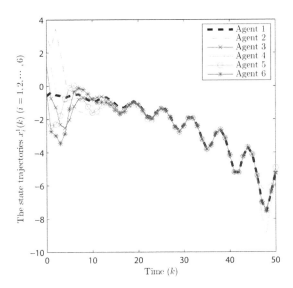

FIGURE 3.6
The trajectories of the first state with $\mathscr{H}_2$.

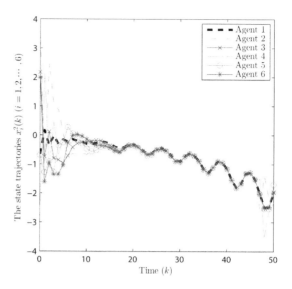

FIGURE 3.7
The trajectories of the second state with $\mathcal{H}_2$.

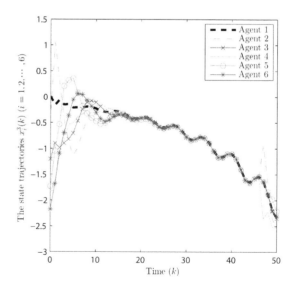

FIGURE 3.8
The trajectories of the third state with $\mathcal{H}_2$.

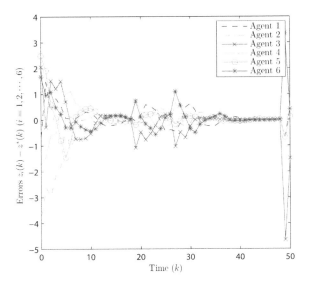

FIGURE 3.9

The errors $z_i(k) - z^*(k)$ with $\mathscr{H}_1$.

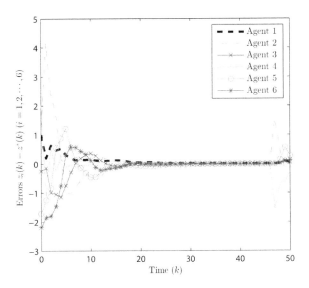

FIGURE 3.10

The errors $z_i(k) - z^*(k)$ with $\mathscr{H}_2$.

4

Finite-Horizon Distributed $\mathcal{H}_\infty$ State Estimation with Stochastic Parameters through Sensor Networks

Sensor networks are composed of small nodes with sensing, computation and wireless communications capabilities. These nodes, also called sensor nodes, are usually distributed spatially to form a wireless ad hoc network and are capable of only a limited amount of processing. Research on sensor networks was initially motivated by military applications such as battlefield surveillance and enemy tracking [4, 111]. Nowadays, due to the rapid developments of technologies in computing and communication, sensor networks have a wide range of applications in areas such as environment and habitat monitoring, health care applications, traffic control, distributed robotics, and industrial and manufacturing automation [28, 132]. The main task of sensor networks is to obtain information on the observed plants.

On the other hand, in network environments, system parameters might be subject to stochastic perturbations within certain intervals. These kinds of systems with stochastic parameters are customarily referred to as stochastic parameter systems, state-dependent or multiplicative noise systems. In this chapter, the distributed $\mathcal{H}_\infty$ state estimation problem is investigated for a class of discrete time-varying stochastic parameter systems in the framework of the backward RDEs developed in Chapter 2 and Chapter 3. The system measurements are collected through sensor networks with sensors distributed according to a given topology. Through available output measurements from not only the individual sensor but also its neighboring sensors, a necessary and sufficient condition is established to achieve the $\mathcal{H}_\infty$ performance constraint, and then the estimator design scheme is proposed via a certain $\mathcal{H}_2$-type criterion. The desired estimator parameters can be obtained by solving coupled backward RDEs. Two numerical simulation examples are provided to demonstrate the effectiveness and applicability of the proposed estimator design approach.

4.1   Modeling and Problem Formulation

In this chapter, it is assumed that the sensor network has n sensor nodes which are distributed in space according to a fixed network topology represented by a directed graph $\mathcal{G} = (\mathcal{V}, \mathcal{E}, \mathcal{H})$ of order n with the set of nodes $\mathcal{V} = \{1, 2, \cdots, n\}$, the set of edges $\mathcal{E} \in \mathcal{V} \times \mathcal{V}$, and the weighted adjacency matrix $\mathcal{H} = [h_{ij}]$ with nonnegative adjacency element h_{ij}. An edge of $\mathcal{G}$ is denoted by the ordered pair (i, j). The adjacency elements associated with the edges of the graph are positive, i.e., $h_{ij} > 0 \iff (i, j) \in \mathcal{E}$, which means that sensor i can obtain information from sensor j. Also, we assume that $h_{ii} = 1$ for all $i \in \mathcal{V}$ and therefore (i, i) can be regarded as an additional edge. The set of neighbors of node $i \in \mathcal{V}$ plus the node itself are denoted by $\mathcal{N}_i = \{j \in \mathcal{V} : (i, j) \in \mathcal{E}\}$.

In this chapter, a target plant is the system whose states are to be estimated through the distributed sensors. Let the target plant be described by the following discrete time-varying nonlinear stochastic system defined on $k \in [0, N-1]$:

$$\begin{cases} x_{k+1} = A_k(\mu_k)x_k + f_k(x_k, \vartheta_k) + B_k w_k, \\ z_k = L_k x_k \end{cases} \tag{4.1}$$

with n sensors modeled by

$$y_k^{(i)} = C_k^{(i)}(\varepsilon_k^{(i)})x_k + D_k^{(i)}v_k^{(i)}, \quad i = 1, 2, \cdots, n \tag{4.2}$$

where $x_k \in \mathbb{R}^{n_x}$ is the state of the target plant that cannot be observed directly, $y_k^{(i)} \in \mathbb{R}^{n_y}$ is the measurement output from sensor i, $z_k \in \mathbb{R}^{n_z}$ is the output to be estimated, and $w_k, v_k^{(i)} \in \ell_2([0, \infty); \mathbb{R})$ are external disturbances. $A_k(\mu_k) \in \mathbb{R}^{n_x \times n_x}$ and $C_k^{(i)}(\varepsilon_k^{(i)}) \in \mathbb{R}^{n_y \times n_x}$ are the stochastic parameter matrices, and B_k, $D_k^{(i)}$ and L_k are known time-varying matrices with compatible dimensions. μ_k, ϑ_k and $\varepsilon_k^{(i)}$ are mutually independent zero mean Gaussian noise sequences with unit variances, and therefore $A_k(\mu_k)$, $C_k^{(i)}(\varepsilon_k^{(i)})$ and $f_k(x_k, \vartheta_k)$ are mutually independent, too.

As in [80], it is assumed that the stochastic parameter matrices $A_k(\mu_k)$ and $C_k^{(i)}(\varepsilon_k^{(i)})$ have the statistical properties as follows:

$$\begin{aligned} \mathbb{E}\{A_k(\mu_k)\} = \bar{A}_k, \quad & \mathbb{E}\{C_k^{(i)}(\varepsilon_k^{(i)})\} = \bar{C}_k^{(i)}, \\ \text{Cov}\{a_{mu}^k, a_{st}^k\} = M_{a_{mu}^k, a_{st}^k}, \quad & \text{Cov}\{c_{mu}^{i,k}, c_{st}^{i,k}\} = M_{c_{mu}^{i,k}, c_{st}^{i,k}}, \end{aligned} \tag{4.3}$$

where a_{mu}^k and $c_{mu}^{i,k}$ are the (m, u)-th entries of matrices $A_k(\mu_k)$ and $C_k^{(i)}(\varepsilon_k^{(i)})$, respectively. They can also be denoted as $a_{mu}^k = A_k(\mu_k)[m, u]$ and $c_{mu}^{i,k} = C_k^{(i)}(\varepsilon_k^{(i)})[m, u]$.

The function $f_k(x_k, \vartheta_k)$ with $f_k(0, \vartheta_k) = 0$ is a stochastic nonlinear function having the following first moment for all x_k:

$$\mathbb{E}\left\{f_k(x_k, \vartheta_k)|x_k\right\} = 0 \tag{4.4}$$

and the covariance given by

$$\mathbb{E}\left\{f_k(x_k, \vartheta_k)f_j^T(x_j, \vartheta_j)|x_k\right\} = 0, \quad k \neq j, \tag{4.5}$$

and

$$\mathbb{E}\left\{f_k(x_k, \vartheta_k)f_k^T(x_k, \vartheta_k)|x_k\right\} = \sum_{i=1}^{s} \Pi_k^{(i)} x_k^T \Gamma_k^{(i)} x_k, \tag{4.6}$$

where s is a known nonnegative integer, and $\Pi_k^{(i)}$ and $\Gamma_k^{(i)}$ ($i = 1, 2, \cdots, s$) are known matrices with appropriate dimensions.

Remark 4.1 *The nonlinearity description satisfying (4.4)-(4.6) is applicable for some of the most investigated stochastic nonlinear models, such as the system with state-dependent multiplicative noises $D_k x_k \vartheta_k$ or $D_k x_k (\vartheta_k)^3$, the system with stochastic vectors whose powers depend on the sector-bound of a nonlinear function of the states $f_k(x_k)\vartheta_k$ with $\|f_k(x_k)\| \leq a_k \|\|x_k\|$ where $a_k > 0$ is a known real scalar, and the systems with a stochastic sequence dependent on the signum of a nonlinear function of the states $\text{sign}[f_k(x_k)]x_k\vartheta_k$ where sign denotes the signum function. Other models that fit this description can be found in [57, 146]. As such, rather than requiring the exact expression of nonlinear functions, we make use of the first- and second-order statistics of the stochastic nonlinearities (see (4.4)-(4.6)) for design of the desired distributed filters so they possess robustness against the unknown (but bounded) nonlinearities. Nevertheless, in the case that the nonlinear function is deterministic and known a priori, the estimate for such an exactly known nonlinear function should be included in the filter structure to reduce possible conservatism.*

In this chapter, the key point in designing distributed estimators for sensor networks is how to fuse the information available for the estimator on the node i both from the sensor i itself and from its neighbors. Keeping such a fact in mind, we propose to construct the following distributed state estimators:

$$\begin{cases} \hat{x}_{k+1}^{(i)} = \bar{A}_k \hat{x}_k^{(i)} + \sum_{j \in \mathcal{N}_i} h_{ij} K_k^{(ij)} (y_k^{(j)} - \bar{C}_k^{(j)} \hat{x}_k^{(j)}), \\ \hat{z}_k^{(i)} = L_k \hat{x}_k^{(i)} \end{cases} \tag{4.7}$$

where $\hat{x}_k^{(i)} \in \mathbb{R}^{n_x}$ and $\hat{z}_k^{(i)} \in \mathbb{R}^{n_z}$ are the estimated state and the estimated output on sensor node i. Here, $K_k^{(ij)}$ is the estimator gain matrix on node i to be determined.

For convenience of later analysis, we rewrite $A_k(\mu_k)$ and $C_k^{(i)}(\varepsilon_k^{(i)})$ as

$$A_k(\mu_k) = \bar{A}_k + \tilde{A}_k(\mu_k), \quad C_k^{(i)}(\varepsilon_k^{(i)}) = \bar{C}_k^{(i)} + \tilde{C}_k^{(i)}(\varepsilon_k^{(i)}).$$

For notational simplicity, in the rest of this chapter, let

$$
\begin{aligned}
e_k &= \mathbf{I}x_k - \hat{x}_k, & \hat{x}_k &= [\hat{x}_k^{(1)T} \quad \hat{x}_k^{(2)T} \quad \cdots \quad \hat{x}_k^{(n)T}]^T, \\
\tilde{z}_k^{(i)} &= z_k - \hat{z}_k^{(i)}, & \tilde{z}_k &= [\tilde{z}_k^{(1)T} \quad \tilde{z}_k^{(2)T} \quad \cdots \quad \tilde{z}_k^{(n)T}]^T, \\
\bar{\mathcal{A}}_k &= \mathrm{diag}_n\{\bar{A}_k\}, & v_k &= [v_k^{(1)} \quad v_k^{(2)} \quad \cdots \quad v_k^{(n)}]^T, \\
\tilde{\mathcal{A}}_k &= \mathrm{diag}_n\{\tilde{A}_k(\mu_k)\}, & \bar{\mathcal{C}}_k^{(i)} &= \mathrm{diag}\{h_{i1}\bar{C}_k^{(1)}, h_{i2}\bar{C}_k^{(2)}, \cdots, h_{in}\bar{C}_k^{(n)}\}, \\
\mathcal{B}_k &= \mathrm{diag}_n\{B_k\}, & \tilde{\mathcal{C}}_k^{(i)} &= \mathrm{diag}\{h_{i1}\tilde{C}_k^{(1)}(\varepsilon_k^{(1)}), \cdots, h_{in}\tilde{C}_k^{(n)}(\varepsilon_k^{(n)})\}, \\
\mathcal{L}_k &= \mathrm{diag}_n\{L_k\}, & \mathcal{D}_k^{(i)} &= \mathrm{diag}\{h_{i1}D_k^{(1)}, h_{i2}D_k^{(2)}, \cdots, h_{in}D_k^{(n)}\}, \\
\bar{\mathcal{L}}_k &= \mathrm{diag}\{0, \mathcal{L}_k\}, & E^{(i)} &= \mathrm{diag}\{\underbrace{0, \cdots, 0}_{i-1}, I, \underbrace{0, \cdots, 0}_{n-i}\}, \\
\mathbf{I} &= [I, I, \cdots, I]_{1\times n}^T, & \mathcal{K}_k &= [\bar{K}_k^{(ij)}]_{n\times n}
\end{aligned}
$$

with $\bar{K}_k^{(ij)} = \begin{cases} K_k^{(ij)}, & h_{ij} \neq 0, \\ 0, & h_{ij} = 0. \end{cases}$

Using the defined notations, the dynamics of the estimation errors can be obtained from (4.1) and (4.7) as follows:

$$
\begin{aligned}
e_{k+1} &= \left(\bar{\mathcal{A}}_k - \sum_{i=1}^n E^{(i)}\mathcal{K}_k\bar{\mathcal{C}}_k^{(i)}\right)e_k + \left(\tilde{\mathcal{A}}_k - \sum_{i=1}^n E^{(i)}\mathcal{K}_k\tilde{\mathcal{C}}_k^{(i)}\right)\mathbf{I}x_k \\
&\quad + \mathbf{I}f_k(x_k, \vartheta_k) + \mathcal{B}_k\mathbf{I}w_k - \sum_{i=1}^n E^{(i)}\mathcal{K}_k\mathcal{D}_k^{(i)}v_k.
\end{aligned} \tag{4.8}
$$

Setting $\eta_k = [x_k^T \quad e_k^T]^T$ and $\xi_k = [w_k \quad v_k^T]^T$, an augmented system can be derived from (4.1) and (4.8) as follows:

$$
\begin{cases} \eta_{k+1} = \mathscr{A}_k\eta_k + \tilde{\mathscr{A}}_k\eta_k + \mathscr{F}_k(\eta_k, \vartheta_k) + \mathscr{B}_k\xi_k, \\ \tilde{z}_k = \bar{\mathcal{L}}_k\eta_k \end{cases} \tag{4.9}
$$

where

$$
\mathscr{A}_k = \begin{pmatrix} \bar{A}_k & 0 \\ 0 & \bar{\mathcal{A}}_k - \sum_{i=1}^n E^{(i)}\mathcal{K}_k\bar{\mathcal{C}}_k^{(i)} \end{pmatrix},
$$

$$
\tilde{\mathscr{A}}_k = \begin{pmatrix} \tilde{A}_k(\mu_k) & 0 \\ \tilde{\mathcal{A}}_k\mathbf{I} - \sum_{i=1}^n E^{(i)}\mathcal{K}_k\tilde{\mathcal{C}}_k^{(i)}\mathbf{I} & 0 \end{pmatrix},
$$

$$
\mathscr{B}_k = \begin{pmatrix} B_k & 0 \\ \mathcal{B}_k\mathbf{I} & -\sum_{i=1}^n E^{(i)}\mathcal{K}_k\mathcal{D}_k^{(i)} \end{pmatrix},
$$

$$
\mathscr{F}_k(\eta_k, \vartheta_k) = \begin{pmatrix} f_k(x_k, \vartheta_k) \\ \mathbf{I}f_k(x_k, \vartheta_k) \end{pmatrix}.
$$

Before proceeding further, we introduce the following definition.

Definition 4.1 *For a given disturbance attenuation level $\gamma > 0$ and a given* *weighted matrix $W > 0$, the estimation error $\tilde{z}_k$ from (4.9) is said to satisfy the $\mathcal{H}_\infty$ performance constraint if the following inequality holds:*

$$\frac{1}{n}\mathbb{E}\left\{||\tilde{z}_k||^2_{[0,N-1]}\right\} < \gamma^2||\xi_k||^2_{[0,N-1]} + \gamma^2\mathbb{E}\left\{\eta_0^T W \eta_0\right\}. \tag{4.10}$$

Remark 4.2 *In terms of (4.9) and (4.10), it can be seen that the value of $\mathbb{E}\{||\tilde{z}_k||^2\}$ would become larger as the number of the nodes increases. Theoretically, the disturbance attenuation level γ for the overall network should account for the average disturbance rejection performance that is insensitive to a change in the number of nodes in the estimator design. For this purpose, the term $1/n$ is used to accommodate the average $\mathcal{H}_\infty$ index over the sensor networks so that the scalar γ reflects the practical significance of the $\mathcal{H}_\infty$ disturbance rejection level.*

Our aim in this chapter is to design an estimator of the form (4.7) on each node i for system (4.1) with stochastic parameter matrices and stochastic nonlinearities. In other words, we are going to determine the estimator gain matrices $K_k^{(ij)}$ $(i = 1, 2, \cdots, n, j \in \mathcal{N}_i)$ such that the estimation error $\tilde{z}_k$ from (4.9) satisfies the $\mathcal{H}_\infty$ performance constraint (4.10) over a finite horizon $[0, N-1]$.

4.2 $\mathcal{H}_\infty$ Performance Analysis

In this section, we are aiming at establishing a sufficient criterion for the estimation error $\tilde{z}_k$ in (4.9) to satisfy the $\mathcal{H}_\infty$ performance constraint. By resorting to a combination of the completing squares method, the backward recursive Riccati difference equation (RDE) approach, and the stochastic analysis technique, verifiable conditions can be derived step by step.

First, let us introduce two lemmas and a partitioning matrix $\mathcal{P}_k$ as follows

$$\mathcal{P}_k = \left(\begin{array}{c|c} P_1^k & P_2^k \\ \hline P_3^k & P_4^k \end{array}\right) = \left(\begin{array}{ccc|c} P_{11}^k & P_{12}^k & \cdots & P_{1,n+1}^k \\ \hline P_{21}^k & P_{22}^k & \cdots & P_{2,n+1}^k \\ \vdots & \vdots & \vdots & \vdots \\ P_{n+1,1}^k & P_{n+1,2}^k & \cdots & P_{n+1,n+1}^k \end{array}\right).$$

Lemma 4.1 *[3] Let matrices G, M and Γ be given with appropriate sizes. Then the following matrix equation*

$$GXM = \Gamma \tag{4.11}$$

has a solution X if and only if $GG^\dagger \Gamma M^\dagger M = \Gamma$ where $G^\dagger$ is the Moore-Penrose pseudoinverse of G. Moreover, a general solution to (4.11) can be given by

$$X = G^\dagger \Gamma M^\dagger + Y - G^\dagger GYMM^\dagger$$

where Y is any matrix with appropriate dimension.

Lemma 4.2 *Let the disturbance attenuation level $\gamma > 0$ and the positive definite matrix W be given. For any nonzero $\xi \in \ell_2$, the augmented system (4.9) satisfies the $\mathcal{H}_\infty$ performance constraint (4.10) if and only if there exist a set of real-valued matrices $\{\mathcal{K}_k\}_{0 \leq k \leq N-1}$ and a family of non-negative definite matrices $\{\mathcal{P}_k\}_{0 \leq k \leq N}$ with the final condition $\mathcal{P}_N = 0$ to the following backward recursive RDE:*

$$
\begin{aligned}
\mathcal{P}_k = {}& \mathscr{A}_k^T \mathcal{P}_{k+1} \mathscr{A}_k + \frac{1}{n} \bar{\mathcal{L}}_k^T \bar{\mathcal{L}}_k \\
& + \Theta_k^{(1)} + \Theta_k^{(2)} + \mathscr{A}_k^T \mathcal{P}_{k+1} \mathscr{B}_k \Phi_k^{-1} \mathscr{B}_k^T \mathcal{P}_{k+1} \mathscr{A}_k
\end{aligned}
\tag{4.12}
$$

subject to

$$
\begin{cases}
\Phi_k = \gamma^2 I - \mathscr{B}_k^T \mathcal{P}_{k+1} \mathscr{B}_k > 0, \\
\mathcal{P}_0 < \gamma^2 W
\end{cases}
\tag{4.13}
$$

where

$$
\Theta_k^{(1)} = diag \left\{ \sum_{p=1}^{n+1} \sum_{q=1}^{n+1} \sum_{i=1}^{s} \Pi_k^{(i)} \, tr \left(P_{pq}^{k+1} \Gamma_k^{(i)} \right), 0, \cdots, 0 \right\},
$$

$$
\Theta_k^{(2)} = diag \left\{ \Theta_k^{(21)}, 0, \cdots, 0 \right\}
$$

with

$$
\begin{aligned}
\Theta_k^{(21)}[s,t] = {}& \sum_{p=1}^{n+1} \sum_{q=1}^{n+1} \sum_{j=1}^{n} \sum_{i=1}^{n} M_{a_{tj}^k, a_{si}^k} (P_{pq}^{k+1})_{[i,j]} - \sum_{j=1}^{n} \sum_{i=1}^{n} M_{a_{tj}^k, a_{si}^k} (P_{11}^{k+1})_{[i,j]} \\
& + \sum_{r=1}^{n} \sum_{p=1}^{n} \sum_{q=1}^{n} \sum_{j=1}^{n} \sum_{i=1}^{n} h_{pr} h_{qr} M_{c_{tj}^k, c_{si}^k} (\bar{K}_k^{(pr)T} P_{p+1,q+1}^{k+1} \bar{K}_k^{(qr)})_{[i,j]}.
\end{aligned}
$$

Proof *Sufficiency*: Define $V_k(\eta_k) = \eta_k^T \mathcal{P}_k \eta_k$. Calculate $V_{k+1}(\eta_{k+1}) - V_k(\eta_k)$ along the system trajectory (4.9) and take conditional expectation as follows

$$
\begin{aligned}
J_k^{(1)} := {}& \mathbb{E} \left\{ \eta_{k+1}^T \mathcal{P}_{k+1} \eta_{k+1} - \eta_k^T \mathcal{P}_k \eta_k | \eta_k \right\} \\
= {}& \eta_k^T \mathscr{A}_k^T \mathcal{P}_{k+1} \mathscr{A}_k \eta_k + \mathbb{E} \left\{ \eta_k^T \tilde{\mathscr{A}}_k^T \mathcal{P}_{k+1} \tilde{\mathscr{A}}_k \eta_k \right\} \\
& + \mathbb{E} \left\{ \mathscr{F}_k^T(\eta_k, \vartheta_k) \mathcal{P}_{k+1} \mathscr{F}_k(\eta_k, \vartheta_k) \right\} \\
& + \xi_k^T \mathscr{B}_k^T \mathcal{P}_{k+1} \mathscr{B}_k \xi_k + 2\eta_k^T \mathscr{A}_k^T \mathcal{P}_{k+1} \mathscr{B}_k \xi_k - \eta_k^T \mathcal{P}_k \eta_k.
\end{aligned}
\tag{4.14}
$$

Adding the following zero term

$$
\frac{1}{n} ||\tilde{z}_k||^2 - \gamma^2 ||\xi_k||^2 - \left(\frac{1}{n} ||\tilde{z}_k||^2 - \gamma^2 ||\xi_k||^2 \right)
$$

to the right-hand side of (4.14) results in

$$J_k^{(1)} = \eta_k^T \left(\mathscr{A}_k^T \mathcal{P}_{k+1} \mathscr{A}_k + \frac{1}{n} \bar{\mathcal{L}}_k^T \bar{\mathcal{L}}_k - \mathcal{P}_k \right) \eta_k + \mathbb{E} \left\{ \eta_k^T \tilde{\mathscr{A}}_k^T \mathcal{P}_{k+1} \tilde{\mathscr{A}}_k \eta_k \right\}$$

$$+ \mathbb{E} \left\{ \mathscr{F}_k^T(\eta_k, \vartheta_k) \mathcal{P}_{k+1} \mathscr{F}_k(\eta_k, \vartheta_k) \right\} + \xi_k^T (\mathscr{B}_k^T \mathcal{P}_{k+1} \mathscr{B}_k - \gamma^2 I) \xi_k$$

$$+ 2\eta_k^T \mathscr{A}_k^T \mathcal{P}_{k+1} \mathscr{B}_k \xi_k - \left(\frac{1}{n} ||\tilde{z}_k||^2 - \gamma^2 ||\xi_k||^2 \right).$$

$$(4.15)$$

For (4.15), we have

$$\mathbb{E} \left\{ \eta_k^T \tilde{\mathscr{A}}_k^T \mathcal{P}_{k+1} \tilde{\mathscr{A}}_k \eta_k \right\} = \eta_k^T \left(\begin{array}{cc} \bar{\Theta}_k^{(2)} + \mathbf{I}^T \bar{\Theta}_k^{(3)} \mathbf{I} & 0 \\ 0 & 0 \end{array} \right) \eta_k,$$

$$\bar{\Theta}_k^{(1)} := \mathbb{E} \left\{ \mathscr{F}_k^T(\eta_k, \vartheta_k) \mathcal{P}_{k+1} \mathscr{F}_k(\eta_k, \vartheta_k) \right\}$$

$$= \mathbb{E} \left\{ \sum_{i=1}^{n+1} \sum_{j=1}^{n+1} f_k^T(x_k, \vartheta_k) P_{ij}^{k+1} f_k(x_k, \vartheta_k) \right\},$$

$$\bar{\Theta}_k^{(2)} := \mathbb{E} \left\{ \tilde{A}_k^T P_1^{k+1} \tilde{A}_k + \tilde{A}_k^T P_2^{k+1} \tilde{\mathcal{A}}_k \mathbf{I} + \mathbf{I}^T \tilde{\mathcal{A}}_k^T P_3^{k+1} \tilde{A}_k + \mathbf{I}^T \tilde{\mathcal{A}}_k^T P_4^{k+1} \tilde{\mathcal{A}}_k \mathbf{I} \right\}$$

$$= \mathbb{E} \left\{ \tilde{A}_k^T P_{11}^{k+1} \tilde{A}_k + 2 \sum_{i=1+1}^{n+1} \tilde{A}_k^T P_{1i}^{k+1} \tilde{A}_k + \sum_{i=1}^{n} \sum_{j=1}^{n} \tilde{A}_k^T P_{i+1,j+1}^{k+1} \tilde{A}_k \right\},$$

$$\bar{\Theta}_k^{(3)} := \mathbb{E} \left\{ \left(\sum_{i=1}^{n} E^{(i)} \mathcal{K}_k \tilde{\mathcal{C}}_k^{(i)} \right)^T P_4^{k+1} \left(\sum_{i=1}^{n} E^{(i)} \mathcal{K}_k \tilde{\mathcal{C}}_k^{(i)} \right) \right\}$$

$$= \mathbb{E} \left\{ \sum_{i=1}^{n} \sum_{j=1}^{n} \tilde{\mathcal{C}}_k^{(i)T} \mathcal{K}_k^T E^{(i)} P_4^{k+1} E^{(j)} \mathcal{K}_k \tilde{\mathcal{C}}_k^{(j)} \right\}$$

$$= \mathbb{E} \sum_{i=1}^{n} \sum_{j=1}^{n} diag \left\{ h_{i1} h_{j1} \tilde{C}_k^{(1)T} \bar{K}_k^{(i1)T} P_{i+1,j+1}^{k+1} \bar{K}_k^{(j1)} \tilde{C}_k^{(1)}, \right.$$

$$\left. \cdots, h_{in} h_{jn} \tilde{C}_k^{(n)T} \bar{K}_k^{(in)T} P_{i+1,j+1}^{k+1} \bar{K}_k^{(jn)} \tilde{C}_k^{(n)} \right\}$$

$$:= \mathbb{E} \left\{ diag \left\{ \bar{\Theta}_k^{(31)}, \bar{\Theta}_k^{(32)}, \cdots, \bar{\Theta}_k^{(3n)} \right\} \right\}.$$

Based on [80] and together with (4.3)-(4.6), one has

$$\bar{\Theta}_k^{(1)} = \eta_k^T \Theta_k^{(1)} \eta_k,$$

$$\bar{\Theta}_k^{(2)}[s,t] = \sum_{p=1}^{n+1} \sum_{q=1}^{n+1} \sum_{j=1}^{n} \sum_{i=1}^{n} M_{a_{tj}^k, a_{si}^k} (P_{pq}^{k+1})_{[i,j]} - \sum_{j=1}^{n} \sum_{i=1}^{n} M_{a_{tj}^k, a_{si}^k} (P_{11}^{k+1})_{[i,j]},$$

$$\mathbb{E} \left\{ \bar{\Theta}_k^{(3r)} \right\}_{[s,t]} = \sum_{p=1}^{n} \sum_{q=1}^{n} \sum_{j=1}^{n} \sum_{i=1}^{n} h_{pr} h_{qr} M_{c_{tj}^k, c_{si}^k} (\bar{K}_k^{(pr)T} P_{p+1,q+1}^{k+1} \bar{K}_k^{(qr)})_{[i,j]}.$$

$$(4.16)$$

Furthermore, it follows from (4.15) and (4.16) that

$$
\begin{aligned}
J_k^{(1)} = \quad & \eta_k^T \left(\mathscr{A}_k^T \mathcal{P}_{k+1} \mathscr{A}_k + \frac{1}{n} \bar{\mathcal{L}}_k^T \bar{\mathcal{L}}_k - \mathcal{P}_k + \Theta_k^{(1)} + \Theta_k^{(2)} \right) \eta_k \\
& - \xi_k^T \Phi_k \xi_k + 2\eta_k^T \mathscr{A}_k^T \mathcal{P}_{k+1} \mathscr{B}_k \xi_k - \left(\frac{1}{n} ||\tilde{z}_k||^2 - \gamma^2 ||\xi_k||^2 \right).
\end{aligned}
\tag{4.17}
$$

By the completing squares method, it is not difficult to see that

$$
\begin{aligned}
J_k^{(1)} = \quad & \eta_k^T \left(\mathscr{A}_k^T \mathcal{P}_{k+1} \mathscr{A}_k + \frac{1}{n} \bar{\mathcal{L}}_k^T \bar{\mathcal{L}}_k - \mathcal{P}_k + \Theta_k^{(1)} \right. \\
& \left. + \Theta_k^{(2)} + \mathscr{A}_k^T \mathcal{P}_{k+1} \mathscr{B}_k \Phi_k^{-1} \mathscr{B}_k^T \mathcal{P}_{k+1} \mathscr{A}_k \right) \eta_k \\
& - (\xi_k - \xi_k^*)^T \Phi_k (\xi_k - \xi_k^*) - \left(\frac{1}{n} ||\tilde{z}_k||^2 - \gamma^2 ||\xi_k||^2 \right)
\end{aligned}
\tag{4.18}
$$

where $\xi_k^* = \Phi_k^{-1} \mathscr{B}_k^T \mathcal{P}_{k+1} \mathscr{A}_k \eta_k$, and then it follows readily that

$$
\begin{aligned}
& \mathbb{E} \sum_{k=0}^{N-1} J_k^{(1)} \\
& = \mathbb{E}\{\eta_N^T \mathcal{P}_N \eta_N - \eta_0^T \mathcal{P}_0 \eta_0\} \\
& = -\sum_{k=0}^{N-1} (\xi_k - \xi_k^*)^T \Phi_k (\xi_k - \xi_k^*) - \sum_{k=0}^{N-1} \mathbb{E}\left\{ \frac{1}{n} ||\tilde{z}_k||^2 - \gamma^2 ||\xi_k||^2 \right\}.
\end{aligned}
\tag{4.19}
$$

By noticing $\Phi_k > 0$, $\mathcal{P}_0 < \gamma^2 W$ and the final condition $\mathcal{P}_N = 0$, we have

$$
\begin{aligned}
\mathcal{J}_1(\eta_0, \xi) \triangleq \quad & \mathbb{E}\left\{ \frac{1}{n} ||\tilde{z}_k||_{[0,N-1]}^2 - \gamma^2 ||\xi_k||_{[0,N-1]}^2 \right\} - \gamma^2 \mathbb{E}\{\eta_0^T W \eta_0\} \\
\leq \quad & \mathbb{E}\left\{ \frac{1}{n} ||\tilde{z}_k||_{[0,N-1]}^2 - \gamma^2 ||\xi_k||_{[0,N-1]}^2 \right\} - \mathbb{E}\{\eta_0^T \mathcal{P}_0 \eta_0\} \\
= \quad & -\sum_{k=0}^{N-1} (\xi_k - \xi_k^*)^T \Phi_k (\xi_k - \xi_k^*) \leq 0.
\end{aligned}
\tag{4.20}
$$

It remains to show that (4.20) holds as a strict inequality. If $\eta_0 \neq 0$, there would be $\mathcal{J}_1(\eta_0, \xi) < 0$ due to $\mathcal{P}_0 < \gamma^2 W$. On the other hand, if $\eta_0 = 0$, substituting $\xi_k^* = \Phi_k^{-1} \mathscr{B}_k^T \mathcal{P}_{k+1} \mathscr{A}_k \eta_k$ into the system dynamics (4.9) leads to $\eta_k \equiv 0$ for $\forall k \in [0, N-1]$, and therefore $\xi_k^* \equiv 0$ for $\forall k \in [0, N-1]$. To this end, it can be deduced that $\mathcal{J}_1(\eta_0, \xi) = 0$ if and only if $\xi_k = \xi_k^* = 0$ for $\forall k \in [0, N-1]$, which contradicts the assumption of $\xi \neq 0$. Hence, at least one ξ_k cannot be equal to ξ_k^* in (4.9) or, equivalently, $\mathcal{J}_1(0, \xi) < 0$ for any nonzero $\xi \in \ell_2$. This means that the pre-specified $\mathcal{H}_\infty$ performance is satisfied.

__Necessity.__ We proceed to show that "if (4.12) is not true, then (4.10) is also not true." For convenience, let us provide an expression for the condition and conclusion of this proposition, respectively.

The IF statement (i.e., the condition that (4.12) is not true):

Due to $\Phi_{N+1} = \gamma^2 I > 0$, $\mathcal{P}_N$ can be calculated from the recursion (4.12). It is easy to see that, by the same procedure, the recursion RDE (4.12) can be solved backward when $|\Phi_{k+1}| \neq 0$ for all $k \in [0, N-1]$. It means that the recursion RDE (4.12) fails if there exists some k_0 satisfying $|\Phi_{k_0+1}| = 0$, which fails without the condition (4.13).

In short, in terms of (4.13) and the backward recursion character of (4.12), the "if statement" can be divided into three cases:

a) *$\Phi_{k+1} > 0$ for all $k \in [0, N-1]$, but the initial condition $\mathcal{P}_0 < \gamma^2 W$ can't be satisfied.*

b) *There exists a k_0 such that $|\Phi_{k_0+1}| = 0$ and $\Phi_{k+1} > 0$ $(k_0 < k < N)$. This means that Φ_{k_0+1} has at least one zero eigenvalue.*

c) *There exists a k_0 such that i) Φ_{k_0+1} is neither positive semi-definite nor positive definite; ii) $\Phi_{k+1} > 0$ $(k_0 < k < N)$, that is, $\mathcal{R}^{22}_{k_0+1}$ has at least one negative eigenvalue.*

Furthermore, combining b) and c), one has that, for some k_0, $\mathcal{R}^{22}_{k+1} > 0$ $(k_0 < k < N)$ and $\mathcal{R}^{22}_{k_0+1}$ has at least one zero or negative eigenvalue denoted as $\lambda_{k_0} \leq 0$.

The THEN statement (i.e., the conclusion that (4.10) is not true):

There exists $(\eta_0, \xi) \neq 0$ such that

$$\frac{1}{n}\mathbb{E}\left\{||\tilde{z}_k||^2_{[0,N-1]}\right\} \geq \gamma^2||\xi_k||^2_{[0,N-1]} + \gamma^2\mathbb{E}\left\{\eta_0^T W \eta_0\right\}. \tag{4.21}$$

First, denote

$$\mathcal{J}_1(\eta_0, \xi) = \mathbb{E}\left\{\frac{1}{n}||\tilde{z}_k||^2_{[0,N-1]} - \gamma^2||\xi_k||^2_{[0,N-1]}\right\} - \mathbb{E}\{\eta_0^T\gamma^2 W \eta_0\}. \tag{4.22}$$

Case a): *We can choose $\xi_k = \xi_k^*$, and then obtain from (4.19) that*

$$\mathcal{J}(\bar{x}_0, \tilde{\eta}) = \mathbb{E}\{\eta_0^T\mathcal{P}_0\eta_0 - \eta_N^T\mathcal{P}_N\eta_N\} - \mathbb{E}\{\eta_0^T\gamma^2 W \eta_0\}$$
$$- \sum_{k=0}^{N-1}(\xi_k - \xi_k^*)^T\Phi_k(\xi_k - \xi_k^*) \tag{4.23}$$
$$= \mathbb{E}\{\eta_0^T(\mathcal{P}_0 - \gamma^2 W)\eta_0\}.$$

Obviously, there always exists a $\eta_0 \neq 0$ satisfying $\mathcal{J}(\bar{x}_0, \tilde{\eta}) \geq 0$, even if $\mathcal{P}_0 - \gamma^2 W$ has at least one non-negative eigenvalue.

Case b) *and* **Case c):** *We assume that there exists a non-positive eigenvalue of Φ_{k+1} at time k_0, and design the special sequence $(\eta_0, \xi) \neq 0$ as follows:*

$$\eta_0 = 0 \quad and \quad \xi_k = \begin{cases} \psi_{k_0}, & k = k_0, \\ \xi_k^*, & k_0 < k < N, \\ 0, & 0 \leq k < k_0 \end{cases} \tag{4.24}$$

where ψ_{k_0} is the eigenvector of Φ_{k_0+1} with respect to λ_{k_0}. For the purpose of simplicity, denote $\xi := \{\xi_k\}_{0 \le k \le N}$.

For this given sequence $(\bar{x}_0, \tilde{\eta})$, it is easy to see from (4.9) that $\eta_k = 0$ for all $0 \le k \le k_0$, and then, one has from (4.19) that

$$
\begin{aligned}
\mathcal{J}(\bar{x}_0, \tilde{\eta}) = \sum_{k=0, k \ne k_0}^{N} & \mathbb{E}\Big\{ \|z_k\|^2 - \gamma^2 \|\tilde{\eta}_k\|^2 + \bar{\alpha}\gamma^2 \|\varepsilon_k \aleph_k \bar{x}_k\|^2 \Big\} \\
& + \mathbb{E}\Big\{ \|z_{k_0}\|^2 - \gamma^2 \|\tilde{\eta}_{k_0}\|^2 + \bar{\alpha}\gamma^2 \|\varepsilon_{k_0} \aleph_{k_0} \bar{x}_{k_0}\|^2 \Big\}.
\end{aligned}
\tag{4.25}
$$

Moreover, one can obviously show that

$$
\sum_{k=0}^{k_0-1} \mathbb{E}\left\{ \frac{1}{n}\|\tilde{z}_k\|^2 - \gamma^2 \|\xi_k\|^2 \right\} = 0,
\tag{4.26}
$$

and

$$
\begin{aligned}
& \mathbb{E}\left\{ \frac{1}{n}\|\tilde{z}_{k_0}\|^2 - \gamma^2 \|\xi_{k_0}\|^2 \right\} \\
& = -\xi_{k_0}^T \Phi_{k_0} \xi_{k_0} - \mathbb{E}J_{k_0}^{(1)} + \mathbb{E}\Big\{ 2\eta_{k_0}^T \mathscr{A}_{k_0}^T \mathcal{P}_{k_0+1} \mathscr{B}_{k_0} \xi_{k_0} \\
& \quad + \eta_{k_0}^T \Big(\mathscr{A}_{k_0}^T \mathcal{P}_{k_0+1} \mathscr{A}_{k_0} + \frac{1}{n}\bar{\mathcal{L}}_{k_0}^T \bar{\mathcal{L}}_{k_0} - \mathcal{P}_{k_0} + \Theta_{k_0}^{(1)} + \Theta_{k_0}^{(2)} \Big) \eta_{k_0} \Big\}.
\end{aligned}
$$

Furthermore, it is easy to see from (4.19) that

$$
\begin{aligned}
& \sum_{k=k_0+1}^{N-1} \mathbb{E}\left\{ \frac{1}{n}\|\tilde{z}_k\|^2 - \gamma^2 \|\xi_k\|^2 \right\} \\
& = -\sum_{k=k_0+1}^{N-1} (\xi_k - \xi_k^*)^T \Phi_k (\xi_k - \xi_k^*) - \sum_{k=k_0+1}^{N-1} \mathbb{E}J_k^{(1)}.
\end{aligned}
\tag{4.27}
$$

Finally, substituting (4.26)-(4.27) into (4.25) yields

$$
\begin{aligned}
\mathcal{J}_1(\eta_0, \xi) & = \mathbb{E}\left\{ \frac{1}{n}\|\tilde{z}_k\|_{[0,N-1]}^2 - \gamma^2 \|\xi_k\|_{[0,N-1]}^2 \right\} - \mathbb{E}\{\eta_0^T \gamma^2 W \eta_0\} \\
& = \sum_{k=0}^{k_0-1} \mathbb{E}\left\{ \frac{1}{n}\|\tilde{z}_k\|^2 - \gamma^2 \|\xi_k\|^2 \right\} + \sum_{k=k_0+1}^{N-1} \mathbb{E}\left\{ \frac{1}{n}\|\tilde{z}_k\|^2 - \gamma^2 \|\xi_k\|^2 \right\} \\
& \quad + \mathbb{E}\left\{ \frac{1}{n}\|\tilde{z}_{k_0}\|^2 - \gamma^2 \|\xi_{k_0}\|^2 \right\} \\
& = -\xi_{k_0}^T \Phi_{k_0} \xi_{k_0} = -\psi_{k_0}^T \Phi_{k_0} \psi_{k_0} = -\lambda_{k_0} \|\psi_{k_0}\|^2 \\
& \ge 0.
\end{aligned}
\tag{4.28}
$$

Obviously, from the above inequality, the conclusion (4.21) is satisfied, i.e. the inverse negative proposition holds, which completes the proof.

Remark 4.3 *As discussed in [17], the "stochastic" version of the bounded real lemma (BRL) provides the key for the resolution of stochastic disturbance attenuation problems. There are basically three kinds of techniques involved in the BRLs, i.e., LMI (see [41, 108]), RDE (see [17, 49, 107]) and Hamilton Jacobi Inequality (HJI) (see [112]). Lemma 4.2 can be viewed as a "stochastic" version of the BRL to solve the distributed $\mathcal{H}_\infty$ state estimation problem for discrete time-varying systems with stochastic parameter matrices and stochastic nonlinearities over sensor networks. Obviously, when the sensor network reduces to a single sensor, Lemma 4.2 can be simplified to the BRL for the traditional $\mathcal{H}_\infty$ state estimation issue.*

4.3 Distributed Filter Design

In what follows, we shall propose an approach to the determination of the estimator gain matrices $K_k^{(ij)}$ ($i \in \mathcal{V}, j \in \mathcal{N}_i$) under the situation of worst-case disturbance. Firstly, a cost functional is defined as

$$\mathcal{J}_2(\mathcal{K}, \xi^*) = \mathbb{E}\left\{ \frac{1}{n}||\tilde{z}_k||^2_{[0,N-1]} + \frac{1}{n}||\varphi_k||^2_{[0,N-1]} \right\} \tag{4.29}$$

with $\varphi_k = -\sum_{i=1}^n E^{(i)} K_k \bar{\mathcal{C}}_k^{(i)} e_k$. Then, the original system (4.9) can be rewritten as follows:

$$\begin{cases} \eta_{k+1} = \left(\bar{\mathscr{A}}_k + \mathscr{B}_k \Phi_k^{-1} \mathscr{B}_k^T \mathcal{P}_{k+1} \mathscr{A}_k \right) \eta_k + \tilde{\mathscr{A}}_k \eta_k + \mathscr{F}_k(\eta_k, \vartheta_k) + \tilde{\varphi}_k, \\ \tilde{z}_k = \bar{\mathcal{L}}_k \eta_k \end{cases} \tag{4.30}$$

where $\bar{\mathscr{A}}_k = \text{diag}_{n+1}\{\bar{A}_k\}$ and $\tilde{\varphi}_k = [0 \quad \varphi_k^T]^T$. Moreover, we define $Y_k(\eta_k) = \eta_k^T \mathcal{Q}_k \eta_k$ and construct the function

$$J_k^{(2)} = \mathbb{E}\{Y_{k+1}(\eta_{k+1}) - Y_k(\eta_k)|\eta_k\}. \tag{4.31}$$

Substituting $\xi_k = \xi_k^*$ into (4.31) and taking mathematical expectation, one has

$$\begin{aligned} J_k^{(2)}\big|_{\xi=\xi^*} &= \mathbb{E}\left\{ \eta_{k+1}^T \mathcal{Q}_{k+1} \eta_{k+1} - \eta_k^T \mathcal{Q}_k \eta_k | \eta_k \right\}\Big|_{\xi=\xi^*} \\ &= \eta_k^T \left(\bar{\mathscr{A}}_k + \mathscr{B}_k \Phi_k^{-1} \mathscr{B}_k^T \mathcal{P}_{k+1} \mathscr{A}_k \right)^T \mathcal{Q}_{k+1} \\ &\quad \times \left(\bar{\mathscr{A}}_k + \mathscr{B}_k \Phi_k^{-1} \mathscr{B}_k^T \mathcal{P}_{k+1} \mathscr{A}_k \right) \eta_k + \mathbb{E}\left\{ \eta_k^T \tilde{\mathscr{A}}_k^T \mathcal{Q}_{k+1} \tilde{\mathscr{A}}_k \eta_k \right\} \\ &\quad + \mathbb{E}\left\{ \mathscr{F}_k^T(\eta_k, \vartheta_k) \mathcal{Q}_{k+1} \mathscr{F}_k(\eta_k, \vartheta_k) \right\} + \tilde{\varphi}_k^T \mathcal{Q}_{k+1} \tilde{\varphi}_k \\ &\quad + 2\eta_k^T \left(\bar{\mathscr{A}}_k + \mathscr{B}_k \Phi_k^{-1} \mathscr{B}_k^T \mathcal{P}_{k+1} \mathscr{A}_k \right)^T \mathcal{Q}_{k+1} \tilde{\varphi}_k - \eta_k^T \mathcal{Q}_k \eta_k. \end{aligned} \tag{4.32}$$

Then, it is easy to see that

$$J_k^{(2)}|_{\xi=\xi^*} = \eta_k^T \Delta_k \eta_k + 2e_k^T \bar{\mathcal{A}}_k^T Q_4^{k+1} \varphi_k + \varphi_k^T Q_4^{k+1} \varphi_k \tag{4.33}$$

where

$$\begin{aligned}
\Delta_k &= \left(\bar{\mathscr{A}}_k + \mathscr{B}_k \Phi_k^{-1} \mathscr{B}_k^T \mathcal{P}_{k+1} \mathscr{A}_k \right)^T \mathcal{Q}_{k+1} \left(\bar{\mathscr{A}}_k + \mathscr{B}_k \Phi_k^{-1} \mathscr{B}_k^T \mathcal{P}_{k+1} \mathscr{A}_k \right) \\
&\quad + \Xi_k^{(1)} + \Xi_k^{(2)} + 2 \left(\mathcal{R}_k + \mathscr{B}_k \Phi_k^{-1} \mathscr{B}_k^T \mathcal{P}_{k+1} \mathscr{A}_k \right)^T \mathcal{Q}_{k+1} \mathcal{S}_k - \mathcal{Q}_k
\end{aligned}$$

with

$$\mathcal{R}_k = \text{diag}\left\{ \bar{A}_k, 0 \right\}, \quad \mathcal{S}_k = \text{diag}\left\{ 0, -\sum_{i=1}^{n} E^{(i)} \mathcal{K}_k \bar{\mathcal{C}}_k^{(i)} \right\},$$

$$\Xi_k^{(1)} = \text{diag}\left\{ \sum_{p=1}^{n+1} \sum_{q=1}^{n+1} \sum_{i=1}^{s} \Pi_k^{(i)} \text{tr}\left(Q_{pq}^{k+1} \Gamma_k^{(i)} \right), 0, \cdots, 0 \right\},$$

$$\Xi_k^{(2)} = \text{diag}\left\{ \Xi_k^{(21)}, 0, \cdots, 0 \right\}, \tag{4.34}$$

$$\begin{aligned}
\Xi_k^{(21)}[s,t] &= \sum_{p=1}^{n+1} \sum_{q=1}^{n+1} \sum_{j=1}^{n} \sum_{i=1}^{n} M_{a_{tj}^k, a_{si}^k} (Q_{pq}^{k+1})_{[i,j]} \\
&\quad - \sum_{j=1}^{n} \sum_{i=1}^{n} M_{a_{tj}^k, a_{si}^k} (Q_{11}^{k+1})_{[i,j]} \\
&\quad + \sum_{r=1}^{n} \sum_{p=1}^{n} \sum_{q=1}^{n} \sum_{j=1}^{n} \sum_{i=1}^{n} h_{pr} h_{qr} M_{c_{tj}^k, c_{si}^k} (\bar{K}_k^{(pr)T} Q_{p+1,q+1}^{k+1} \bar{K}_k^{(qr)})_{[i,j]}.
\end{aligned}$$

Here, Q_4^{k+1} and Q_{ij}^{k+1} $(i, j = 1, 2, \cdots, n+1)$ are block elements of the matrix $\mathcal{Q}_{k+1}$ (same as the form $\mathcal{P}_{k+1}$). Furthermore, it follows that

$$\begin{aligned}
J_k^{(2)}|_{\xi=\xi^*} &= \eta_k^T \Delta_k \eta_k + 2e_k^T \bar{\mathcal{A}}_k^T Q_4^{k+1} \varphi_k + \varphi_k^T Q_4^{k+1} \varphi_k \\
&\quad + \frac{1}{n} ||\tilde{z}_k||^2 + \frac{1}{n} ||\varphi_k||^2 - \left(\frac{1}{n} ||\tilde{z}_k||^2 + \frac{1}{n} ||\varphi_k||^2 \right) \\
&= \eta_k^T \left(\Delta_k + \frac{1}{n} \bar{\mathcal{L}}_k^T \bar{\mathcal{L}}_k \right) \eta_k + 2e_k^T \bar{\mathcal{A}}_k^T Q_4^{k+1} \varphi_k \\
&\quad + \varphi_k^T \Psi_k \varphi_k - \left(\frac{1}{n} ||\tilde{z}_k||^2 + \frac{1}{n} ||\varphi_k||^2 \right)
\end{aligned} \tag{4.35}$$

where $\Psi_k = Q_4^{k+1} + \frac{1}{n} I$.

By the completing squares method again, we have

$$\begin{aligned}
J_k^{(2)}|_{\xi=\xi^*} &= \eta_k^T \left(\Delta_k + \frac{1}{n} \bar{\mathcal{L}}_k^T \bar{\mathcal{L}}_k - (\bar{\mathscr{A}}_k - \mathcal{R}_k)^T \mathcal{U}_{k+1} \Psi_k^{-1} \mathcal{U}_{k+1} (\bar{\mathscr{A}}_k - \mathcal{R}_k) \right) \eta_k \\
&\quad + (\varphi_k - \varphi_k^*)^T \Psi_k (\varphi_k - \varphi_k^*) - \left(\frac{1}{n} ||\tilde{z}_k||^2 + \frac{1}{n} ||\varphi_k||^2 \right)
\end{aligned} \tag{4.36}$$

with $\varphi_k^* = -\Psi_k^{-1} Q_4^{k+1} \bar{\mathscr{A}}_k e_k$ and $\mathcal{U}_{k+1} = [0 \; Q_4^{k+1T}]^T$. Then, it is easily shown that

$$
\begin{aligned}
&\mathcal{J}_2(\mathcal{K}, \xi^*) \\
&= \mathbb{E}\left\{ \frac{1}{n}\|\tilde{z}_k\|_{[0,N-1]}^2 + \frac{1}{n}\|\varphi_k\|_{[0,N-1]}^2 \right\} \\
&= \sum_{k=0}^{N-1} (\varphi_k - \varphi_k^*)^T \Psi_k (\varphi_k - \varphi_k^*) \\
&\quad + \mathbb{E}\left\{ \eta_0^T \mathcal{Q}_0 \eta_0 - \eta_N^T \mathcal{Q}_N \eta_N \right\} + \sum_{k=0}^{N-1} \mathbb{E}\left\{ \eta_k^T \left(\Delta_k + \frac{1}{n}\bar{\mathcal{L}}_k^T \bar{\mathcal{L}}_k \right. \right. \\
&\quad \left. \left. - (\bar{\mathscr{A}}_k - \mathcal{R}_k)^T \mathcal{U}_{k+1} \Psi_k^{-1} \mathcal{U}_{k+1} (\bar{\mathscr{A}}_k - \mathcal{R}_k) \right) \eta_k \right\}.
\end{aligned}
\tag{4.37}
$$

Under the final condition $\mathcal{Q}_N = 0$, in order to suppress the cost of $\mathcal{J}_2(\mathcal{K}, \xi^*)$, the best choice of $\mathcal{K}_k$ is that satisfying

$$
\begin{cases}
\mathcal{Q}_k = \left(\bar{\mathscr{A}}_k + \mathscr{B}_k \Phi_k^{-1} \mathscr{B}_k^T \mathcal{P}_{k+1} \mathscr{A}_k \right)^T \mathcal{Q}_{k+1} \\
\quad \times \left(\bar{\mathscr{A}}_k + \mathscr{B}_k \Phi_k^{-1} \mathscr{B}_k^T \mathcal{P}_{k+1} \mathscr{A}_k \right) + \frac{1}{n}\bar{\mathcal{L}}_k^T \bar{\mathcal{L}}_k + \Xi_k^{(1)} + \Xi_k^{(2)} \\
\quad + \left(\mathcal{R}_k + \mathscr{B}_k \Phi_k^{-1} \mathscr{B}_k^T \mathcal{P}_{k+1} \mathscr{A}_k \right)^T \mathcal{Q}_{k+1} \mathcal{S}_k \\
\quad + \mathcal{S}_k^T \mathcal{Q}_{k+1} \left(\mathcal{R}_k + \mathscr{B}_k \Phi_k^{-1} \mathscr{B}_k^T \mathcal{P}_{k+1} \mathscr{A}_k \right) \\
\quad - (\bar{\mathscr{A}}_k - \mathcal{R}_k)^T \mathcal{U}_{k+1} \Psi_k^{-1} \mathcal{U}_{k+1} (\bar{\mathscr{A}}_k - \mathcal{R}_k), \\
\mathcal{K}_k = \arg\min_{\mathcal{K}_k} \text{norm} \left(\sum_{i=1}^{n} E^{(i)} \mathcal{K}_k \bar{\mathcal{C}}_k^{(i)} - \Psi_k^{-1} Q_4^{k+1} \bar{\mathscr{A}}_k \right)
\end{cases}
\tag{4.38}
$$

where "arg" is short for "argument", which means the value of $\mathcal{K}_k$ minimizing the norm.

Subsequently, denoting

$$
\mathcal{K}_k^{(i)} \triangleq [\bar{K}_k^{(i1)} \; \bar{K}_k^{(i2)} \; \cdots \; \bar{K}_k^{(in)}]
$$

and

$$
\Psi_k^{-1} Q_4^{k+1} \bar{\mathscr{A}}_k \triangleq \begin{pmatrix} \Pi_{11}^k & \Pi_{12}^k & \cdots & \Pi_{1,n}^k \\ \Pi_{21}^k & \Pi_{22}^k & \cdots & \Pi_{2,n}^k \\ \vdots & \vdots & \vdots & \vdots \\ \Pi_{n,1}^k & \Pi_{n,2}^k & \cdots & \Pi_{n,n}^k \end{pmatrix},
$$

one has $\sum_{i=1}^{n} E^{(i)} \mathcal{K}_k \bar{\mathcal{C}}_k^{(i)} = [\; \bar{\mathcal{C}}_k^{(1)T} \mathcal{K}_k^{(1)T} \quad \bar{\mathcal{C}}_k^{(2)T} \mathcal{K}_k^{(2)T} \quad \cdots \quad \bar{\mathcal{C}}_k^{(n)T} \mathcal{K}_k^{(n)T} \;]^T$.
Furthermore, it is easy to obtain that

$$
\begin{aligned}
[\bar{K}_k^{(i1)} \; \bar{K}_k^{(i2)} \; \cdots \; \bar{K}_k^{(in)}] &= [\Pi_{i1}^k \; \Pi_{i2}^k \; \cdots \; \Pi_{i,n}^k] \bar{\mathcal{C}}_k^{(i)\dagger} \\
&\Longrightarrow \bar{K}_k^{(ij)} = \begin{cases} \Pi_{i,j}^k h_{ij}^{-1} \bar{C}_k^{(j)\dagger}, & h_{ij} \neq 0, \\ 0, & h_{ij} = 0. \end{cases}
\end{aligned}
\tag{4.39}
$$

Therefore, we can easily find from (4.39) that the calculated value has the same structure as specified for K_k in the distributed state estimators (4.7).

Remark 4.4 *It can be seen that the first functional* $\mathcal{J}_1(\eta_0, \xi^*)$ *is associated with the* $\mathcal{H}_\infty$*-constraint criterion, whereas the second functional* $\mathcal{J}_2(\mathcal{K}, \xi^*)$ *is related to the sum of input and output energies of the system (4.30) (or* $\mathcal{H}_2$*-type criterion). Due to the given topology of sensor networks and the existence of both stochastic parameters and stochastic nonlinearity, we are unable to design a globally optimal estimator via minimizing the cost* $\mathcal{J}_2(\mathcal{K}, \xi^*)$*. Instead, a more realistic way is to seek a sub-optimal estimator in terms of (4.38) to realize the* $\mathcal{H}_\infty$*-constraint criterion.*

The following result can be easily accessible from Lemma 4.2 and the analysis conducted above, and the corresponding proof is therefore omitted.

Theorem 4.1 *Let the disturbance attenuation level* $\gamma > 0$ *and the positive definite matrix* W *be given. There exist parameters* $\{K_k^{(ij)}\}_{0 \le k \le N-1}$ *(i =* $\mathcal{V}, j \in \mathcal{N}_i$*) for the distributed state estimator (4.7) such that the augmented system (4.9) satisfies the* $\mathcal{H}_\infty$ *performance constraint (4.10) for any nonzero* $\xi \in \ell_2$*, if the following coupled backward recursive RDEs have solutions* $(\mathcal{P}_k, \mathcal{K}_k)$ *and* $(\mathcal{Q}_k, \mathcal{K}_k)$ *(0 ≤ k ≤ N − 1):*

$$
\begin{cases}
\mathcal{P}_k = \mathscr{A}_k^T \mathcal{P}_{k+1} \mathscr{A}_k + \dfrac{1}{n} \bar{\mathcal{L}}_k^T \bar{\mathcal{L}}_k + \Theta_k^{(1)} \\[2mm]
\quad + \Theta_k^{(2)} + \mathscr{A}_k^T \mathcal{P}_{k+1} \mathscr{B}_k \Phi_k^{-1} \mathscr{B}_k^T \mathcal{P}_{k+1} \mathscr{A}_k, \\[2mm]
\mathcal{P}_k \ge 0, \quad \mathcal{P}_N = 0,
\end{cases} \tag{4.40}
$$

$$
\begin{cases}
\mathcal{Q}_k = \left(\bar{\mathscr{A}}_k + \mathscr{B}_k \Phi_k^{-1} \mathscr{B}_k^T \mathcal{P}_{k+1} \mathscr{A}_k \right)^T \mathcal{Q}_{k+1} \left(\bar{\mathscr{A}}_k + \mathscr{B}_k \Phi_k^{-1} \mathscr{B}_k^T \mathcal{P}_{k+1} \mathscr{A}_k \right) \\[2mm]
\quad + \dfrac{1}{n} \bar{\mathcal{L}}_k^T \bar{\mathcal{L}}_k + \left(\mathcal{R}_k + \mathscr{B}_k \Phi_k^{-1} \mathscr{B}_k^T \mathcal{P}_{k+1} \mathscr{A}_k \right)^T \mathcal{Q}_{k+1} \mathcal{S}_k \\[2mm]
\quad + \mathcal{S}_k^T \mathcal{Q}_{k+1} \left(\mathcal{R}_k + \mathscr{B}_k \Phi_k^{-1} \mathscr{B}_k^T \mathcal{P}_{k+1} \mathscr{A}_k \right) + \Xi_k^{(1)} + \Xi_k^{(2)} \\[2mm]
\quad - \left(\bar{\mathscr{A}}_k - \mathcal{R}_k \right)^T \mathcal{U}_{k+1} \Psi_k^{-1} \mathcal{U}_{k+1} \left(\bar{\mathscr{A}}_k - \mathcal{R}_k \right), \\[2mm]
\mathcal{Q}_k \ge 0, \quad \mathcal{Q}_N = 0,
\end{cases} \tag{4.41}
$$

such that

$$
\begin{cases}
\Phi_k = \gamma^2 I - \mathscr{B}_k^T \mathcal{P}_{k+1} \mathscr{B}_k > 0, & (4.42\text{a}) \\[2mm]
\Psi_k = Q_4^{k+1} + \dfrac{1}{n} I, & (4.42\text{b}) \\[2mm]
\mathcal{P}_0 < \gamma^2 W, & (4.42\text{c}) \\[2mm]
\mathcal{K}_k = \arg\min_{\mathcal{K}_k} norm \left(\displaystyle\sum_{i=1}^n E^{(i)} \mathcal{K}_k \bar{\mathcal{C}}_k^{(i)} - \Psi_k^{-1} Q_4^{k+1} \bar{A}_k \right) & (4.42\text{d})
\end{cases}
$$

where the corresponding matrix parameters are defined as in (4.30), (4.34) and Lemma 4.2.

Noticing that the estimator gain matrices are involved in the proposed coupled RDEs, by means of Theorem 4.1, we can summarize the distributed estimator design algorithm as follows.

Algorithm:

Step 1. Set $k = N - 1$. Then $\mathcal{P}_N = \mathcal{Q}_N = 0$ are available.

Step 2. Compute Φ_k and Ψ_k via (4.42a) and (4.42b), and then obtain the estimator gain matrices $K_k^{(ij)}$ $(i = 1, 2, \cdots, n, j \in \mathcal{N}_i)$ by (4.42d) and the matrix's generalized inverse.

Step 3. Compute $\Theta_k^{(1)}$, $\Theta_k^{(2)}$, $\Xi_k^{(1)}$ and $\Xi_k^{(2)}$ in (4.34) and Lemma 4.2, respectively.

Step 4. Solve the first equations of (4.40) and (4.41) to get $\mathcal{P}_k$ and $\mathcal{Q}_k$, respectively.

Step 5. If $\mathcal{P}_k \geq 0$ and $\mathcal{Q}_k \geq 0$, then $K_k^{(ij)}$ are suitable estimator parameters, and go to the next step, else go to *Step 7*.

Step 6. If $k \neq 0$, set $k = k - 1$ and go to *Step 2*, else go to the next step.

Step 7. If $\mathcal{P}_k < 0$, $\mathcal{Q}_k < 0$ $(\forall k \neq 0)$ or $\mathcal{P}_0 \geq \gamma^2 W$, this algorithm is infeasible. Stop.

In what follows, we show that our main results can be easily specialized to the sensor networks with certain special topology. In the case of $h_{ij} \neq 0$, we have the following corollary readily.

Corollary 4.1 *Let the disturbance attenuation level $\gamma > 0$ and the positive definite matrix W be given. Let us assume that $h_{ij} \neq 0$, and $\bar{C}_k^{(i)}$ is row full rank for $\forall i, j \in \mathcal{V}$, $k \in [0, N - 1]$. If there exist two families of non-negative definite real-symmetric matrices $\{\mathcal{P}_k\}_{0 \leq k < N}$ and $\{\mathcal{Q}_k\}_{0 \leq k < N}$ to the coupled backward recursive RDEs (4.40) and (4.41) together with*

$$
\begin{cases}
\Phi_k = \gamma^2 I - \mathscr{B}_k^T \mathcal{P}_{k+1} \mathscr{B}_k > 0, \\
\Psi_k = Q_4^{k+1} + \dfrac{1}{n} I, \\
\mathcal{P}_0 < \gamma^2 W,
\end{cases}
\tag{4.43}
$$

then,

(i) the worst-case disturbance ξ_k^* and the estimator gain matrices $K_k^{(ij)*}$ are given by

$$
\xi_k^* = \Phi_k^{-1} \mathscr{B}_k^T \mathcal{P}_{k+1} \mathscr{A}_k \eta_k, \quad K_k^{(ij)*} = \Pi_{i,j}^k h_{ij}^{-1} \bar{C}_k^{(j)\dagger};
\tag{4.44}
$$

(ii) the augmented system (4.9) satisfies the $\mathcal{H}_\infty$ performance constraint (4.10) for any nonzero $\xi \in \ell_2$;

(iii) the costs or performance objectives for $\mathcal{J}_1(\eta_0, \xi)$ and $\mathcal{J}_2(K, \xi)$ are

$$
\mathcal{J}_1(\eta_0, \xi^*; K_k^*) = \eta_0^T (\mathcal{P}_0 - \gamma^2 W) \eta_0, \quad \mathcal{J}_2(K_k^*, \xi^*; \eta_0) = \eta_0^T \mathcal{Q}_0 \eta_0.
$$

Proof *First, it can be shown from Lemma 4.2 that when there exist solutions $\mathcal{P}_k$ to (4.40) such that $\Phi_k > 0$ and $\mathcal{P}_0 < \gamma^2 W$, the system (4.9) satisfies the pre-specified $\mathcal{H}_\infty$ performance. Subsequently, because of $h_{ij} \neq 0$ (for $\forall i, j \in \mathscr{V}$) and according to Lemma 4.1 as well as (4.42d) in Theorem 4.1, we have $\mathcal{K}_k^{(i)} = \mathscr{K}_k^{(i)} \bar{\mathcal{C}}_k^{(i)\dagger}$ and further obtain the worst-case disturbance $\xi_k^* = \Phi_k^{-1} \mathscr{B}_k^T \mathcal{P}_{k+1} \mathscr{A}_k \eta_k$. Furthermore, substituting ξ_k^* and $\mathcal{K}_k^*$ into (4.37) gives*

$$\mathcal{J}_1(\eta_0, \xi; \mathcal{K}_k^*) \leq \mathcal{J}_1(\eta_0, \xi^*; \mathcal{K}_k^*) = \eta_0^T (\mathcal{P}_0 - \gamma^2 W) \eta_0,$$

and

$$\mathcal{J}_2(\mathcal{K}_k^*, \xi^*; \eta_0) = \eta_0^T \mathcal{Q}_0 \eta_0.$$

The proof is complete.

Remark 4.5 *In this chapter, the distributed estimation problem is considered for time-varying systems with stochastic parameters and nonlinearities. Due to such a complicated time-varying nature, our research is focused on the finite horizon case and we are only interested in the transient property over the finite horizon $k \in [0 \ N-1]$, i.e., the estimation error is bounded at every time-instant $k \in [0 \ N-1]$, and the average $\mathcal{H}_\infty$-norm constraint is satisfied over the finite horizon $k \in [0 \ N-1]$.*

Remark 4.6 *To deal with the computational complexity of the developed RDE-based algorithm, we recall that the sensor network size is n, the length of finite time horizon is N, and the variable dimensions can be seen from $x_k \in \mathbb{R}^{n_x}$, $z_k \in \mathbb{R}^{n_z}$, $x_k^{(i)} \in \mathbb{R}^{n_x}$, $y_k^{(i)} \in \mathbb{R}^{n_y}$, w_k and $v_k^{(i)} \in \mathbb{R}$ ($i = 1, 2, \cdots, n$). It follows from Theorem 4.1 that the proposed algorithm is implemented recursively for N steps and each step consists of $n + 2$ instances of the matrix inversion operation and $4n^5 + 4(n+1)^2 s + 5n + 38$ instances of the matrix multiplication operation. Therefore, it is not difficult to calculate the overall computational complexity of the given algorithm as $O(Nn^5 n_x^3)$, which depends linearly on the length of the finite time horizon and polynomially on the sensor network size and the variable dimensions. Note that the computational burden is mainly caused by the basic mathematical operations. Fortunately, research on improving the computational efficiency of mathematical operations is a very active area in computational mathematics, optimization, and operational research, and substantial progress can be expected in the future.*

Remark 4.7 *In this chapter, we examine how the stochastic parameters and stochastic nonlinearities influence the performance of the distributed estimation over a sensor network with a given topology. In Theorem 4.1, all the system parameters as well as the occurrence probabilities for the stochastic parameters and stochastic nonlinearities are reflected in the backward recursive RDEs. Also, an efficient algorithm DED is proposed to solve the addressed distributed estimator design problem. Due to the recursive nature of the algorithm, the proposed design procedure is suitable for online application without the need of increasing the problem size. This is particularly attractive* *for relatively complex systems involving severe nonlinearities and strong coupling of the sensor nodes. Compared to existing literature, the target plant considered in this chapter is more general and covers both the stochastic parameters and the stochastic nonlinearities. One of the main contributions is the development of a new "stochastic version" of the bounded real lemma specifically for handling the $\mathcal{H}_\infty$ estimation performance of the sensor networks.*

4.4 Simulation Examples

In this section, we present a simulation example to illustrate the effectiveness of the proposed distributed estimator design scheme for nonlinear systems with both stochastic parameters and stochastic nonlinearities through sensor networks.

The target plant considered is modeled by (4.1) with the following parameters:

$$B_k = [\ 0.16 \quad 0.18\ ]^T, \quad L_k = [\ 0.40 \quad -0.60\ ],$$
$$A_k(\mu_k) = \bar{A}_k + \tilde{A}_k(\mu_k)$$
$$= \begin{bmatrix} 0.90 + 0.22\cos(0.12k) & 0.4 \\ 0.12 & -0.75 \end{bmatrix} + \mu_k \begin{bmatrix} 0.02 & 0 \\ 0 & 0.03 \end{bmatrix},$$

and the stochastic nonlinear function $f_k(x_k, \vartheta_k)$ is chosen as

$$f_k(x_k, \vartheta_k) = \left(0.1\mathrm{sign}(x_k^{(1)})x_k^{(1)}\vartheta_k^{(1)} + 0.2\mathrm{sign}(x_k^{(2)})x_k^{(2)}\sin(\vartheta_k^{(2)})\right) \begin{bmatrix} 0.06 \\ 0.09 \end{bmatrix}$$

where $x_k^{(i)}$ $(i = 1, 2)$ denotes the i-th element of the system state, and μ_k, $\vartheta_k^{(1)}$ and $\vartheta_k^{(2)}$ are zero mean, uncorrelated Gaussian white noise sequences with unity covariances. By using MATLAB® software, it is not difficult to verify that the above stochastic nonlinear function satisfies

$$\mathbb{E}\left\{f_k(x_k, \vartheta_k)|x_k\right\} = 0,$$
$$\mathbb{E}\left\{f_k(x_k, \vartheta_k)f_k^T(x_k, \vartheta_k)|x_k\right\} = \begin{bmatrix} 0.06 \\ 0.09 \end{bmatrix}\begin{bmatrix} 0.06 \\ 0.09 \end{bmatrix}^T x_k^T \begin{bmatrix} 0.01 & 0 \\ 0 & 0.0173 \end{bmatrix} x_k.$$

The sensor network shown in Fig. 4.1 is represented by a directed graph $\mathscr{G} = (\mathscr{V}, \mathscr{E}, \mathscr{H})$ with the set of nodes $\mathscr{V} = \{1, 2, 3, 4\}$, the set of edges

$$\mathscr{E} = \{(1, 1), (1, 2), (2, 1), (2, 2), (2, 3), (3, 1), (3, 3), (4, 1), (4, 4)\}$$

and the following adjacency matrix

$$\mathscr{H} = \begin{bmatrix} 1 & 1 & 0 & 0 \\ 1 & 1 & 1 & 0 \\ 1 & 0 & 1 & 0 \\ 1 & 0 & 0 & 1 \end{bmatrix}.$$

The dynamics of the sensor nodes are described by (4.2) with parameters as

$$C_k^{(i)}(\varepsilon_k^{(i)}) = \bar{C}_k^{(i)} + \tilde{C}_k^{(i)}(\varepsilon_k^{(i)}) = \bar{C}_k^{(i)} + \varepsilon_k^{(i)}[\ 0.01 \quad 0\ ],$$

$$\bar{C}_k^{(1)} = [\ 0.82 \quad 0.62\ ], \quad \bar{C}_k^{(2)} = [\ 0.75 \quad 0.80\ ],$$

$$\bar{C}_k^{(3)} = [\ 0.74 \quad 0.75\ ], \quad \bar{C}_k^{(4)} = [\ 0.75 \quad 0.65\ ],$$

$$D_k^{(1)} = 0.18, \quad D_k^{(2)} = 0.12, \quad D_k^{(3)} = 0.16, \quad D_k^{(4)} = 0.14$$

where $\varepsilon_k^{(i)}(i = 1, 2, 3, 4)$ are mutually independent zero mean Gaussian noise sequences with unity covariances.

The $\mathcal{H}_\infty$ performance level, positive definite matrix W, and time-horizon N are taken as $\gamma = 0.95$, $\text{diag}_{10}\{0.3\}$, and 91, respectively. Using the developed computational algorithm and MATLAB (with YALMIP 3.0), we can check the feasibility of the coupled recursive RDEs and obtain the desired estimator parameters as shown in Table 4.1.

In the simulation, the exogenous disturbance inputs are selected as

$$\begin{aligned} w_k &= 0.15\cos(0.2k), \\ v_k^{(1)} &= 2\sin(10k)/(3k), \\ v_k^{(2)} &= 4k/(3k^2 + 4), \\ v_k^{(3)} &= -0.12\cos(0.2k), \\ v_k^{(4)} &= 0.10\sin(5k)\exp(-2k). \end{aligned}$$

The initial conditions are set as $x_0 = [3.0 \ -5.4]^T$ and $\hat{x}_0^{(i)} = [0\ 0]^T$ ($i = 1, 2, 3, 4$). Simulation results are shown in Figs. 4.2-4.4, where Fig. 4.2 and Fig. 4.3 depict the trajectories for the states and estimates, and Fig. 4.4 plots the estimation errors $z_k - \hat{z}_k^{(i)}$ ($i = 1, 2, 3, 4$). The simulation results show that estimators have a satisfactory tracking performance even though the state trajectories of the target plant are divergent. Therefore, it has been confirmed that the distributed estimation technology presented in this chapter is indeed effective.

Remark 4.8 *It is well known that the traditional Kalman filter (KF) serves as an optimal filter in the least mean square sense for linear systems with the assumption that the system model is exactly known. In the case that the system model is nonlinear and/or uncertain, many alternative filtering*

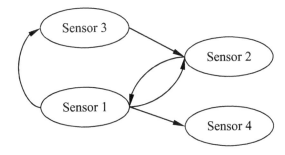

FIGURE 4.1
Topological structure of the sensor network.

schemes have been reported in the literature including the extended Kalman filter (EKF) and $\mathcal{H}_\infty$ filter. EKF has been shown to be an effective way of tackling estimation problems of the nonlinear system. EKF requires that the noise statistics are accurate and the deterministic nonlinear functions are continuously differentiable, while the $\mathcal{H}_\infty$ filtering method provides a bound for the worst-case estimation error without the need for knowledge of noise statistics. Very recently, the RLMI technique has been developed to effectively solve the finite-horizon $\mathcal{H}_\infty$ filtering problems in a recursive manner [109], where the nonlinearity and/or uncertainty need to satisfy certain sector-like conditions. It should be pointed out that, for the problem addressed in this chapter, 1) the considered external disturbances do not possess known statistical properties, 2) the nonlinearities do not explicitly satisfy any bounded conditions, and 3) the stochastic parameters and nonlinearities bring in some complicated nonlinear terms such as $\Theta_k^{(1)}$, $\Theta_k^{(2)}$, $\Xi_k^{(1)}$ and $\Xi_k^{(2)}$. Unfortunately, these factors prevent the existing methods (KF, EKF, $\mathcal{H}_\infty$-filtering and RLMI) from being applied to the distributed $\mathcal{H}_\infty$ state estimation problem for the underlying system in this chapter. In summary, the proposed RDE algorithm provides yet another approach that complements the existing techniques for handling certain types of complex systems.

4.5 Summary

In this chapter, we have dealt with the distributed $\mathcal{H}_\infty$ state estimation problem for a class of discrete time-varying nonlinear systems in sensor networks. To reflect more realistic situations, we have considered the stochastic parameters in both the system and output matrices. Also, the stochastic nonlinearities have been taken into account. By employing the completing

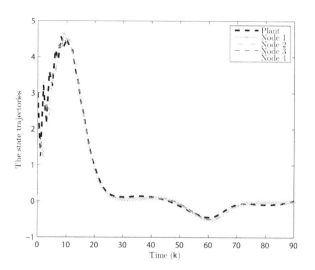

FIGURE 4.2
The trajectories of $x_k^{(1)}$ and $\hat{x}_k^{(i,1)}(i = 1, 2, 3, 4)$.

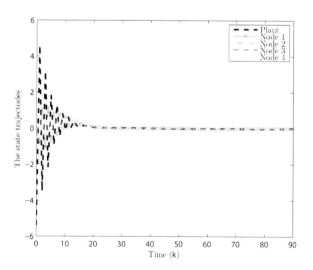

FIGURE 4.3
The trajectories of $x_k^{(2)}$ and $\hat{x}_k^{(i,2)}(i = 1, 2, 3, 4)$.

TABLE 4.1
The distributed state estimator gain matrices.

k	0	1	2	3
K_{11}	$\begin{bmatrix} 0.4329 \\ -0.1725 \end{bmatrix}$	$\begin{bmatrix} 0.4244 \\ -0.1741 \end{bmatrix}$	$\begin{bmatrix} 0.4131 \\ -0.1761 \end{bmatrix}$	$\begin{bmatrix} 0.3996 \\ -0.1785 \end{bmatrix}$
$K_{12}(10^{-3})$	$\begin{bmatrix} 1.1000 \\ -0.7628 \end{bmatrix}$	$\begin{bmatrix} 1.0000 \\ -0.7468 \end{bmatrix}$	$\begin{bmatrix} 0.9818 \\ -0.7279 \end{bmatrix}$	$\begin{bmatrix} 0.9241 \\ -0.7071 \end{bmatrix}$
$K_{21}(10^{-3})$	$\begin{bmatrix} 1.2000 \\ -0.5701 \end{bmatrix}$	$\begin{bmatrix} 1.2000 \\ -0.5618 \end{bmatrix}$	$\begin{bmatrix} 1.1000 \\ -0.5518 \end{bmatrix}$	$\begin{bmatrix} 1.1000 \\ -0.5409 \end{bmatrix}$
K_{22}	$\begin{bmatrix} 0.3824 \\ -0.1987 \end{bmatrix}$	$\begin{bmatrix} 0.3752 \\ -0.2001 \end{bmatrix}$	$\begin{bmatrix} 0.3658 \\ -0.2039 \end{bmatrix}$	$\begin{bmatrix} 0.2062 \\ -0.3544 \end{bmatrix}$
$K_{23}(10^{-3})$	$\begin{bmatrix} 1.3000 \\ -0.5920 \end{bmatrix}$	$\begin{bmatrix} 1.2000 \\ -0.5815 \end{bmatrix}$	$\begin{bmatrix} 1.2000 \\ -0.5691 \end{bmatrix}$	$\begin{bmatrix} 1.1000 \\ -0.5534 \end{bmatrix}$
$K_{31}(10^{-3})$	$\begin{bmatrix} 1.2000 \\ -0.5981 \end{bmatrix}$	$\begin{bmatrix} 1.2000 \\ -0.5889 \end{bmatrix}$	$\begin{bmatrix} 1.1000 \\ -0.5778 \end{bmatrix}$	$\begin{bmatrix} 1.000 \\ -0.5657 \end{bmatrix}$
K_{33}	$\begin{bmatrix} 0.4025 \\ -0.2014 \end{bmatrix}$	$\begin{bmatrix} 0.3948 \\ -0.2029 \end{bmatrix}$	$\begin{bmatrix} 0.3848 \\ -0.2047 \end{bmatrix}$	$\begin{bmatrix} 0.3727 \\ -0.2069 \end{bmatrix}$
$K_{41}(10^{-3})$	$\begin{bmatrix} 1.1000 \\ -0.6831 \end{bmatrix}$	$\begin{bmatrix} 1.000 \\ -0.6716 \end{bmatrix}$	$\begin{bmatrix} 1.0000 \\ -0.6579 \end{bmatrix}$	$\begin{bmatrix} 0.9914 \\ -0.6429 \end{bmatrix}$
K_{44}	$\begin{bmatrix} 0.4398 \\ -0.1954 \end{bmatrix}$	$\begin{bmatrix} 0.4312 \\ -0.1970 \end{bmatrix}$	$\begin{bmatrix} 0.4200 \\ -0.1991 \end{bmatrix}$	$\begin{bmatrix} 0.4065 \\ -0.2014 \end{bmatrix}$

squares method and the stochastic analysis technique, a necessary and sufficient condition has been established to ensure the dynamics of the estimation error to satisfy the $\mathcal{H}_\infty$ performance constraint. Furthermore, the estimator gains have been explicitly characterized by means of the solutions to two coupled backward recursive Riccati difference equations (RDEs). Finally, an illustrative example has been provided that highlights the usefulness of the developed state estimation approach. Further research topics include the extension of the main results to the consensus control problem for multi-agent networks with both stochastic parameters and communication noises, to the mobile robot localization problem with missing measurements and also to the $\mathcal{H}_\infty$ control/filtering problem for network control systems with missing measurements, and so on.

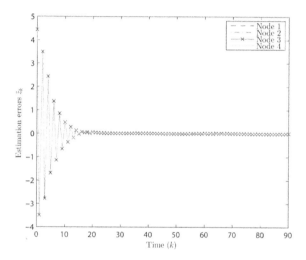

FIGURE 4.4
Estimation errors $\tilde{z}_k^{(i)} (i = 1, 2, 3, 4)$.

5

Finite-Horizon Dissipative Control for State-Saturated Discrete Time-Varying Systems with Missing Measurements

In some practical mechanical systems and motor propulsion systems, the state variables are forced to stay within a given bounded set due to the physical constraints of devices or the safety settings. These kinds of systems are usually called state-saturated systems, and the inherent saturation gives rise to nonlinear characteristics that might severely restrict the application of the existing control or filtering scheme. Specifically, the developed backward RDE method in Chapter 2 cannot be utilized to analyze the system performance because the saturation function is usually differentiable.

This chapter investigates the dissipative control problem for a class of discrete time-varying systems with the simultaneous presence of state saturations, randomly occurring nonlinearities, as well as multiple missing measurements. In order to render more practical significance of the system model, some Bernoulli distributed white sequences with known conditional probabilities are adopted to describe the phenomena of the randomly occurring nonlinearities and the multiple missing measurements. By introducing a free matrix with its infinity norm less than or equal to 1, the system state is bounded by a convex hull so that some sufficient conditions can be obtained in the form of recursive nonlinear matrix inequalities. A novel controller design algorithm is then developed to deal with the recursive nonlinear matrix inequalities. Furthermore, the obtained results are extended to the case when the state saturation is partial. Two numerical simulation examples are provided to demonstrate the effectiveness and applicability of the proposed controller design approach.

5.1 Modeling and Problem Formulation

Let a finite time horizon be denoted by $[0, N] \triangleq \{0, 1, 2, \cdots, N\}$, on which we consider the following class of discrete time-varying nonlinear systems with state saturations:

$$\begin{cases} x_{k+1} = \varphi\Big(A_1^k x_k + A_2^k x_k \varepsilon_k + \alpha_k f(k, x_k) + B_k u_k + D_k w_k\Big), \\ z_k = L_k x_k \end{cases} \quad (5.1)$$

where $x_k \in \mathbb{R}^{n_x}$ is the state vector, $z_k \in \mathbb{R}^{n_z}$ is the system output, $u_k \in \mathbb{R}^{n_u}$ is the control input, and w_k is the disturbance input belonging to $\ell_2([0, N]; \mathbb{R}^p)$. The matrices A_1^k, A_2^k, B_k, D_k and L_k are known real matrices with appropriate dimensions. The stochastic variable ε_k is a zero mean Gaussian white noise sequence with variance σ_ε. The stochastic variable α_k, which is independent of ε_k, is a Bernoulli distributed white sequence taking values of 0 or 1 with:

$$\begin{cases} \text{Prob}\{\alpha_k = 1\} = \bar{\alpha}, \\ \text{Prob}\{\alpha_k = 0\} = 1 - \bar{\alpha} \end{cases}$$

where $\bar{\alpha} \in [0, 1]$ is a known constant.

The nonlinear vector-valued function $f : [0, N] \times \mathbb{R}^{n_x} \mapsto \mathbb{R}^{n_x}$ is assumed to be continuous and satisfies the following sector-bounded condition:

$$[f(k, x) - \Phi_k x]^T [f(k, x) - \Psi_k x] \le 0 \quad (5.2)$$

where Φ_k and Ψ_k are real matrices with appropriate dimensions.

The saturation function $\varphi : \mathbb{R}^{n_x} \to \mathbb{R}^{n_x}$ is defined as

$$\varphi(x) = \begin{bmatrix} \varphi_1(x_1) & \varphi_2(x_2) & \cdots & \varphi_{n_x}(x_{n_x}) \end{bmatrix}^T \quad (5.3)$$

with $\varphi_i(x_i) = \text{sign}(x_i)\min\{\delta_i, |x_i|\}$, where x_i is the ith element of the vector x, and $\delta_i > 0$ is the saturation level. Here, the notation of "sign" means the signum function. For the purpose of simplicity, we denote $\Theta \triangleq \text{diag}\{\delta_1, \delta_2, \cdots, \delta_{n_x}\}$. From this definition, we have $x_k \in \mathbb{D}^{n_x} \triangleq \{x | -\delta_i \le x_i \le \delta_i, \ i = 1, 2, \cdots, n_x\}$ for $\forall k \in [0, N]$.

In this chapter, the measurement with probabilistic data missing is described by

$$y_k = \Xi_k C_k x_k + E_k v_k \quad (5.4)$$

where $y_k \in \mathbb{R}^m$ is the actual signal received by the controller, C_k and E_k are known real matrices with appropriate dimensions, and v_k is the disturbance input belonging to $\ell_2([0, N]; \mathbb{R}^q)$. $\Xi_k \triangleq \text{diag}\{\vartheta_1^k, \vartheta_2^k, \cdots, \vartheta_m^k\}$ with ϑ_i^k ($i = 1, 2, \cdots, m$) being m mutually independent stochastic variables which are also independent of α_k and ε_k. It is assumed that ϑ_i^k is a Bernoulli distributed white sequence taking values of 0 or 1 with

$$\begin{cases} \text{Prob}\{\vartheta_i^k = 1\} = \bar{\vartheta}_i, \\ \text{Prob}\{\vartheta_i^k = 0\} = 1 - \bar{\vartheta}_i. \end{cases}$$

Remark 5.1 *In networked environments, many practical systems are influenced by randomly occurring nonlinear disturbances while the system* states may suffer from saturation constraints. In this chapter, the model (5.1) is put forward to specifically provide for the network-induced phenomena coupled with state saturations and state-dependent noises, which typically emerge in mechanical systems with position and speed limits, electoral systems with limited power supply for the actuators (motors), digital filters implemented in finite word-length format, and implementable neural networks. Note that the control and filtering problems for state-saturated linear time-invariant (LTI) systems with neither RONs nor state-dependent noises have been dealt with in [58, 73, 74, 122] and the references therein. Furthermore, in real-world networked systems, the measurement data may be collected and then transmitted through multiple sensors. Due to different networked environments and/or sampling rates, the data missing probability for each sensor may be different if there exist packet losses [30]. In this chapter, the model (5.4) is employed to describe this kind of measurement phenomenon with multiple sensors, in which the diagonal matrix Ξ_k and its element ϑ_i^k represent the data missing status for the whole sensor system and the ith sensor, respectively.

Definition 5.1 *[14] Consider the discrete time-varying stochastic system (5.1) defined on $[0, N]$ and let $\mathcal{W} : \mathbb{R}^{q+p} \times \mathbb{R}^{n_z} \to \mathbb{R}$ be a Borel measurable function. Denote $\sum_{i=j}^{k-1} \mathcal{W}(\xi_i, z_i) = 0$ when $k = j$. Then, the controlled system (5.1) is said to be dissipative with respect to $\mathcal{W}$ if, for some admissible control sequences $\{u_k\}_{k\in[0,N]}$, there exists a family of functions $V_k : \mathbb{R}^{n_x} \to \mathbb{R}$ with $V_k(x_k) \geq 0$ ($\forall x_k \in \mathbb{R}^{n_x}$, $k \in [0, N]$) so that $V_k(0) = 0$, $\mathbb{E}\{V_0(x_0)\} < \infty$ and*

$$\mathbb{E}\{V_k(x_k)\} \leq \mathbb{E}\{V_j(x_j)\} - \mathbb{E}\left\{\sum_{i=j}^{k-1} \mathcal{W}(\xi_i, z_i)\right\}, \quad k \geq j \geq 0 \qquad (5.5)$$

for all nonzero $\xi \triangleq \{\xi_k = (v_k^T, w_k^T)^T\}_{k\geq 0} \in \ell_2([0, N]; \mathbb{R}^{q+p})$. Here, $-\mathcal{W}$ is called a supply rate and the family $V = \{V_k\}_{k\in[0,N]}$ is said to be a storage function of the controlled system (5.1).

Definition 5.2 *Let the supply rate be defined by*

$$\mathcal{W}(\xi, z; k, N) = \langle z, \mathcal{Q}z \rangle_{k,N} + 2\langle z, \mathcal{S}\xi \rangle_{k,N} + \langle \xi, \mathcal{R}\xi \rangle_{k,N} \qquad (5.6)$$

where $\mathcal{Q}$, $\mathcal{S}$ and $\mathcal{R}$ are real matrices with $\mathcal{Q}, \mathcal{R}$ symmetric, $N \geq 0$ is an integer, and $\langle a, b \rangle_{k,N} = \mathbb{E}\{\sum_{i=k}^{N} a_i^T b_i\}$. The controlled system (5.1) is said to be $(\mathcal{Q}, \mathcal{S}, \mathcal{R})$-dissipative if, for some scalar $\rho > 0$, the following holds

$$\mathcal{W}(\xi, z; 0, N) \leq \rho \mathbb{E}\{x_0^T x_0\}, \quad \forall N \geq 0. \qquad (5.7)$$

Remark 5.2 *In the literature concerning dissipativity, the concept of the quadratic supply rate has come to play an important role. As pointed out in [23], the notion of $(\mathcal{Q}, \mathcal{S}, \mathcal{R})$-dissipativity implies both H_∞ performance and positive realness. Special cases of dissipativity can be covered by choosing different values for $\mathcal{Q}$, $\mathcal{S}$ and $\mathcal{R}$. For example, (a) if $\mathcal{Q} = I$, $\mathcal{S} = 0$ and*

$\mathcal{R} = -\gamma^2 I$, the inequality (5.7) reduces to the H_∞ performance constraint; (b) if $\mathcal{Q} = 0$, $\mathcal{S} = -I$ and $\mathcal{R} = 0$, the inequality (5.7) reduces to positive realness; (c) if $\mathcal{Q} = \theta I$, $\mathcal{S} = (\theta - 1)I$, and $\mathcal{R} = -\gamma^2 \theta I$ $(\theta \in (0,1))$, the inequality (5.7) reflects a mixed H_∞ and positive real performance, where θ represents a weighting parameter that defines the trade-off between H_∞ and positive real performance. It should be pointed out that the purpose of H_∞ control is to design a controller which stabilizes the plant and also reduces the norm of the closed-loop transfer function from the disturbance input to the controlled output to a prescribed level. On the other hand, the purpose of positive-real control is to guarantee that the plant is stable and the closed-loop transfer function is ensured to be strictly positive real, i.e., the phase-lag is less than 180° of the loop transfer function with a negative feedback connecting two (strictly) positive real systems. It is obvious that both the H_∞ control and the positive real control guarantee the overall stability, but they ignore the phase information and the loop gain, respectively. When it comes to the system analysis, although similar to H_∞ control, the dissipative control problem with the quadratic supply rate could reflect more system characteristics simultaneously (such as stability, phase character, loop gain, etc.).

The purpose of this chapter is to design an output-feedback controller $u_k = K_k y_k$ for the discrete time-varying system with RONs, multiple missing measurements as well as state saturations. More specifically, we are interested in looking for the controller parameter K_k so that the closed-loop system (5.1) is $(\mathcal{Q}, \mathcal{S}, \mathcal{R})$-dissipative.

5.2   Dissipative Control for Full State Saturation Case

In this section, a sufficient condition is provided for the controlled system (5.1) to satisfy the dissipativity constraint (5.7), and an algorithm is then proposed to deal with the addressed controller design problem. Furthermore, the corresponding corollary is provided for the time-invariant system.

First, let $\mathscr{S}_{n_x}$ be the set of $n_x \times n_x$ diagonal matrices whose diagonal elements are either 1 or 0. It is easy to see that there are 2^{n_x} elements in $\mathscr{S}_{n_x}$, whose ith element is denoted as S_i, $i \in [1, 2^{n_x}]$. For example, when $n_x = 2$, one has

$$\mathscr{S}_{n_x} = \left\{ \begin{bmatrix} 1 & 0 \\ 0 & 1 \end{bmatrix}, \begin{bmatrix} 0 & 0 \\ 0 & 1 \end{bmatrix}, \begin{bmatrix} 1 & 0 \\ 0 & 0 \end{bmatrix}, \begin{bmatrix} 0 & 0 \\ 0 & 0 \end{bmatrix} \right\}.$$

For convenience of later analysis, we let $S_1 = I$. Furthermore, define $S_i^- = I - S_i$, and hence $S_i^- \in \mathscr{S}_{n_x}$. Before proceeding, we introduce the following lemmas that will be used in deriving our main results.

Lemma 5.1 *[58] For any symmetric positive definite matrix $\mathcal{P} \in \mathbb{R}^{n_x \times n_x}$, the map $\varrho \rightarrow \varrho^T \mathcal{P} \varrho$ is convex, i.e.*

$$\left(\sum_{i=1}^{n} \delta_i \varrho_i \right)^T \mathcal{P} \left(\sum_{i=1}^{n} \delta_i \varrho_i \right) \leq \sum_{i=1}^{n} \delta_i \varrho_i^T \mathcal{P} \varrho_i, \quad \forall \varrho_i \in \mathbb{R}^{n_x} \tag{5.8}$$

where $\sum_{i=1}^{n} \delta_i = 1$.

Lemma 5.2 *Let $G_k = [g_{ij}] \in \mathbb{R}^{n_x \times n_x}$ satisfy $\|G_k\|_\infty \leq 1$. For any vector $v \in \mathbb{R}^{n_x}$,*

$$\begin{aligned}
&\Theta^{-1}\varphi(A_k x + v) \\
&\in co\{S_i \Theta^{-1}(A_k x + v) + S_i^- G_k \Theta^{-1} x : \ \forall x \in \mathbb{D}^{n_x}, \ i \in [1, 2^{n_x}]\}
\end{aligned} \tag{5.9}$$

where $co\{\cdot\}$ denotes the convex hull of a set.

Proof *Let A_i^k and G_i^k represent the i-th row vectors of A_k and G_k, respectively, and v_i be the ith element of the vector v. Taking $\|G_k\|_\infty \leq 1$ into consideration, one has*

$$|G_l^k \Theta^{-1} x| \leq \sum_{j=1}^{n_x} |g_{lj} \delta_j^{-1} x_j| \leq \sum_{j=1}^{n_x} |g_{lj}| \leq 1, \ \forall x \in \mathbb{D}^{n_x}, \ l \in [1, \ n_x].$$

In the case that $\varphi(A_l^k x + v_l)$ is free of saturation, it is easy to see that

$$\begin{aligned}
\delta_l^{-1} \varphi(A_l^k x + v_l) &= \delta_l^{-1}(A_l^k x + v_l) \\
&\in co\{\delta_l^{-1}(A_l^k x + v_l), \ G_l^k \Theta^{-1} x\}, \ l \in [1, n_x].
\end{aligned} \tag{5.10}$$

Furthermore, when the state does incur saturation, it follows that

$$\begin{cases}
G_l^k \Theta^{-1} x \leq \delta_l^{-1} \varphi(A_l^k x + v_l) = 1 \leq \delta_l^{-1}(A_l^k x + v_l), \ \text{if } A_l^k x + v_l > \delta_l, \\
\delta_l^{-1}(A_l^k x + v_l) \leq \delta_l^{-1} \varphi(A_l^k x + v_l) = -1 \leq G_l^k \Theta^{-1} x, \ \text{if } A_l^k x + v_l < -\delta_l,
\end{cases} \tag{5.11}$$

which means that

$$\delta_l^{-1} \varphi(A_l^k x + v_l) = \pm 1 \in co\{\delta_l^{-1}(A_l^k x + v_l), \ G_l^k \Theta^{-1} x\}, \ l \in [1, n_x]. \tag{5.12}$$

By Lemma 1 in [52], the desired result can be obtained immediately, and the proof is complete.

Lemma 5.3 *For given matrices $\mathcal{Q}$, $\mathcal{S}$ and $\mathcal{R}$ with $\mathcal{Q}$, $\mathcal{R}$ symmetric, the controlled system (5.1) satisfies the following dissipativity constraint*

$$\mathcal{W}(\xi, z; j, k-1) \leq \mathbb{E}\{V_j(x_j)\} \tag{5.13}$$

for all $0 \leq j \leq k$ and $\xi \in \ell_2([0, N]; \mathbb{R}^{q+p})$ if and only if there exists a family *of positive real-valued functions $V_k : \mathbb{R}^{n_x} \to \mathbb{R}$ ($V_k(0) = 0$ for all $k \in [0, N]$) satisfying the following Hamilton-Jacobi inequality (HJI)*

$$V_k(x) \geq \sup_{\xi \in \ell_2} \left\{ \mathcal{W}(\xi, z; k, k) + \mathbb{E}_{(\varepsilon_k, \alpha_k, \Xi_k)} V_{k+1}\left(\phi(k, x, \xi_k, \varepsilon_k, \alpha_k, \Xi_k)\right) \right\} \quad (5.14)$$

for all $x \in \mathbb{R}^{n_x}$, where

$$\phi(k, x, \xi_k, \varepsilon_k, \alpha_k, \Xi_k)$$
$$= \varphi\left(A_1^k x + A_2^k x \varepsilon_k + B_k K_k \Xi_k C_k x + \alpha_k f(k, x) + B_k K_k E_k v_k + D_k w_k\right).$$

Proof *Along a similar line of the proof of Theorem 2 in [14], the result is easily accessible and the proof is therefore omitted.*

Remark 5.3 *Lemma 5.3 serves as a bounded real lemma (BRL) for dissipative problems of general systems with stochastic variables $\varepsilon_k, \alpha_k, \Xi_k$. Similar results can be found in [14, 112] for H_∞ control/filtering problems of the stochastic system. Such "stochastic" version of the BRL provides the key for the resolution of stochastic disturbance attenuation problems and dissipative problems. Basically, three kinds of techniques have been involved in the BRLs, namely, LMIs (see [41]), Riccati difference equations (RDEs) (see [17, 107]), as well as HJIs (see [14, 112]).*

Theorem 5.1 *For given matrices $\mathcal{Q}$, $\mathcal{S}$ and $\mathcal{R}$ with $\mathcal{Q}$, $\mathcal{R}$ symmetric, the controlled system (5.1) is ($\mathcal{Q}$, $\mathcal{S}$, $\mathcal{R}$)-dissipative if there exist a family of positive-definite matrices $\{\mathcal{P}_k\}_{0 \leq k \leq N+1}$, a set of positive scalars $\{\lambda_k\}_{0 \leq k \leq N}$, and families of real-valued matrices $\{G_k\}_{0 \leq k \leq N}$, $\{K_k\}_{0 \leq k \leq N}$ satisfying the following recursive matrix inequalities*

$$\|G_k\|_\infty \leq 1, \quad (5.15)$$

$$\Pi_1^{k,i} = \begin{bmatrix} \Pi_{11}^{k,i} & \Pi_{12}^{k,i} & \Pi_{13}^{k,i} & \Pi_{14}^{k,i} & \Pi_{15}^{k,i} \\ * & \Pi_{22}^{k,i} & \Pi_{23}^{k,i} & 0 & 0 \\ * & * & \mathcal{R} & \Pi_{34}^{k,i} & 0 \\ * & * & * & -\mathcal{P}_{k+1} & 0 \\ * & * & * & * & -I_m \otimes \mathcal{P}_{k+1} \end{bmatrix} < 0,$$

$$i \in [1, 2^{n_x}] \quad (5.16)$$

where

$$\bar{\Psi}_k = \frac{\Psi_k^T \Phi_k + \Phi_k^T \Psi_k}{2}, \quad \bar{\Phi}_k = \frac{\Psi_k^T + \Phi_k^T}{2}, \quad \tilde{\vartheta}_i = \sqrt{\bar{\vartheta}_i(1 - \bar{\vartheta}_i)},$$

$$\mathcal{A}_k = A_1^k + B_k K_k \bar{\Xi} C_k, \quad \mathcal{D}_k = [B_k K_k E_k \quad D_k],$$

$$\Pi_{11}^{k,i} = L_k^T \mathcal{Q} L_k - \lambda_k \bar{\Psi}_k - \mathcal{P}_k + \sigma_\varepsilon A_2^{kT} S_i \mathcal{P}_{k+1} S_i A_2^k,$$

$$\Pi_{12}^{k,i} = \bar{\alpha}(S_i \mathcal{A}_k + \Theta S_i^- G_k \Theta^{-1})^T \mathcal{P}_{k+1} S_i + \lambda_k \bar{\Phi}_k,$$

$$\Pi_{13}^{k,i} = L_k^T \mathcal{S}, \quad \Pi_{14}^{k,i} = (S_i \mathcal{A}_k + \Theta S_i^- G_k \Theta^{-1})^T \mathcal{P}_{k+1},$$

$$
\begin{aligned}
\Pi_{15}^{k,i} &= [\ \tilde{\vartheta}_1 C_k^T M_1 K_k^T B_k^T S_i \mathcal{P}_{k+1} \quad \cdots \quad \tilde{\vartheta}_m C_k^T M_m K_k^T B_k^T S_i \mathcal{P}_{k+1}\ ], \\
\Pi_{22}^{k,i} &= \bar{\alpha} S_i \mathcal{P}_{k+1} S_i - \lambda_k I, \quad \Pi_{23}^{k,i} = \bar{\alpha} \mathcal{D}_k^T S_i \mathcal{P}_{k+1} S_i, \quad \Pi_{34}^k = \mathcal{D}_k^T S_i \mathcal{P}_{k+1}, \\
M_l &= diag\{\underbrace{0, \cdots, 0}_{l-1}, 1, \underbrace{0, \cdots, 0}_{m-l}\}, \quad \bar{\Xi} = diag\{\bar{\vartheta}_1, \bar{\vartheta}_2, \cdots, \bar{\vartheta}_m\}.
\end{aligned}
$$

Furthermore, $u_k = K_k y_k$ is a desired output feedback control law.

Proof *For the purpose of simplicity, we denote*

$$
\begin{aligned}
\mathscr{A}_k(x_k) &= (S_i \mathcal{A}_k + \Theta S_i^- G_k \Theta^{-1}) x_k + \bar{\alpha} S_i f(k, x_k), \\
\mathscr{B}_k(x_k) &= S_i B_k K_k (\Xi_k - \bar{\Xi}) C_k x_k + S_i A_2^k x_k \varepsilon_k, \\
\mathscr{C}_k(x_k) &= (\alpha_k - \bar{\alpha}) S_i f(k, x_k), \\
\mathscr{D}_k(\xi_k) &= S_i B_k K_k E_k v_k + S_i D_k w_k.
\end{aligned}
$$

Let $V_k(x) = x^T \mathcal{P}_k x$, where $\{\mathcal{P}_k\}_{0 \le k \le N+1}$ are the solutions of the time-varying matrix inequalities (5.16). It can be calculated from Lemma 5.2 that

$$
\sup_{\xi \in \ell_2} \Big\{ \mathcal{W}(\xi, z; k, k) + \mathbb{E}_{(\varepsilon_k, \alpha_k, \Xi_k)} V_{k+1}\Big(\phi(k, x_k, \xi_k, \varepsilon_k, \alpha_k, \Xi_k) \Big) \Big\}
$$

$$
= \sup_{\xi \in \ell_2} \Big\{ \mathcal{W}(\xi, z; k, k) + \mathbb{E}_{(\varepsilon_k, \alpha_k, \Xi_k)} \varphi^T\Big(A_1^k x_k + A_2^k x_k \varepsilon_k + \alpha_k f(k, x_k)
$$
$$
+ B_k K_k \Xi_k C_k x_k + B_k K_k E_k v_k + D_k w_k \Big) \mathcal{P}_{k+1} \varphi\Big(A_1^k x_k + A_2^k x_k \varepsilon_k
$$
$$
+ \alpha_k f(k, x_k) + B_k K_k \Xi_k C_k x_k + B_k K_k E_k v_k + D_k w_k \Big) \Big\}
$$

$$
= \sup_{\xi \in \ell_2} \Big\{ \mathcal{W}(\xi, z; k, k) + \mathbb{E}_{(\varepsilon_k, \alpha_k, \Xi_k)} \Big\{ \sum_{i=1}^{2^{n_x}} \theta_i^k \Big(\mathscr{A}_k^T(x_k) + \mathscr{B}_k^T(x_k) + \mathscr{C}_k^T(x_k)
$$
$$
+ \mathscr{D}_k^T(\xi_k) \Big) \Big\} \mathcal{P}_{k+1} \Big\{ \sum_{i=1}^{2^{n_x}} \theta_i^k \Big(\mathscr{A}_k(x_k) + \mathscr{B}_k(x_k) + \mathscr{C}_k(x_k) + \mathscr{D}_k(\xi_k) \Big) \Big\} \Big\}
$$

$$
\le \sup_{\xi \in \ell_2} \Big\{ \mathcal{W}(\xi, z; k, k) + \mathbb{E}_{(\varepsilon_k, \alpha_k, \Xi_k)} \sum_{i=1}^{2^{n_x}} \theta_i^k \Big(\mathscr{A}_k^T(x_k) + \mathscr{B}_k^T(x_k) + \mathscr{C}_k^T(x_k)
$$
$$
+ \mathscr{D}_k^T(\xi_k) \Big) \mathcal{P}_{k+1} \Big(\mathscr{A}_k(x_k) + \mathscr{B}_k(x_k) + \mathscr{C}_k(x_k) + \mathscr{D}_k(\xi_k) \Big) \Big\}
$$

$$
\le \sup_{\xi \in \ell_2} \Big\{ \mathcal{W}(\xi, z; k, k) + \mathbb{E}_{(\varepsilon_k, \alpha_k, \Xi_k)} \sum_{i=1}^{2^{n_x}} \theta_i^k \Big\{ \mathscr{A}_k^T(x_k) \mathcal{P}_{k+1} \mathscr{A}_k(x_k)
$$
$$
+ 2 \mathscr{A}_k^T(x_k) \mathcal{P}_{k+1} \mathscr{B}_k(x_k) + 2 \mathscr{A}_k^T(x_k) \mathcal{P}_{k+1} \mathscr{C}_k(x_k)
$$
$$
+ 2 \mathscr{A}_k^T(x_k) \mathcal{P}_{k+1} \mathscr{D}_k(\xi_k) + \mathscr{B}_k^T(x_k) \mathcal{P}_{k+1} \mathscr{B}_k(x_k)
$$
$$
+ 2 \mathscr{B}_k^T(x_k) \mathcal{P}_{k+1} \mathscr{C}_k(x_k) + 2 \mathscr{B}_k^T(x_k) \mathcal{P}_{k+1} \mathscr{D}_k(\xi_k) + \mathscr{C}_k^T(x_k) \mathcal{P}_{k+1} \mathscr{C}_k(x_k)
$$
$$
+ \mathscr{C}_k^T(\varepsilon_k, \zeta_k) \mathcal{P}_{k+1} \mathscr{D}_k(\xi_k) + \mathscr{D}_k^T(\xi_k) \mathcal{P}_{k+1} \mathscr{D}_k(\xi_k) \Big\} \Big\}
$$

$$
\begin{aligned}
&\leq \sup_{\xi \in \ell_2} \Big\{ \mathcal{W}(\xi, z; k, k) + \sum_{i=i}^{2^{n_x}} \theta_i^k \Big\{ \mathscr{A}_k^T(x_k)\mathcal{P}_{k+1}\mathscr{A}_k(x_k) + 2\mathscr{A}_k^T(x_k)\mathcal{P}_{k+1}\mathscr{D}_k(\xi_k) \\
&\quad + \sum_{l=1}^{m} \tilde{\vartheta}_l^2 x_k^T C_k^T M_l K_k^T B_k^T S_i \mathcal{P}_{k+1} S_i B_k K_k M_l C_k x_k + \sigma_\varepsilon x_k^T A_2^{kT} S_i \mathcal{P}_{k+1} S_i \\
&\quad \times A_2^k x_k + \bar{\alpha}(1-\bar{\alpha}) f^T(k, x_k) S_i \mathcal{P}_{k+1} S_i f(k, x_k) + \mathscr{D}_k^T(\xi_k)\mathcal{P}_{k+1}\mathscr{D}_k(\xi_k) \Big\} \Big\} \\
&\leq \sup_{\xi \in \ell_2} \sum_{i=1}^{2^{n_x}} \theta_i^k \Big\{ \mathscr{A}_k^T(x_k)\mathcal{P}_{k+1}\mathscr{A}_k(x_k) + 2\mathscr{A}_k^T(x_k)\mathcal{P}_{k+1}\mathscr{D}_k(\xi_k) + x_k^T L_k^T Q L_k x_k \\
&\quad + 2x_k^T L_k^T \mathcal{S}\xi_k + \sum_{l=1}^{m} \tilde{\vartheta}_l^2 x_k^T C_k^T M_l K_k^T B_k^T S_i \mathcal{P}_{k+1} S_i B_k K_k M_l C_k x_k \\
&\quad + \sigma_\varepsilon x_k^T A_2^{kT} S_i \mathcal{P}_{k+1} S_i A_2^k x_k + \bar{\alpha}(1-\bar{\alpha}) f^T(k, x_k) S_i \mathcal{P}_{k+1} S_i f(k, x_k) \\
&\quad + \xi_k^T \mathcal{R}\xi_k + \mathscr{D}_k^T(\xi_k)\mathcal{P}_{k+1}\mathscr{D}_k(\xi_k) \Big\} \quad\quad\quad\quad\quad\quad\quad\quad\quad\quad\quad (5.17)
\end{aligned}
$$

where $\sum_{i=1}^{2^{n_x}} \theta_i^k = 1$.

By denoting $\tilde{f}(k, x_k) = [x_k^T \ f^T(k, x_k)]^T$ and applying the completing squares method, it is not difficult to see that

$$
\begin{aligned}
&\sup_{\xi \in \ell_2} \Big\{ \mathcal{W}(\xi, z; k, k) + \mathbb{E}_{(\varepsilon_k, \alpha_k, \Xi_k)} V_{k+1}\Big(\phi(k, x_k, \xi_k, \varepsilon_k, \alpha_k, \Xi_k) \Big) \Big\} \\
&\leq \sup_{\xi \in \ell_2} \Big\{ \sum_{i=1}^{2^{n_x}} \theta_i^k \Big\{ \mathscr{A}_k^T(x_k)\mathcal{P}_{k+1}\mathscr{A}_k(x_k) + x_k^T L_k^T Q L_k x_k \\
&\quad + \bar{\alpha}(1-\bar{\alpha}) f^T(k, x_k) S_i \mathcal{P}_{k+1} S_i f(k, x_k) + \sigma_\varepsilon x_k^T A_2^{kT} S_i \mathcal{P}_{k+1} S_i A_2^k x_k \\
&\quad + \sum_{l=1}^{m} \tilde{\vartheta}_l^2 x_k^T C_k^T M_l K_k^T B_k^T S_i \mathcal{P}_{k+1} S_i B_k K_k M_l C_k x_k \Big\} \\
&\quad + 2\Big\{ \sum_{i=1}^{2^{n_x}} \theta_i^k \mathscr{A}_k^T(x_k)\mathcal{P}_{k+1} S_i \mathcal{D}_k + x_k^T L_k^T \mathcal{S} \Big\}\xi_k \\
&\quad + \xi_k^T \Big\{ \mathcal{R} + \sum_{i=1}^{2^{n_x}} \theta_i^k \mathcal{D}_k^T S_i \mathcal{P}_{k+1} S_i \mathcal{D}_k \Big\}\xi_k \Big\} \quad\quad\quad\quad\quad\quad (5.18) \\
&\leq \tilde{f}^T(k, x_k) \left(\sum_{i=1}^{2^{n_x}} \theta_i^k \begin{bmatrix} \Lambda_1 & \Lambda_2 \\ * & \Lambda_3 \end{bmatrix} + \begin{bmatrix} \Lambda_4^T \\ \Lambda_5^T \end{bmatrix} \Lambda_6^{-1} \begin{bmatrix} \Lambda_4 \\ \Lambda_5 \end{bmatrix} \right) \tilde{f}(k, x_k)
\end{aligned}
$$

where the maximum of $\xi_k = \Lambda_6^{-1}(\Lambda_4 x_k + \Lambda_5 f(k, x_k))$ is achieved with

$$
\begin{aligned}
\Lambda_1 &= (S_i \mathcal{A}_k + \Theta S_i^- G_k \Theta^{-1})^T \mathcal{P}_{k+1}(S_i \mathcal{A}_k + \Theta S_i^- G_k \Theta^{-1}) \\
&\quad + \sum_{l=1}^{m} \tilde{\vartheta}_l^2 C_k^T M_l K_k^T B_k^T S_i \mathcal{P}_{k+1} S_i B_k K_k M_l C_k
\end{aligned}
$$

$$+ L_k^T Q L_k + \sigma_\varepsilon A_2^{kT} S_i \mathcal{P}_{k+1} S_i A_2^k,$$

$$\Lambda_2 = \bar{a}(S_i \mathcal{A}_k + \Theta S_i^- G_k \Theta^{-1})^T \mathcal{P}_{k+1} S_i, \quad \Lambda_3 = \bar{a} S_i \mathcal{P}_{k+1} S_i,$$

$$\Lambda_4 = \sum_{i=1}^{2^{n_x}} \theta_i^k \mathcal{D}_k^T S_i \mathcal{P}_{k+1} (S_i \mathcal{A}_k + \Theta S_i^- G_k \Theta^{-1}) + \mathcal{S}^T L_k,$$

$$\Lambda_5 = \bar{a} \sum_{i=1}^{2^{n_x}} \theta_i^k \mathcal{D}_k^T S_i \mathcal{P}_{k+1} S_i, \quad \Lambda_6 = -\mathcal{R} - \sum_{i=1}^{2^{n_x}} \theta_i^k \mathcal{D}_k^T S_i \mathcal{P}_{k+1} S_i \mathcal{D}_k > 0.$$

Subsequently, taking (5.2) into consideration, one has

$$\sup_{\xi \in \ell_2} \left\{ \mathcal{W}(\xi, z; k, k) + \mathbb{E}_{(\varepsilon_k, \alpha_k, \Xi_k)} V_{k+1} \Big(\phi(k, x_k, \xi_k, \varepsilon_k, \alpha_k, \Xi_k) \Big) \right\} - V_k(x_k)$$

$$\leq - \sum_{i=1}^{2^{n_x}} \theta_i^k \lambda_k (f(k, x_k) - \Psi_k x_k)^T (f(k, x_k) - \Phi_k x_k)$$

$$+ \tilde{f}^T(k, x_k) \left(\sum_{i=1}^{2^{n_x}} \theta_i^k \begin{bmatrix} \Lambda_1 - \mathcal{P}_k & \Lambda_2 \\ * & \Lambda_3 \end{bmatrix} + \begin{bmatrix} \Lambda_4^T \\ \Lambda_5^T \end{bmatrix} \Lambda_6^{-1} \begin{bmatrix} \Lambda_4 \\ \Lambda_5 \end{bmatrix} \right) \tilde{f}(k, x_k)$$

$$\leq \tilde{f}^T(k, x_k) \left(\begin{bmatrix} \sum_{i=1}^{2^{n_x}} \theta_i^k \Lambda_1 - \lambda_k \bar{\Psi}_k - \mathcal{P}_k & \sum_{i=1}^{2^{n_x}} \theta_i^k \Lambda_2 + \lambda_k \bar{\Phi}_k \\ * & \sum_{i=1}^{2^{n_x}} \theta_i^k \Lambda_3 - \lambda_k I \end{bmatrix} \right.$$

$$\left. + \begin{bmatrix} \Lambda_4^T \\ \Lambda_5^T \end{bmatrix} \Lambda_6^{-1} \begin{bmatrix} \Lambda_4 \\ \Lambda_5 \end{bmatrix} \right) \tilde{f}(k, x_k)$$

$$(5.19)$$

for some positive scalar λ_k. Furthermore, it follows from (5.16) and the Schur complement Lemma that

$$\sup_{\xi \in \ell_2} \Big\{ \mathcal{W}(\xi, z; k, k)$$
$$+ \mathbb{E}_{(\varepsilon_k, \alpha_k, \Xi_k)} V_{k+1} \Big(\phi(k, x_k, \xi_k, \varepsilon_k, \alpha_k, \Xi_k) \Big) \Big\} - V_k(x) \leq 0.$$
$$(5.20)$$

Then, taking $j = 0$ and $k = N + 1$, we obtain directly from Lemma 5.3 that

$$\mathcal{W}(\xi, z; 0, N) \leq \mathbb{E}\{V_0(x_0)\} = \mathbb{E}\{x_0^T \mathcal{P}_0 x_0\} \leq \lambda_{\max}(\mathcal{P}_0) \mathbb{E}\{x_0^T x_0\}.$$

Taking $\rho = \lambda_{\max}(\mathcal{P}_0)$, it can be concluded that the controlled system (5.1) is $(\mathcal{Q}, \mathcal{S}, \mathcal{R})$-dissipative, and the proof is now complete.

Remark 5.4 *It can be seen from Theorem 5.1 that the desired performance can be guaranteed for the discrete time-varying system (5.1) with state saturation, RONs and multiple missing measurements if there exist two sets of parameters $\{\mathcal{P}_k\}_{0 \leq k \leq N+1}$ and $\{G_k, K_k, \lambda_k\}_{0 \leq k \leq N}$ satisfying the recursive* *matrix inequalities (5.15) and (5.16). Note that the inequalities (5.15) and (5.16) are nonlinear because of $K_k^T B_k^T S_i \mathcal{P}_{k+1}$ and $G_k^T (S_i^-)^T \Theta \mathcal{P}_{k+1}$ in $\Pi_{12}^{k,i}$, $\Pi_{14}^{k,i}$, $\Pi_{15}^{k,i}$, $\Pi_{23}^{k,i}$ and $\Pi_{34}^{k,i}$, and therefore cannot be solved directly. In the next subsection, a computationally tractable algorithm is developed for the controller design so that the conditions in Theorem 5.1 can be satisfied.*

In the case that the system under consideration is time-invariant, Theorem 5.1 can be readily specialized to the following corollary where the controller parameter is also time-invariant.

Corollary 5.1 *For given matrices $\mathcal{Q}$, $\mathcal{S}$ and $\mathcal{R}$ with $\mathcal{Q}$, $\mathcal{R}$ symmetric, the controlled discrete time-invariant system with the same structure as (5.1) is $(\mathcal{Q}, \mathcal{S}, \mathcal{R})$-dissipative if there exist a positive-definite matrix $\mathcal{P}$, a positive scalar λ, and two real-valued matrices G and K satisfying the following matrix inequities*

$$\|G\|_\infty \leq 1, \tag{5.21}$$

$$\Pi_1 = \begin{bmatrix} \Pi_{11}^i & \Pi_{12}^i & \Pi_{13}^i & \Pi_{14}^i & \Pi_{15}^i \\ * & \Pi_{22}^i & \Pi_{23}^i & 0 & 0 \\ * & * & \mathcal{R} & \Pi_{34}^i & 0 \\ * & * & * & -\mathcal{P} & 0 \\ * & * & * & * & -I_m \otimes \mathcal{P} \end{bmatrix} < 0, \quad i \in [1,\ 2^{n_x}] \tag{5.22}$$

where

$$\begin{aligned}
\bar{\Psi} &= \frac{\Psi^T \Phi + \Phi^T \Psi}{2}, \quad \bar{\Phi} = \frac{\Psi^T + \Phi^T}{2}, \\
\mathcal{A} &= A_1 + BK\bar{\Xi}C, \quad \mathcal{D} = [BKE \;\; D], \\
\Pi_{11}^i &= L^T \mathcal{Q}L - \lambda\bar{\Psi} - \mathcal{P} + \sigma_\varepsilon A_2^T S_i \mathcal{P} S_i A_2, \\
\Pi_{12}^i &= \bar{\alpha}(S_i\mathcal{A} + \Theta S_i^- GO^{-1})^T \mathcal{P}S_i + \lambda\bar{\Phi}, \\
\Pi_{13}^i &= L^T \mathcal{S}, \quad \Pi_{14}^i = (S_i\mathcal{A} + \Theta S_i^- GO^{-1})^T \mathcal{P}, \\
\Pi_{15}^i &= [\;\tilde{\vartheta}_1 C^T M_1 K^T B^T S_i \mathcal{P} \;\; \cdots \;\; \tilde{\vartheta}_m C^T M_m K^T B^T S_i \mathcal{P}\;], \\
\Pi_{22}^i &= \bar{\alpha}S_i\mathcal{P}S_i - \lambda I, \quad \Pi_{23}^i = \bar{\alpha}\mathcal{D}^T S_i\mathcal{P}S_i, \quad \Pi_{34} = \mathcal{D}^T S_i \mathcal{P}.
\end{aligned}$$

Furthermore, $u_k = Ky_k$ is a desired output feedback control law.

In what follows, similar to [47, 58], we introduce the controller design algorithm via the iterative LMIs to solve inequalities (5.15) and (5.16). Let Υ be the set of n_x-dimensional row vectors which has only one nonzero element 1. Denote the ith element of Υ as η_i whose ith element is 1. $\mathcal{Y}$ denotes the set of n_x-dimensional column vectors whose elements are 1 or -1. For 2^{n_x} elements in $\mathcal{Y}$, let its sth element be denoted as ψ_s. Therefore, $\|G_k\|_\infty \leq 1$ can be guaranteed by $\eta_i G_k \psi_s \leq 1$, for all $i \in [1, n_x]$, $s \in [1, 2^{n_x}]$.

Let us first consider the inequality in (5.16) for $i = 1$. Denoting $Y_k =$

$\mathcal{P}_{k+1}S_1 B_k K_k$ and taking $S_1 = I$ into consideration, we have the following LMIs

$$\begin{bmatrix} \Pi_{11}^{k,1} & \Pi_{12}^{k,1} & L_k^T \mathcal{S} & \Pi_{14}^{k,1} & \Pi_{15}^{k,1} \\ * & \bar{\alpha}\mathcal{P}_{k+1} - \lambda_k I & \Pi_{23}^{k,1} & 0 & 0 \\ * & * & \mathcal{R} & \Pi_{34}^{k,1} & 0 \\ * & * & * & -\mathcal{P}_{k+1} & 0 \\ * & * & * & * & -I_m \otimes \mathcal{P}_{k+1} \end{bmatrix} < 0 \qquad (5.23)$$

with

$$\Pi_{11}^{k,1} = L_k^T \mathcal{Q} L_k - \lambda_k \bar{\Psi}_k - \mathcal{P}_k + \sigma_\varepsilon A_2^{kT} \mathcal{P}_{k+1} A_2^k,$$
$$\Pi_{12}^{k,1} = \bar{\alpha} A_1^{kT} \mathcal{P}_{k+1} + \bar{\alpha} C_k^T \bar{\Xi} Y_k^T + \lambda_k \bar{\Phi}_k, \quad \Pi_{14}^{k,1} = A_1^{kT} \mathcal{P}_{k+1} + C_k^T \bar{\Xi} Y_k^T,$$
$$\Pi_{15}^{k,1} = [\, \tilde{\vartheta}_1 C_k^T M_1 Y_k^T \quad \tilde{\vartheta}_2 C_k^T M_2 Y_k^T \quad \cdots \quad \tilde{\vartheta}_m C_k^T M_m Y_k^T \,],$$
$$\Pi_{23}^{k,1} = \bar{\alpha}[Y_k E_k \quad \mathcal{P}_{k+1} D_k]^T, \quad \Pi_{34}^{k,1} = [Y_k E_k \quad \mathcal{P}_{k+1} D_k]^T.$$

The detailed algorithm for solving (5.23) is described as follows and its the flow chart is shown in Fig. 5.1.

Remark 5.5 *By attempting to solve (5.23), if we cannot obtain a feasible solution $\mathcal{P}_{k+1}$, the inequalities (5.15) and (5.16) in Theorem 5.1 are infeasible. When a feasible solution $\mathcal{P}_{k+1}$ is obtained, it is natural to get other parameters via fixing $\mathcal{P}_{k+1}$ in the inequalities (5.15) and (5.16). Therefore, a convex optimization problem (**Op1**) in Algorithm 1, which is an improvement of (5.15) and (5.16), is constructed for minimizing μ. If $\mu < 0$, the solution must satisfy the inequalities (5.15) and (5.16), otherwise the parameter $\mathcal{P}_{k+1}$ needs to be adjusted by using the obtained G_k and K_k. So, in the following, another new convex optimization problem (**Op2**) in Algorithm 1 is constituted for minimizing π. Similarly, if $\pi < 0$, the solution must satisfy the inequalities (5.15) and (5.16), otherwise we need to adjust the parameters G_k, K_k and λ_k again, i.e. solve the optimization problem (**Op1**).*

Remark 5.6 *Note that, in Algorithm 1, μ and π are the additional variables in optimization problems (**Op1**) and (**Op2**), τ is also an additional variable for counting the number of inner loops, $\bar{\mu}$ is an intermediate variable for saving the latest μ, and ϖ is a given positive scalar to avoid endless loops in the algorithm. Obviously, ϖ is also linked to the convergence rate of the iterations. More specifically, when ϖ increases, the computational complexity will decrease and the possibility for finding a feasible solution reduces too. Therefore, ϖ can be seen as a regulatory factor between the computational speed and the solvability of inequalities. Note that, as shown in Algorithm 1, reducing the computational complexity and increasing the solvability of inequalities are two of the most important questions in further research. Furthermore, the introduced free matrix G_k, which is employed to describe the saturation phenomenon, results in a nonlinear matrix inequality. So, in*

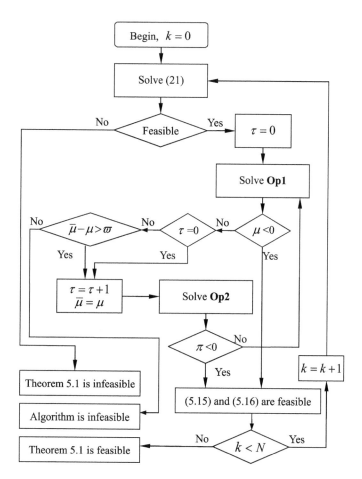

FIGURE 5.1
The flow chart of the proposed algorithm.

The controller design algorithm

Step 1. Given matrices $\mathcal{Q}$, $\mathcal{S}$, $\mathcal{R}$ and a prescribed tolerance $\varpi > 0$, and set $k = 0$.

Step 2. Given $S_1 = I$. Solve LMI (5.23) for $\mathcal{P}_{k+1}$, Y_k, λ_k when $k \neq 0$, or for $\mathcal{P}_1$, $\mathcal{P}_0$, Y_0, λ_0 when $k = 0$. If successful in finding a feasible solution, set $\tau = 0$, $\bar{\mu} = 0$ and go to the next step, else go to *Step 7*.

Step 3. Using $\mathcal{P}_{k+1}$ obtained previously, solve the following optimization problem for $U_k = \{\ G_k,\ K_k,\ \lambda_k,\ \mu\ \}$ when $k \neq 0$, or for $U_0 = \{\ \mathcal{P}_0,\ G_0,\ K_0\ \lambda_0,\ \mu\ \}$ when $k = 0$

$$\mathbf{Op1}: \inf_{U_k}\ \mu,\ \text{s.t.} \begin{cases} \Pi_1^{k,i} - \mu I < 0, \\ \eta_s G_k \psi_i \leq 1, \end{cases} i \in [1, 2^{n_x}],\ s \in [1, n_x]. \quad (5.24)$$

- If $\mu < 0$, then go to *Step 5*.

- If $\tau = 0$ and $\mu > 0$, then set $\tau = \tau + 1$, $\bar{\mu} = \mu$, and go to the next step.

- If $\tau > 0$ and $\bar{\mu} - \mu > \varpi$, then set $\tau = \tau + 1$, $\bar{\mu} = \mu$, and go to the next step, else go to *Step 5*.

Step 4. Using G_k and K_k obtained previously, solve the following optimization problem for $U_k^* = \{\ \mathcal{P}_{k+1},\ \lambda_k,\ \pi\ \}$ when $k \neq 0$, or for $U_0^* = \{\ \mathcal{P}_0,\ \mathcal{P}_1,\ \lambda_0,\ \pi\ \}$ when $k = 0$

$$\mathbf{Op2}: \inf_{U_k^*}\ \pi,\ \text{s.t.}\ \Pi_1^{k,i} - \pi I < 0,\ i \in [1, 2^{n_x}]. \quad (5.25)$$

If $\pi < 0$, then go to the next step, else go to *Step 3*.

Step 5. If $\mu < 0$ or $\pi < 0$, then K_k is a suitable controller, and go to the next step, else no condition can be drawn, and go to *Step 7*.

Step 6. If $k < N$, $k = k + 1$, then go to *Step 2*, else go to the next step.

Step 7. Stop.

order to avoid the difficulty of solving matrix inequality, how to design a new method to represent saturations is also a further topic to be investigated. A more detailed discussion of the similar iterative procedure as well as the convergence can be found in [21].

In the following section, the results obtained so far will be extended to the discrete time-varying systems with *partial* state saturations, RONs and multiple missing measurements.

5.3 Dissipative Control for Partial State Saturation Case

In many practical systems, it is quite common that only *partial* states are subjected to saturation constraints. For example, in the case of the dynamics of a car, variables such as speed and steering angle have finite physical limits and they saturate when reaching these limits, whereas variables such as yaw velocity and roll velocity are usually assumed to have no constraints (see e.g. [74]). Similar to the full state saturations, if not adequately handled, partial states saturations may also pose significant side effects in the controller design. Specifically, saturated nonlinear functions are likely to cause undesirable oscillatory behaviors especially at the switching between saturation and unsaturation. Then, the saturated and unsaturated states may affect each other to make the system dynamics even more complicated. The aim of this section is to extend the previously obtained results to the discrete time-varying systems with partial state saturations, RONs, as well as multiple missing measurements.

First, let us decompose the discrete time-varying system (5.1) as follows

$$
\begin{cases}
x_{k+1}^c = A_{11}^{1k}x_k^c + A_{12}^{1k}x_k^s + A_{11}^{2k}x_k^c\varepsilon_k + A_{12}^{2k}x_k^s\varepsilon_k \\
\qquad\quad + \alpha_k f_1(k, x_k^c, x_k^s) + B_1^k u_k + D_1^k w_k, \\
x_{k+1}^s = \varphi\big(A_{21}^{1k}x_k^c + A_{22}^{1k}x_k^s + A_{21}^{2k}x_k^c\varepsilon_k + A_{22}^{2k}x_k^s\varepsilon_k \\
\qquad\quad + \alpha_k f_2(k, x_k^c, x_k^s) + B_2^k u_k + D_2^k w_k\big), \\
y_k = \Xi_k(C_1^k x_k^c + C_2^k x_k^s) + E_k v_k, \\
z_k = L_1^k x_k^c + L_2^k x_k^s
\end{cases}
\tag{5.26}
$$

where $x_k^c \in \mathbb{R}^{n_x-s}$ and $x_k^s \in \mathbb{R}^s$ are, respectively, the vectors of saturation-free and saturated states, and all matrices in model (5.26) are time-varying matrices with appropriate dimensions. Denote $x_k \triangleq [x_k^{cT}\ x_k^{sT}]^T$ and $f(k, x_k) \triangleq [\ f_1^T(k, x_k^c, x_k^s)\ \ f_2^T(k, x_k^c, x_k^s)\ ]^T$ satisfying (5.2). According to Lemma 5.2, we have

$$
x_{k+1}^s = \sum_{i=1}^{2^s} \theta_i^k \Big(\Theta S_i\Theta^{-1}(A_{21}^{1k}x_k^c + A_{22}^{1k}x_k^s + A_{21}^{2k}x_k^c\varepsilon_k + A_{22}^{2k}x_k^s\varepsilon_k
$$
$$
+ \alpha_k f_2(k, x_k^c, x_k^s) + B_2^k u_k + D_2^k w_k) + \Theta S_i^- G_k\Theta^{-1}x_k^s\Big)
\tag{5.27}
$$

where $\theta_i^k > 0$ and $\sum_{i=1}^{2^s} \theta_i^k = 1$. Furthermore, the following system is obtained

$$
\begin{cases}
x_{k+1} = \sum_{i=1}^{2^s} \theta_i^k \left(\mathcal{A}_i^{1k}x_k + \mathcal{A}_i^{2k}x_k\varepsilon_k + \mathcal{B}_i^k \Xi_k \mathcal{C}_k x_k + \alpha_k \mathcal{F}_i f(k, x_k) + \mathcal{D}_i^k \xi_k\right), \\
z_k = \mathcal{L}_k x_k
\end{cases}
$$

$$
\tag{5.28}
$$

with

$$\mathcal{A}_i^{1k} = \begin{bmatrix} A_{11}^{1k} & A_{12}^{1k} \\ S_i A_{21}^{1k} & S_i A_{22}^{1k} + \Theta S_i^- G_k \Theta^{-1} \end{bmatrix}, \quad \mathcal{B}_i^k = \begin{bmatrix} B_1^k K_k \\ S_i B_2^k K_k \end{bmatrix},$$

$$\mathcal{A}_i^{2k} = \begin{bmatrix} A_{11}^{2k} & A_{12}^{2k} \\ S_i A_{21}^{2k} & S_i A_{22}^{2k} \end{bmatrix}, \quad \mathcal{D}_i^k = \begin{bmatrix} B_1^k K_k E_k & D_1^k \\ S_i B_2^k K_k E_k & S_i D_2^k \end{bmatrix},$$

$$\mathcal{C}_k = \begin{bmatrix} C_1^k & C_2^k \end{bmatrix}, \quad \mathcal{F}_i = \text{diag}\{I, S_i\}, \quad \mathcal{L}_k = \begin{bmatrix} L_1^k & L_2^k \end{bmatrix}.$$

Following the same lines as in Theorem 5.1 and Corollary 5.1, we have the following two results where only the sketches of the proofs are given in order to ensure conciseness.

Theorem 5.2 *For given matrices $\mathcal{Q}$, $\mathcal{S}$ and $\mathcal{R}$ with $\mathcal{Q}$, $\mathcal{R}$ symmetric, the controlled system (5.26) is $(\mathcal{Q}, \mathcal{S}, \mathcal{R})$-dissipative if there exist a family of positive-definite matrices $\{\mathcal{P}_k\}_{0 \le k \le N+1}$, a set of positive scalars $\{\lambda_k\}_{0 \le k \le N}$, and families of real-valued matrices $\{G_k\}_{0 \le k \le N}$, $\{K_k\}_{0 \le k \le N}$ satisfying the following recursive matrix inequities*

$$\|G_k\|_\infty \le 1, \tag{5.29}$$

$$\Omega_1^k = \begin{bmatrix} \Omega_{11}^{k,i} & \Omega_{12}^{k,i} & \Omega_{13}^{k,i} & \Omega_{14}^{k,i} & \Omega_{15}^{k,i} \\ * & \Omega_{22}^{k,i} & \Omega_{23}^{k,i} & 0 & 0 \\ * & * & \mathcal{R} & \Omega_{34}^{k,i} & 0 \\ * & * & * & -\mathcal{P}_{k+1} & 0 \\ * & * & * & * & -I_m \otimes \mathcal{P}_{k+1} \end{bmatrix} < 0, \; i \in [1, \, 2^s] \tag{5.30}$$

where

$$\Omega_{11}^{k,i} = \mathcal{L}_k^T \mathcal{Q} \mathcal{L}_k - \lambda_k \bar{\Psi}_k - \mathcal{P}_k + \sigma_\varepsilon \mathcal{A}_i^{2kT} \mathcal{P}_{k+1} \mathcal{A}_i^{2k}, \quad \Omega_{13}^{k,i} = \mathcal{L}_k^T \mathcal{S},$$

$$\Omega_{12}^{k,i} = \bar{\alpha}(\mathcal{A}_i^{1k} + \mathcal{B}_i^k \bar{\Xi} \mathcal{C}_k)^T \mathcal{P}_{k+1} \mathcal{F}_i + \lambda_k \bar{\Phi}_k, \quad \Omega_{14}^{k,i} = (\mathcal{A}_i^{1k} + \mathcal{B}_i^k \bar{\Xi} \mathcal{C}_k)^T \mathcal{P}_{k+1},$$

$$\Omega_{15}^{k,i} = \begin{bmatrix} \tilde{\vartheta}_1 \mathcal{C}_k^T M_1 \mathcal{B}_i^{kT} \mathcal{P}_{k+1} & \tilde{\vartheta}_2 \mathcal{C}_k^T M_2 \mathcal{B}_i^{kT} \mathcal{P}_{k+1} & \cdots & \tilde{\vartheta}_m \mathcal{C}_k^T M_m \mathcal{B}_i^{kT} \mathcal{P}_{k+1} \end{bmatrix},$$

$$\Omega_{22}^{k,i} = \bar{\alpha} \mathcal{F}_i \mathcal{P}_{k+1} \mathcal{F}_i - \lambda_k I, \quad \Omega_{23}^{k,i} = \bar{\alpha} \mathcal{D}_i^{kT} \mathcal{P}_{k+1} \mathcal{F}_i, \quad \Omega_{34}^k = \mathcal{D}_i^{kT} \mathcal{P}_{k+1},$$

and other parameters are defined as in Theorem 5.1. Furthermore, $u_k = K_k y_k$ is a desired output feedback control law.

Proof *Firstly, denote*

$$\bar{\mathscr{A}}_k(x_k) = (\mathcal{A}_i^{1k} + \mathcal{B}_i^k \bar{\Xi} \mathcal{C}_k)x_k + \bar{\alpha} \mathcal{F}_i f(k, x_k), \quad \bar{\mathscr{D}}_k(\xi_k) = \mathcal{D}_i^k \xi_k,$$

$$\bar{\mathscr{B}}_k(x_k) = \mathcal{B}_i^k (\Xi_k - \bar{\Xi}) \mathcal{C}_k x_k + \mathcal{A}_i^{2k} x_k \varepsilon_k, \quad \bar{\mathscr{C}}_k(x_k) = (\alpha_k - \bar{\alpha}) \mathcal{F}_i f(k, x_k).$$

It can be calculated that

$$\sup_{\xi \in \ell_2} \left\{ \mathcal{W}(\xi, z; k, k) + \mathbb{E}_{(\varepsilon_k, \alpha_k, \Xi_k)} V_{k+1} \Big(\phi(k, x_k, \xi_k, \varepsilon_k, \alpha_k, \Xi_k) \Big) \right\}$$

$$= \sup_{\xi \in \ell_2} \left\{ \mathcal{W}(\xi, z; k, k) + \mathbb{E}_{(\varepsilon_k, \alpha_k, \Xi_k)} \left\{ \sum_{i=1}^{2^s} \theta_i^k \left(\bar{\mathscr{A}}_k^T(x_k) + \bar{\mathscr{B}}_k^T(x_k) + \bar{\mathscr{C}}_k^T(x_k) \right. \right. \right.$$

$$\left. \left. \left. + \bar{\mathscr{D}}_k^T(\xi_k) \right) \right\} \mathcal{P}_{k+1} \left\{ \sum_{i=1}^{2^s} \theta_i^k \left(\bar{\mathscr{A}}_k(x_k) + \bar{\mathscr{B}}_k(x_k) + \bar{\mathscr{C}}_k(x_k) + \bar{\mathscr{D}}_k(\xi_k) \right) \right\} \right\}$$

$$\leq \sup_{\xi \in \ell_2} \left\{ \mathcal{W}(\xi, z; k, k) + \mathbb{E}_{(\varepsilon_k, \zeta_k, \alpha_k, \Xi_k)} \sum_{i=1}^{2^s} \theta_i^k \left(\bar{\mathscr{A}}_k^T(x_k) + \bar{\mathscr{B}}_k^T(x_k) + \bar{\mathscr{C}}_k^T(\varepsilon_k, \zeta_k) \right. \right.$$

$$\left. \left. + \bar{\mathscr{D}}_k^T(\xi_k) \right) \mathcal{P}_{k+1} \left(\bar{\mathscr{A}}_k(x_k) + \bar{\mathscr{B}}_k(x_k) + \bar{\mathscr{C}}_k(\varepsilon_k, \zeta_k) + \bar{\mathscr{D}}_k(\xi_k) \right) \right\}$$

$$\leq \sup_{\xi \in \ell_2} \left\{ \sum_{i=1}^{2^s} \theta_i^k \left\{ \bar{\mathscr{A}}_k^T(x_k) \mathcal{P}_{k+1} \bar{\mathscr{A}}_k(x_k) + 2 \bar{\mathscr{A}}_k^T(x_k) \mathcal{P}_{k+1} \bar{\mathscr{D}}_k(\xi_k) \right. \right.$$

$$+ \sigma_\varepsilon x_k^T \mathcal{A}_i^{2kT} \mathcal{P}_{k+1} \mathcal{A}_i^{2k} x_k + \sum_{l=1}^m \tilde{\vartheta}_l^2 x_k^T \mathcal{C}_k^T M_l \mathcal{B}_i^{kT} \mathcal{P}_{k+1} \mathcal{B}_i^k M_l \mathcal{C}_k x_k$$

$$+ x_k^T \mathcal{L}_k^T Q \mathcal{L}_k x_k + 2 x_k^T \mathcal{L}_k^T S \xi_k + \xi_k^T \mathcal{R} \xi_k$$

$$\left. \left. + \bar{\alpha}(1-\bar{\alpha}) f^T(k, x_k) \mathcal{F}_i^T \mathcal{P}_{k+1} \mathcal{F}_i f(k, x_k) + \bar{\mathscr{D}}_k^T(\xi_k) \mathcal{P}_{k+1} \bar{\mathscr{D}}_k(\xi_k) \right\} \right\}. \quad (5.31)$$

By applying the completing squares method again, it is not difficult to see that

$$\sup_{\xi \in \ell_2} \left\{ \mathcal{W}(\xi, z; k, k) + \mathbb{E}_{(\varepsilon_k, \alpha_k, \Xi_k)} V_{k+1} \left(\phi(k, x_k, \xi_k, \varepsilon_k, \alpha_k, \Xi_k) \right) \right\}$$

$$\leq \tilde{f}^T(k, x_k) \left(\sum_{i=1}^{2^s} \theta_i^k \begin{bmatrix} \bar{\Lambda}_1 & \bar{\Lambda}_2 \\ * & \bar{\Lambda}_3 \end{bmatrix} + \begin{bmatrix} \bar{\Lambda}_4^T \\ \bar{\Lambda}_5^T \end{bmatrix} \bar{\Lambda}_6^{-1} \begin{bmatrix} \bar{\Lambda}_4 \\ \bar{\Lambda}_5 \end{bmatrix} \right) \tilde{f}(k, x_k) \qquad (5.32)$$

where the maximum of $\xi_k = \bar{\Lambda}_6^{-1}(\bar{\Lambda}_4 x_k + \bar{\Lambda}_5 f(k, x_k))$ can be achieved with

$$\begin{aligned}
\bar{\Lambda}_1 &= (\mathcal{A}_i^{1k} + \mathcal{B}_i^k \bar{\Xi} \mathcal{C}_k)^T \mathcal{P}_{k+1}(\mathcal{A}_i^{1k} + \mathcal{B}_i^k \bar{\Xi} \mathcal{C}_k) + \sigma_\varepsilon \mathcal{A}_i^{2kT} \mathcal{P}_{k+1} \mathcal{A}_i^{2k} \\
&\quad + \sum_{l=1}^m \tilde{\vartheta}_l^2 \mathcal{C}_k^T M_l \mathcal{B}_i^{kT} \mathcal{P}_{k+1} \mathcal{B}_i^k M_l \mathcal{C}_k + \mathcal{L}_k^T Q \mathcal{L}_k, \\
\bar{\Lambda}_2 &= \bar{\alpha}(\mathcal{A}_i^{1k} + \mathcal{B}_i^k \bar{\Xi} \mathcal{C}_k)^T \mathcal{P}_{k+1} \mathcal{F}_i, \quad \bar{\Lambda}_3 = \bar{\alpha} \mathcal{F}_i^T \mathcal{P}_{k+1} \mathcal{F}_i, \\
\bar{\Lambda}_4 &= \sum_{i=1}^{2^s} \theta_i^k \mathcal{D}_i^{kT} \mathcal{P}_{k+1}(\mathcal{A}_i^{1k} + \mathcal{B}_i^k \bar{\Xi} \mathcal{C}_k) + \mathcal{S}^T \mathcal{L}_k, \\
\bar{\Lambda}_5 &= \bar{\alpha} \sum_{i=1}^{2^s} \theta_i^k \mathcal{D}_i^{kT} \mathcal{P}_{k+1} \mathcal{F}_i, \quad \bar{\Lambda}_6 = -\mathcal{R} - \sum_{i=1}^{2^s} \theta_i^k \mathcal{D}_i^{kT} \mathcal{P}_{k+1} \mathcal{D}_i^k.
\end{aligned}$$

Furthermore, it follows from (5.2) that

$$\sup_{\xi \in \ell_2} \left\{ \mathcal{W}(\xi, z; k, k) + \mathbb{E}_{(\varepsilon_k, \alpha_k, \Xi_k)} V_{k+1} \left(\phi(k, x_k, \xi_k, \varepsilon_k, \alpha_k, \Xi_k) \right) \right\} - V_k(x_k)$$

$$\leq \tilde{f}^T(k,x_k) \left(\begin{bmatrix} \sum\limits_{i=1}^{2^s} \theta_i^k \bar{\Lambda}_1 - \lambda_k \bar{\Psi}_k - \mathcal{P}_k & \sum\limits_{i=1}^{2^s} \theta_i^k \bar{\Lambda}_2 + \lambda_k \bar{\Phi}_k \\ * & \sum\limits_{i=1}^{2^s} \theta_i^k \bar{\Lambda}_3 - \lambda_k I \end{bmatrix} \right.$$

$$\left. + \begin{bmatrix} \bar{\Lambda}_4^T \\ \bar{\Lambda}_5^T \end{bmatrix} \bar{\Lambda}_6^{-1} \begin{bmatrix} \bar{\Lambda}_4 \\ \bar{\Lambda}_5 \end{bmatrix} \right) \tilde{f}(k,x_k) \tag{5.33}$$

holds for any positive scalar λ_k. Then, along the same line of the proof for Theorem 5.1, the rest of the proof follows immediately.

Similar to Corollary 5.1, when the systems are subject to partial state saturations, we have the following corollary for dissipative control problems of time-invariant systems.

Corollary 5.2 *For given matrices $\mathcal{Q}$, $\mathcal{S}$ and $\mathcal{R}$ with $\mathcal{Q}$, $\mathcal{R}$ symmetric, the controlled discrete time-invariant system with the same structure as (5.26) is $(\mathcal{Q}, \mathcal{S}, \mathcal{R})$-dissipative if there exist a positive-definite matrix $\mathcal{P}$, a positive scalar λ, and two real-valued matrices G and K satisfying the following matrix inequities*

$$\|G\|_\infty \leq 1, \tag{5.34}$$

$$\begin{bmatrix} \Omega_{11}^i & \Omega_{12}^i & \Omega_{13}^i & \Omega_{14}^i & \Omega_{15}^i \\ * & \Omega_{22}^i & \Omega_{23}^i & 0 & 0 \\ * & * & \mathcal{R} & \Omega_{34}^i & 0 \\ * & * & * & -\mathcal{P} & 0 \\ * & * & * & * & -I_m \otimes \mathcal{P} \end{bmatrix} < 0, \quad i \in [1, 2^s] \tag{5.35}$$

where

$$\mathcal{A}_i^1 = \begin{bmatrix} A_{11}^1 & A_{12}^1 \\ S_i A_{21}^1 & S_i A_{22}^1 + \Theta S_i^- G\Theta^{-1} \end{bmatrix}, \quad \mathcal{B}_i = \begin{bmatrix} B_1 K \\ S_i B_2 K \end{bmatrix},$$

$$\mathcal{A}_i^2 = \begin{bmatrix} A_{11}^2 & A_{12}^2 \\ S_i A_{21}^2 & S_i A_{22}^2 \end{bmatrix}, \quad \mathcal{D}_i = \begin{bmatrix} B_1 KE & D_1 \\ S_i B_2 KE & S_i D_2 \end{bmatrix},$$

$$\mathcal{C} = [\ C_1 \quad C_2\ ], \quad \mathcal{L} = [\ L_1 \quad L_2\ ],$$

$$\Omega_{11}^i = \mathcal{L}^T \mathcal{Q}\mathcal{L} - \lambda\bar{\Psi} - \mathcal{P} + \sigma_\varepsilon \mathcal{A}_i^{2T} \mathcal{P} \mathcal{A}_i^2, \quad \Omega_{13}^i = \mathcal{L}^T \mathcal{S},$$

$$\Omega_{12}^i = \bar{\alpha}(\mathcal{A}_i^1 + \mathcal{B}_i \bar{\Xi}\mathcal{C})^T \mathcal{P}\mathcal{F}_i + \lambda\bar{\Phi}, \quad \Omega_{14}^i = (\mathcal{A}_i^1 + \mathcal{B}_i \bar{\Xi}\mathcal{C})^T \mathcal{P},$$

$$\Omega_{15}^i = [\ \tilde{\vartheta}_1 \mathcal{C}^T M_1 \mathcal{B}_i^T \mathcal{P} \quad \tilde{\vartheta}_2 \mathcal{C}^T M_2 \mathcal{B}_i^T \mathcal{P} \quad \cdots \quad \tilde{\vartheta}_m \mathcal{C}^T M_m \mathcal{B}_i^T \mathcal{P}\ ],$$

$$\Omega_{22}^i = \bar{\alpha}\mathcal{F}_i \mathcal{P}\mathcal{F}_i - \lambda I, \quad \Omega_{23}^i = \bar{\alpha}\mathcal{D}_i^T \mathcal{P}\mathcal{F}_i, \quad \Omega_{34} = \mathcal{D}_i^T \mathcal{P},$$

and other parameters are defined as in Corollary 5.1. Furthermore, $u_k = K y_k$ is a desired output feedback control law.

Remark 5.7 *In this chapter, instead of the commonly used Lipschitz-type functions, the more general sector-like nonlinear functions are employed to describe the nonlinearities existing in the system. Furthermore, we have* *examined how the state saturations, RONs, and missing measurements influence the performance of the controlled systems in terms of dissipativity. In Theorem 5.1 and Theorem 5.2, all the system parameters, as well as the occurrence probabilities for the RONs and the missing measurements, are reflected in the recursive nonlinear matrix inequalities. Also, an efficient iterative algorithm is proposed to solve the addressed dissipative controller design problem. In the next section, two simulation examples are provided to show the usefulness of the proposed dissipative control conditions.*

5.4 Simulation Examples

To illustrate the effectiveness of the proposed methods, two numerical examples are given in this section.

Example 5.1 *The first example concerns the discrete time-varying system (5.1) with state saturation whose parameters are as follows:*

$$A_1^k = \begin{bmatrix} 1.06 - 0.4\sin(k+2) & 0.24 \\ 0.1 & -0.95 \end{bmatrix}, \quad B_k = \begin{bmatrix} -0.50 & 0.75 \\ 1.2 & 0.75 \end{bmatrix},$$

$$A_2^k = \begin{bmatrix} -0.06 & -0.01 \\ 0 & 0.10 \end{bmatrix}, \quad C_k = \begin{bmatrix} 1.10 & 0.95 \\ 0.25 & 0.85 \end{bmatrix},$$

$$D_k = [-0.03 \ \ 0.04]^T, \quad E_k = [0.02 \ \ -0.03]^T, \quad L_k = [0.80 \ \ 0.80],$$

$$f(k, x_k) = \begin{cases} \begin{bmatrix} -0.48x_{k,1} + 0.24x_{k,2} + tanh(0.24x_{k,1}) \\ 0.48x_{k,2} - tanh(0.16x_{k,2}) \end{bmatrix}, & 0 \le k < 6, \\ \begin{bmatrix} -0.45x_{k,1} + 0.20x_{k,2} + tanh(0.25x_{k,1}) \\ 0.45x_{k,2} - tanh(0.20x_{k,2}) \end{bmatrix}, & 6 \le k < 21. \end{cases}$$

The time horizon is taken as $N = 21$. It is easy to see that the constraint (5.2) can be met with

$$\Phi_k = \begin{cases} \begin{bmatrix} -0.48 & 0.24 \\ 0 & 0.32 \end{bmatrix}, & 0 \le k < 6, \\ \begin{bmatrix} -0.45 & 0.25 \\ 0 & 0.45 \end{bmatrix}, & 6 \le k < 21, \end{cases}$$

$$\Psi_k = \begin{cases} \begin{bmatrix} -0.24 & 0.24 \\ 0 & 0.48 \end{bmatrix}, & 0 \le k < 6, \\ \begin{bmatrix} -0.25 & 0.25 \\ 0 & 0.25 \end{bmatrix}, & 6 \le k < 21. \end{cases}$$

Considering both the RONs and the multiple missing measurements, the probabilities are taken as $\bar{\alpha} = 0.12$, $\bar{\vartheta}_1 = 0.95$ and $\bar{\vartheta}_2 = 0.92$, respectively. Moreover, the variance σ_ε, the matrices Θ, $\mathcal{Q}$, $\mathcal{S}$ and $\mathcal{R}$ are chosen as

$$\sigma_\varepsilon = 0.5, \quad \Theta = diag\{1.6, 1.4\}, \quad \mathcal{Q} = 1, \quad \mathcal{S} = [0.1 \ \ 0.1], \quad \mathcal{R} = -I.$$

Using the given algorithm and MATLAB® (with YALMIP 3.0), the set of solutions to recursive matrix inequalities (5.15) and (5.16) in Theorem 5.1 are obtained and shown in Table 5.1.

Example 5.2 *The second example considers the discrete time-varying system (5.26) with partial state saturation, where the second state $x_{2,k}$ is under saturation constraint with the saturation level $\Theta = 1.4$. Consider the system with*

$$\begin{pmatrix} A_{11}^{1k} & A_{12}^{1k} \\ A_{21}^{1k} & A_{21}^{1k} \end{pmatrix} = A_1^k, \quad \begin{pmatrix} A_{11}^{2k} & A_{12}^{2k} \\ A_{21}^{2k} & A_{21}^{2k} \end{pmatrix} = A_2^k, \quad \begin{pmatrix} B_1^k \\ B_2^k \end{pmatrix} = B_k,$$

$$\begin{pmatrix} C_1^k & C_2^k \end{pmatrix} = C_k, \quad \begin{pmatrix} D_1^{kT} & D_2^{kT} \end{pmatrix}^T = D_k, \quad \begin{pmatrix} L_1^k & L_2^k \end{pmatrix} = L_k$$

where A_1^k, A_2^k, B_k, C_k, D_k, L_k and other parameters are the same as the above example. The set of solutions to recursive matrix inequalities (5.29) and (5.30) in Theorem 5.2 are shown in Table 5.2.

In the simulations, the initial values of the system states in both examples are set as $[1.5 \; -1.35]^T$. The exogenous disturbance inputs are selected as $w_k = 5\exp(-2k)$ and $v_k = \frac{k}{k+2}\cos(k)$, respectively. The simulation results for Example 1 and Example 2 are shown in Fig. 5.2 and Fig. 5.3, respectively. Fig. 5.2 plots the system outputs with state saturation, and Fig. 5.3 depicts the system outputs with partial state saturation. The simulation results have confirmed that the designed dissipative control system performs very well.

5.5 Summary

In this chapter, we have addressed the dissipative control problem for a class of discrete time-varying systems with state saturations, RONs, as well as multiple missing measurements over a finite-horizon. Some sequences of mutually independent random variables that obey the Bernoulli distribution are introduced to describe the addressed randomness. With a free matrix whose infinity norm is less than or equal to 1, some sufficient conditions have been derived for the finite horizon controller to satisfy the prescribed dissipativity performance requirement. These conditions are expressed in terms of recursive nonlinear matrix inequalities that can be checked by the presented controller design algorithm. Two numerical simulation examples have been provided to demonstrate the effectiveness and applicability of the proposed design approach. It should be pointed out that the obtained result could be applied to mechanical systems, electric power systems with hard limits on power inputs, field voltages and currents, digitally controlled systems with sampled data, and so on. Further research topics include the extension of the main results to suspension systems, and also to general network control systems with both state saturations and incomplete information.

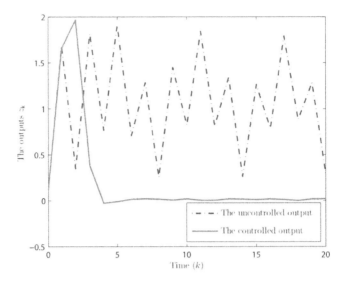

FIGURE 5.2
The trajectory for the full state saturation.

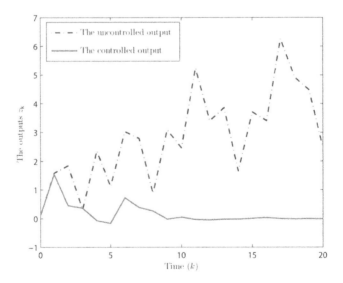

FIGURE 5.3
The trajectory for the partial state saturation.

TABLE 5.1
Variables and controller parameters.

k	P_k	G_k	K_k	λ_k
0	$\begin{bmatrix} 748.79 & -56.55 \\ -56.55 & 728.17 \end{bmatrix}$	$\begin{bmatrix} -0.1121 & 0.3231 \\ -0.9864 & 0.0131 \end{bmatrix}$	$\begin{bmatrix} 0.2244 & 0.2096 \\ -0.1266 & -0.0174 \end{bmatrix}$	3287.4
1	$\begin{bmatrix} 779.79 & 377.20 \\ 377.20 & 386.24 \end{bmatrix}$	$\begin{bmatrix} -0.2889 & 0.1013 \\ -0.2241 & 0.1112 \end{bmatrix}$	$\begin{bmatrix} 0.4605 & 0.3222 \\ 0.2844 & 0.1378 \end{bmatrix}$	4253.0
2	$\begin{bmatrix} 212.83 & 39.02 \\ 39.02 & 143.99 \end{bmatrix}$	$\begin{bmatrix} 0.0556 & -0.0401 \\ 0.0500 & -0.0340 \end{bmatrix}$	$\begin{bmatrix} 0.5234 & 0.3104 \\ -0.4329 & 0.1396 \end{bmatrix}$	2404.6
3	$\begin{bmatrix} 119.71 & -7.27 \\ -7.27 & 124.96 \end{bmatrix}$	$\begin{bmatrix} 0.0448 & -0.0503 \\ 0.0306 & -0.0316 \end{bmatrix}$	$\begin{bmatrix} 0.5873 & 0.3015 \\ -0.7932 & 0.1158 \end{bmatrix}$	4008.0
4	$\begin{bmatrix} 64.20 & -5.57 \\ -5.57 & 72.28 \end{bmatrix}$	$\begin{bmatrix} 0.0622 & -0.0563 \\ 0.0323 & -0.0284 \end{bmatrix}$	$\begin{bmatrix} 0.4237 & 0.4338 \\ -0.4008 & -0.1986 \end{bmatrix}$	339.6
$\cdots$				$\cdots$
20	$\begin{bmatrix} 15.08 & -0.46 \\ -0.46 & 14.52 \end{bmatrix}$	$\begin{bmatrix} 0.0501 & -0.0400 \\ 0.0191 & -0.0147 \end{bmatrix}$	$\begin{bmatrix} 0.3381 & 0.4310 \\ -0.3312 & -0.0585 \end{bmatrix}$	77.1
21	$\begin{bmatrix} 14.37 & -0.76 \\ -0.76 & 14.51 \end{bmatrix}$			

TABLE 5.2
Variables and controller parameters.

k	P_k	G_k	K_k	λ_k
0	$\begin{bmatrix} 726.97 & -82.65 \\ -82.65 & 698.96 \end{bmatrix}$	0.9995	$\begin{bmatrix} 0.5587 & 0.5126 \\ -0.3495 & -0.1077 \end{bmatrix}$	1572.4
1	$\begin{bmatrix} 779.79 & 377.20 \\ 377.20 & 386.24 \end{bmatrix}$	0.3437	$\begin{bmatrix} 0.4607 & 0.3233 \\ 0.2611 & 0.1448 \end{bmatrix}$	4473.6
2	$\begin{bmatrix} 212.83 & 39.02 \\ 39.02 & 143.99 \end{bmatrix}$	-0.0845	$\begin{bmatrix} 0.5209 & 0.3164 \\ -0.4092 & 0.0866 \end{bmatrix}$	2498.9
3	$\begin{bmatrix} 119.71 & -7.27 \\ -7.27 & 124.96 \end{bmatrix}$	-0.0407	$\begin{bmatrix} 0.5844 & 0.3123 \\ -0.7738 & 0.0443 \end{bmatrix}$	4159.5
4	$\begin{bmatrix} 64.20 & -5.57 \\ -5.57 & 72.28 \end{bmatrix}$	-0.0767	$\begin{bmatrix} 0.4122 & 0.4582 \\ -0.3528 & -0.2977 \end{bmatrix}$	253.0
...	...	...	...	...
20	$\begin{bmatrix} 15.08 & -0.46 \\ -0.46 & 14.52 \end{bmatrix}$	-0.0322	$\begin{bmatrix} 0.3335 & 0.4432 \\ -0.3137 & -0.1039 \end{bmatrix}$	81.58
21	$\begin{bmatrix} 14.37 & -0.76 \\ -0.76 & 14.51 \end{bmatrix}$			

6

Finite-Horizon $\mathcal{H}_\infty$ Filtering for State-Saturated Discrete Time-Varying Systems with Packet Dropouts

The previous chapter was concerned with the dissipative control issue for state-saturated discrete time-varying systems with missing measurements. A novel design of dissipative controller is developed to ensure the desired performance of closed-loop systems subject to state-saturated nonlinearity and missing measurements. It is worth noting that the control or filtering with more complex network-induced phenomena should be further discussed in theory. This chapter deals with the $\mathcal{H}_\infty$ filtering problem for a class of discrete time-varying systems with state saturations, randomly occurring nonlinearities, as well as successive packet dropouts. Two mutually independent sequences of random variables that obey the Bernoulli distribution are employed to describe the random occurrence of the nonlinearities and packet dropouts. Similar to the method adopted in Chapter 5, i.e., by introducing a free matrix with its infinity norm less than or equal to 1, the error state is bounded by a convex hull. In addition, some sufficient conditions are obtained such that the $\mathcal{H}_\infty$ disturbance attenuation level is guaranteed, over a given finite horizon, for the filtering error dynamics in the presence of saturated states, randomly occurring nonlinearities, and successive packet dropouts. Furthermore, the obtained results are extended to the case when state saturations are partial. Two numerical simulation examples are provided to demonstrate the effectiveness and applicability of the proposed filter design approach.

6.1 Modeling and Problem Formulation

Let a finite time horizon be denoted by $[0, N] := \{0, 1, 2, \cdots, N\}$, on which we consider the following class of discrete time-varying nonlinear systems with state saturations:

$$\begin{cases} x_{k+1} = \hbar \left(A_k x_k + \alpha_k f(k, x_k) + B_k w_k \right), \\ y_{ck} = C_k x_k + D_k w_k, \\ z_k = E_k x_k \end{cases} \quad (6.1)$$

where $x_k \in \mathbb{R}^n$ is the state vector, $y_{ck} \in \mathbb{R}^r$ represents the measurement output, $z_k \in \mathbb{R}^p$ is the signal to be estimated, and w_k is the disturbance input belonging to $l_2([0,\infty); \mathbb{R}^q)$. The matrices A_k, B_k, C_k, D_k, and E_k are known time-varying matrices with appropriate dimensions. The stochastic variable α_k is a Bernoulli distributed white sequence taking values of 0 and 1 with the following probability:

$$\begin{cases} \text{Prob}\{\alpha_k = 1\} = \bar{\alpha}, \\ \text{Prob}\{\alpha_k = 0\} = 1 - \bar{\alpha} \end{cases}$$

where $\bar{\alpha} \in [0,1]$ is a known constant. It is easy to see that α_k is employed to describe the random occurrence of the nonlinearities $f(k, x_k)$.

The saturation function $\hbar : \mathbb{R}^n \mapsto \mathbb{R}^n$ is defined as

$$\hbar(v) = \begin{bmatrix} \hbar_1(v_1) & \hbar_2(v_2) & \cdots & \hbar_n(v_n) \end{bmatrix}^T \tag{6.2}$$

with $\hbar_i(v_i) = \text{sign}(v_i)\min\{\theta_i, |v_i|\}$, where v_i is the ith element of the vector v, and $\theta_i > 0$ is the saturation level. Here, the notation of "sign" means the signum function. For the purpose of simplicity, we denote $\Theta :=$ $\text{diag}\{\theta_1, \theta_2, \cdots, \theta_n\}$. From this definition, we have $x_k \in \{v| -\theta_i \leq v_i \leq \theta_i, \ i = 1, 2, \cdots, n\}$ for any k.

The nonlinear vector-valued function $f : [0, N] \times \mathbb{R}^n \mapsto \mathbb{R}^n$ is assumed to be continuous and satisfies the following sector-bounded condition

$$[f(k, x) - \Phi_k x]^T [f(k, x) - \Psi_k x] \leq 0 \tag{6.3}$$

for all $k \in [0, N]$, where Φ_k and Ψ_k are real matrices of appropriate dimensions.

Bearing in mind the successive packet dropouts to be addressed, the actual input signal y_k for the filter to be designed is described by

$$y_k = \beta_k y_{ck} + (1 - \beta_k)y_{k-1} = \beta_k(C_k x_k + D_k w_k) + (1 - \beta_k)y_{k-1} \tag{6.4}$$

where, similar to α_k, the stochastic variable β_k is also a Bernoulli distributed white sequence taking values of 0 and 1 with known mathematical expectation $\bar{\beta}$ and variance $\bar{\beta}(1 - \bar{\beta})$. It is assumed that β_k is unrelated to α_k.

Remark 6.1 *In networked environments, many practical systems are influenced by additive randomly occurring nonlinear disturbances with their states suffering from saturation constraint. The model (6.1) is introduced to specifically address the issues of saturations and RONs that may arise in real-time networked dynamic systems, such as mechanical systems with position and speed limits, electoral systems with limited power supply for the actuators (motors), digital filters implemented in finite word-length format, and implementable neural networks. Note that the control and filtering problems for state-saturated linear time-invariant (LTI) systems without stochastic nonlinearities have been dealt with in [58, 62, 73, 74, 122] and the references therein.*

Remark 6.2 *The model (6.4) has been introduced in [103] to describe possible successive packet dropouts (SPDs). It can be seen clearly from (6.4) that the latest measurement received will be used if the current measurement is lost during transmissions, i.e., $\beta_k = 0$. Moreover, when the value of β_k is zero successively for k, it means that the phenomenon of SPDs appears, and the probability of the event of s ($s \geq 2$) SPDs to occur is $\bar{\beta}(1 - \bar{\beta})^{s-1}$, which decreases exponentially when s increases. This property is consistent with the real network circumstance in which it is unlikely for a large number of packet dropouts to occur successively.*

In this chapter, we consider the following time-varying filter for system (6.1)

$$\begin{cases} \hat{x}_{k+1} = A_k^f \hat{x}_k + B_k^f (C_k \hat{x}_k - y_k), \\ \hat{z}_k = E_k \hat{x}_k \end{cases} \tag{6.5}$$

where $\hat{x}_k$ represents the state estimate and $\hat{z}_k$ is the estimated output. A_k^f and B_k^f are the filter parameters to be determined.

Letting $\eta_k = [\ x_k^T\ \ \hat{x}_k^T\ \ y_{k-1}^T\ ]^T$, $\tilde{z}_k = z_k - \hat{z}_k$, and $\xi_k = A_k x_k(k) + \alpha_k f(k, x_k) + B_k w_k$, the following augmented system is obtained

$$\begin{cases} \eta_{k+1} = \mathcal{A}_k \eta_k + \mathcal{B}_k w_k + \varphi_k + (\beta_k - \bar{\beta})\mathcal{H}_k \eta_k + (\beta_k - \bar{\beta})\mathcal{G}_k w_k, \\ \tilde{z}_k = \mathcal{E}_k \eta_k \end{cases} \tag{6.6}$$

where

$$\mathcal{A}_k = \begin{bmatrix} 0 & 0 & 0 \\ -\bar{\beta}B_k^f C_k & A_k^f + B_k^f C_k & -(1-\bar{\beta})B_k^f \\ \bar{\beta}C_k & 0 & (1-\bar{\beta})I \end{bmatrix},$$

$$\mathcal{H}_k = \begin{bmatrix} 0 & 0 & 0 \\ -B_k^f C_k & 0 & B_k^f \\ C_k & 0 & -I \end{bmatrix}, \quad \mathcal{B}_k = \begin{bmatrix} 0 \\ -\bar{\beta}B_k^f D_k \\ \bar{\beta}D_k \end{bmatrix},$$

$$\mathcal{E}_k = \begin{bmatrix} E_k^T \\ -E_k^T \\ 0 \end{bmatrix}^T, \quad \mathcal{G}_k = \begin{bmatrix} 0 \\ -B_k^f D_k \\ D_k \end{bmatrix}, \quad \varphi_k = \begin{bmatrix} \hbar(\xi_k) \\ 0 \\ 0 \end{bmatrix}.$$

Our aim in this chapter is to design a finite-horizon filter in the form of (6.5) such that, under zero initial condition, for the given disturbance attenuation level $\gamma > 0$ and all nonzero $w(k)$, the filtering error $\tilde{z}(k)$ from (6.6) satisfies the following condition

$$\sum_{k=0}^{N-1} \mathbb{E}\left\{||\tilde{z}_k||^2\right\} \leq \gamma^2 \sum_{k=0}^{N-1} ||w_k||^2. \tag{6.7}$$

6.2 $\mathcal{H}_\infty$ Filtering for Full State Saturation Case

In this section, a sufficient condition is given for the augmented system (6.6) to satisfy the $\mathcal{H}_\infty$ performance constraint (6.7) over a finite horizon, and an algorithm is then proposed to deal with the addressed filter design problem for the discrete time-varying systems (6.1) with state saturations, RONs, and SPDs. It is shown that the filter parameters can be obtained by solving a certain set of recursive nonlinear matrix inequalities.

Let $\mathcal{S}_n$ be the set of $n \times n$ diagonal matrices whose diagonal elements are either 1 or 0. It is easy to see that there are 2^n elements in $\mathcal{S}_n$ and one can denote its ith element as S_i, $i \in [1, 2^n]$. Furthermore, define $S_i^- = I - S_i$, and hence $S_i^- \in \mathcal{S}_n$.

Theorem 6.1 *Let the filter parameters A_k^f, B_k^f in (6.5) and the positive scalar $\gamma > 0$ be given. Then, under the zero initial condition, the augmented system (6.6) satisfies the $\mathcal{H}_\infty$ performance constraint (6.7) for all nonzero w_k if there exist three families of positive definite matrices $\{P_{1,k}\}_{0 \le k \le N}$, $\{P_{2,k}\}_{0 \le k \le N}$, $\{P_{3,k}\}_{0 \le k \le N}$, a set of real-valued matrices $\{G_k\}_{0 \le k \le N}$, and a family of positive scalars $\{\lambda_k\}_{0 \le k \le N}$ satisfying the following recursive matrix inequalities:*

$$\|G_k\|_\infty \le 1, \tag{6.8}$$

$$\Xi^{(i)} = \begin{bmatrix} \Xi_{11}^{(i)} & \Xi_{12} & 0 & \Xi_{14}^{(i)} & \Xi_{15}^{(i)} \\ * & \Xi_{22} & \Xi_{23} & 0 & \Xi_{25} \\ * & * & \Xi_{33} & 0 & 0 \\ * & * & * & \Xi_{44}^{(i)} & \Xi_{45}^{(i)} \\ * & * & * & * & \Xi_{55}^{(i)} \end{bmatrix} \le 0, \quad i \in [1, 2^n] \tag{6.9}$$

where

$$\Delta_1 = \frac{\Phi_k^T \Psi_k + \Psi_k^T \Phi_k}{2}, \quad \Delta_2 = \frac{\Phi_k + \Psi_k}{2},$$

$$\Xi_{11}^{(i)} = (S_i \Theta^{-1} A_k + S_i^- G_k \Theta^{-1})^T \Theta P_{1,k+1} \Theta (S_i \Theta^{-1} A_k + S_i^- G_k \Theta^{-1})$$
$$+ \bar\beta C_k^T B_k^{fT} P_{2,k+1} B_k^f C_k + \bar\beta C_k^T P_{3,k+1} C_k + E_k^T E_k - \lambda_k \Delta_1 - P_{1,k},$$

$$\Xi_{12} = -\bar\beta C_k^T B_k^{fT} P_{2,k+1}(A_k^f + B_k^f C_k) - E_k^T E_k,$$

$$\Xi_{14}^{(i)} = \bar\alpha (S_i \Theta^{-1} A_k + S_i^- G_k \Theta^{-1})^T \Theta P_{1,k+1} S_i + \lambda_k \Delta_2,$$

$$\Xi_{15}^{(i)} = (S_i \Theta^{-1} A_k + S_i^- G_k \Theta^{-1})^T \Theta P_{1,k+1} S_i B_k$$
$$+ \bar\beta C_k^T B_k^{fT} P_{2,k+1} B_k^f D_k + \bar\beta C_k^T P_{3,k+1} D_k,$$

$$\Xi_{22} = (A_k^f + B_k^f C_k)^T P_{2,k+1}(A_k^f + B_k^f C_k) + E_k^T E_k - P_{2,k},$$

$$\Xi_{23} = -(1 - \bar\beta)(A_k^f + B_k^f C_k)^T P_{2,k+1} B_k^f,$$

$$\Xi_{25} = -\bar{\beta}(A_k^f + B_k^f C_k)^T P_{2,k+1} B_k^f D_k,$$

$$\Xi_{33} = (1-\bar{\beta})B_k^{fT}P_{2,k+1}B_k^f + (1-\bar{\beta})P_{3,k+1} - P_{3,k},$$

$$\Xi_{44}^{(i)} = \bar{\alpha}S_i^T P_{1,k+1}S_i - \lambda_k I, \quad \Xi_{45}^{(i)} = \bar{\alpha}S_i^T P_{1,k+1}S_i B_k,$$

$$\Xi_{55}^{(i)} = B_k^T S_i^T P_{1,k+1}S_i B_k + \bar{\beta}D_k^T B_k^{fT}P_{2,k+1}B_k^f D_k + \bar{\beta}D_k^T P_{3,k+1}D_k - \gamma^2 I.$$

Proof *Denote $R = [\; I \quad 0 \quad 0 \;]$ and define*

$$J_k = \eta_{k+1}^T P_{k+1}\eta_{k+1} - \eta_k^T P_k \eta_k \tag{6.10}$$

where $P_k = diag\{P_{1,k}, P_{2,k}, P_{3,k}\}$.

Taking (6.6) into consideration, one has

$$
\begin{aligned}
\mathbb{E}\{J_k\} =& \mathbb{E}\{\eta_{k+1}^T P_{k+1}\eta_{k+1} - \eta_k^T P_k \eta_k\} \\
=& \mathbb{E}\Big\{\varphi_k^T P_{k+1}\varphi_k\Big\} + \eta_k^T \mathcal{A}_k^T P_{k+1}\mathcal{A}_k \eta_k + w_k^T \mathcal{B}_k^T P_{k+1}\mathcal{B}_k w_k \\
& + \bar{\beta}(1-\bar{\beta})\eta_k^T \mathcal{H}_k^T P_{k+1}\mathcal{H}_k \eta_k + \bar{\beta}(1-\bar{\beta})w_k^T \mathcal{G}_k^T P_{k+1}\mathcal{G}_k w_k \\
& + 2\eta_k^T \mathcal{A}_k^T P_{k+1}\mathcal{B}_k w_k + 2\bar{\beta}(1-\bar{\beta})\eta_k^T \mathcal{H}_k^T P_{k+1}\mathcal{G}_k w_k - \eta_k^T P_k \eta_k \\
=& \mathbb{E}\Big\{\Big(\Big(\sum_{i=1}^{2^n}\delta_i\big(S_i\Theta^{-1}(A_k R\eta_k + \alpha_k f(k, R\eta_k) + B_k w_k) \\
& + S_i^- G_k\Theta^{-1}R\eta_k\big)^T\Big)\Theta P_{1,k+1}\Theta\Big(\sum_{i=1}^{2^n}\delta_i\big(S_i\Theta^{-1}(A_k R\eta_k \\
& + \alpha_k f(k, R\eta_k) + B_k w_k) + S_i^- G_k\Theta^{-1}R\eta_k\big)\Big)\Big\} \\
& + \eta_k^T \mathcal{A}_k^T P_{k+1}\mathcal{A}_k \eta_k + w_k^T \mathcal{B}_k^T P_{k+1}\mathcal{B}_k w_k - \eta_k^T P_k \eta_k \\
& + \bar{\beta}(1-\bar{\beta})\eta_k^T \mathcal{H}_k^T P_{k+1}\mathcal{H}_k \eta_k + \bar{\beta}(1-\bar{\beta})w_k^T \mathcal{G}_k^T P_{k+1}\mathcal{G}_k w_k \\
& + 2\eta_k^T \mathcal{A}_k^T P_{k+1}\mathcal{B}_k w_k + 2\bar{\beta}(1-\bar{\beta})\eta_k^T \mathcal{H}_k^T P_{k+1}\mathcal{G}_k w_k \tag{6.11}
\end{aligned}
$$

where $\delta_i > 0$ and $\sum_{i=1}^{2^n}\delta_i = 1$.

Denoting $\tilde{\eta}_k = [\; \eta_k^T \quad f(k, R\eta_k)^T \quad w_k^T \;]^T$ and according to Lemma 5.1, we can obtain

$$
\begin{aligned}
\mathbb{E}\{J_k\} \leq & \max_{i\in[1,2^n]} \mathbb{E}\Big\{\big[S_i\Theta^{-1}(A_k R\eta_k + \alpha_k f(k, R\eta_k) + B_k w_k) \\
& + S_i^- G_k\Theta^{-1}R\eta_k\big]^T \Theta P_{1,k+1}\Theta\big[S_i\Theta^{-1}(A_k R\eta_k \\
& + \alpha_k f(k, R\eta_k) + B_k w_k) + S_i^- G_k\Theta^{-1}R\eta_k\big]\Big\} \\
& + \eta_k^T \mathcal{A}_k^T P_{k+1}\mathcal{A}_k \eta_k + w_k^T \mathcal{B}_k^T P_{k+1}\mathcal{B}_k w_k \\
& + \bar{\beta}(1-\bar{\beta})\eta_k^T \mathcal{H}_k^T P_{k+1}\mathcal{H}_k \eta_k \\
& - \eta_k^T P_k \eta_k + \bar{\beta}(1-\bar{\beta})w_k^T \mathcal{G}_k^T P_{k+1}\mathcal{G}_k w_k + 2\eta_k^T \mathcal{A}_k^T P_{k+1}\mathcal{B}_k w_k \\
& + 2\bar{\beta}(1-\bar{\beta})\eta_k^T \mathcal{H}_k^T P_{k+1}\mathcal{G}_k w_k
\end{aligned}
$$

$$
\begin{aligned}
&\leq \max_{i\in[1,2^n]}\Big\{\eta_k^T R^T (S_i\Theta^{-1}A_k + S_i^- G_k\Theta^{-1})^T\Theta P_{1,k+1}\Theta \\
&\quad \times (S_i\Theta^{-1}A_k + S_i^- G_k\Theta^{-1})R\eta_k \\
&\quad + \bar{\alpha}f^T(k,R\eta_k)S_i^T P_{1,k+1}S_i f(k,R\eta_k) \\
&\quad + 2\bar{\alpha}\eta_k^T R^T (S_i\Theta^{-1}A_k + S_i^- G_k\Theta^{-1})^T\Theta P_{1,k+1}S_i f(k,R\eta_k) \\
&\quad + 2\eta_k^T R^T (S_i\Theta^{-1}A_k + S_i^- G_k\Theta^{-1})^T\Theta P_{1,k+1}S_i B_k w_k \\
&\quad + 2\bar{\alpha}f(k,R\eta_k)^T S_i^T P_{1,k+1}S_i B_k w_k + w_k^T B_k^T S_i^T P_{1,k+1}S_i B_k w_k \\
&\quad + \eta_k^T \mathcal{A}_k^T P_{k+1}\mathcal{A}_k\eta_k + w_k^T \mathcal{B}_k^T P_{k+1}\mathcal{B}_k w_k \\
&\quad + \bar{\beta}(1-\bar{\beta})\eta_k^T \mathcal{H}_k^T P_{k+1}\mathcal{H}_k\eta_k \\
&\quad - \eta_k^T P_k\eta_k + \bar{\beta}(1-\bar{\beta})w_k^T \mathcal{G}_k^T P_{k+1}\mathcal{G}_k w_k + 2\eta_k^T \mathcal{A}_k^T P_{k+1}\mathcal{B}_k w_k \\
&\quad + 2\bar{\beta}(1-\bar{\beta})\eta_k^T \mathcal{H}_k^T P_{k+1}\mathcal{G}_k w_k\Big\} \\
&\leq \max_{i\in[1,2^n]}\tilde{\eta}_k^T \Xi_1^{(i)}\tilde{\eta}_k
\end{aligned}
\tag{6.12}
$$

where

$$
\tilde{\Xi}_{11}^{(i)} = \Xi_{11}^{(i)} - E_k^T E_k + \lambda_k\Delta_1,
$$

$$
\Xi_1^{(i)} =
\begin{bmatrix}
\tilde{\Xi}_{11}^{(i)} & \Xi_{12}+E_k^T E_k & 0 & \Xi_{14}^{(i)}-\lambda_k\Delta_2 & \Xi_{15}^{(i)} \\
* & \Xi_{22}-E_k^T E_k & \Xi_{23} & 0 & \Xi_{25} \\
* & * & \Xi_{33} & 0 & 0 \\
* & * & * & \Xi_{44}^{(i)}+\lambda_k I & \Xi_{45}^{(i)} \\
* & * & * & * & \Xi_{55}^{(i)}+\gamma^2 I
\end{bmatrix}.
$$

From (6.3), it is not difficult to see that

$$
\begin{aligned}
&\mathbb{E}\{J_k\} \\
&\leq \max_{i\in[1,2^n]}\tilde{\eta}_k^T \Xi_1^{(i)}\tilde{\eta}_k - \lambda_k[f(k,R\eta_k)-\Phi_k R\eta_k]^T[f(k,R\eta_k)-\Psi_k R\eta_k] \\
&\leq \max_{i\in[1,2^n]}\tilde{\eta}_k^T \Xi_2^{(i)}\tilde{\eta}_k
\end{aligned}
\tag{6.13}
$$

with

$$
\Xi_2^{(i)} =
\begin{bmatrix}
\Xi_{11}^{(i)}-E_k^T E_k & \Xi_{12}+E_k^T E_k & 0 & \Xi_{14}^{(i)} & \Xi_{15}^{(i)} \\
* & \Xi_{22}-E_k^T E_k & \Xi_{23} & 0 & \Xi_{25} \\
* & * & \Xi_{33} & 0 & 0 \\
* & * & * & \Xi_{44}^{(i)} & \Xi_{45}^{(i)} \\
* & * & * & * & \Xi_{55}^{(i)}+\gamma^2 I
\end{bmatrix}.
$$

Adding the zero term $\tilde{z}_k^T\tilde{z}_k - \gamma^2 w_k^T w_k - \tilde{z}_k^T\tilde{z}_k + \gamma^2 w_k^T w_k$ to $\mathbb{E}\{J_k\}$ results in

$$
\mathbb{E}\{J_k\} \leq \max_{i\in[1,2^n]}\tilde{\eta}_k^T \Xi^{(i)}\tilde{\eta}_k - \tilde{z}_k^T\tilde{z}_k + \gamma^2 w_k^T w_k.
\tag{6.14}
$$

Summing up (6.14) on both sides from 0 to $N - 1$ with respect to k, one has

$$\sum_{k=0}^{N-1} \mathbb{E}\{J_k\} \leq \sum_{k=0}^{N-1} \max_{i \in [1,2^n]} \tilde{\eta}_k^T \Xi^{(i)} \tilde{\eta}_k - \sum_{k=0}^{N-1} \left(\tilde{z}_k^T \tilde{z}_k - \gamma^2 w_k^T w_k \right). \tag{6.15}$$

Noting that $\Xi^{(i)} < 0$ $(i \in [1,2^n])$, the $\mathcal{H}_\infty$ performance index defined in (6.7) is satisfied and the proof is now complete.

Up to now, the analysis problem has been dealt with for $\mathcal{H}_\infty$ filtering for a class of discrete time-varying systems with state saturation, RONs, and SDPs. In the following, we proceed to solve the filter design problem using the recursive matrix inequalities approach.

Theorem 6.2 *Let the disturbance attenuation level $\gamma > 0$ be given. The finite-horizon $\mathcal{H}_\infty$ filtering problem is solvable for the discrete time-varying systems (1) with state saturation, RONs, and SDPs if there exist three families of positive definite matrices $\{P_{1,k}\}_{0 \leq k \leq N}$, $\{P_{2,k}\}_{0 \leq k \leq N}$, $\{P_{3,k}\}_{0 \leq k \leq N}$, a family of positive scalars $\{\lambda_k\}_{0 \leq k \leq N}$, and families of real-valued matrices $\{G_k\}_{0 \leq k \leq N-1}$, $\{\tilde{A}_k^f\}_{0 \leq k \leq N-1}$, $\{\tilde{B}_k^f\}_{0 \leq k \leq N-1}$ satisfying the following recursive matrix inequalities:*

$$\|G_k\|_\infty \leq 1, \tag{6.16}$$

$$\Xi^{*(i)} = \begin{bmatrix} \bar{\Xi}^{(i)} & \tilde{S}_i^T P_{1,k+1} & Q_1^T & Q_2^T \\ * & -P_{1,k+1} & 0 & 0 \\ * & * & -P_{2,k+1} & 0 \\ * & * & * & -P_{2,k+1} \end{bmatrix} < 0, i \in [1,2^n] \tag{6.17}$$

where

$$\bar{\Xi}^{(i)} = \begin{bmatrix} \Xi_{11}^* & \Xi_{12}^* & 0 & \Xi_{14}^* & \Xi_{15}^* \\ * & \Xi_{22}^* & 0 & 0 & 0 \\ * & * & \Xi_{33}^* & 0 & 0 \\ * & * & * & \Xi_{44}^{*(i)} & 0 \\ * & * & * & * & \Xi_{55}^* \end{bmatrix},$$

$$\Xi_{11}^* = E_k^T E_k + \bar{\beta} C_k^T P_{3,k+1} C_k - \lambda_k \Delta_1 - P_{1,k},$$

$$\Xi_{12}^* = -E_k^T E_k, \quad \Xi_{14}^* = \lambda_k \Delta_2, \quad \Xi_{15}^* = \bar{\beta} C_k^T P_{3,k+1} D_k,$$

$$\Xi_{22}^* = E_k^T E_k - P_{2,k}, \quad \Xi_{33}^* = (1 - \bar{\beta}) P_{3,k+1} - P_{3,k},$$

$$\Xi_{44}^{*(i)} = \bar{\alpha}(1 - \bar{\alpha}) S_i^T P_{1,k+1} S_i - \lambda_k I, \quad \Xi_{55}^* = \bar{\beta} D_k^T P_{3,k+1} D_k - \gamma^2 I,$$

$$\Delta_1 = \frac{\Phi_k^T \Psi_k + \Psi_k^T \Phi_k}{2}, \quad \Delta_2 = \frac{\Phi_k + \Psi_k}{2}, \quad \bar{\beta}^* = \sqrt{\bar{\beta}(1 - \bar{\beta})},$$

$$\tilde{S}_i = \begin{bmatrix} S_i A_k + \Theta S_i^- G_k \Theta^{-1} & 0 & 0 & \bar{\alpha} S_i & S_i B_k \end{bmatrix},$$

$$Q_1 = \begin{bmatrix} -\bar{\beta} \tilde{B}_k^f C_k & \tilde{A}_k^f + \tilde{B}_k^f C_k & -(1 - \bar{\beta}) \tilde{B}_k^f & 0 & -\bar{\beta} \tilde{B}_k^f D_k \end{bmatrix},$$

$$Q_2 = \begin{bmatrix} \bar{\beta}^* \tilde{B}_k^f C_k & 0 & -\bar{\beta}^* \tilde{B}_k^f & 0 & \bar{\beta}^* \tilde{B}_k^f D_k \end{bmatrix}.$$

Furthermore, the desired filter parameters can be determined by

$$A_k^f = P_{2,k+1}^{-1} \tilde{A}_k^f, \quad B_k^f = P_{2,k+1}^{-1} \tilde{B}_k^f \tag{6.18}$$

for all $0 \le k \le N - 1$.

Proof *It is easily shown that*

$$\Xi^{(i)} \;=\; \bar{\Xi}^{(i)} + \tilde{S}_i^T P_{1,k+1} \tilde{S}_i + \tilde{Q}_1^T P_{2,k+1} \tilde{Q}_1 + \tilde{Q}_2^T P_{2,k+1} \tilde{Q}_2 \tag{6.19}$$

where

$$\begin{aligned}
\tilde{Q}_1 &= [\; -\bar{\beta} B_k^f C_k \quad A_k^f + B_k^f C_k \quad -(1-\bar{\beta}) B_k^f \quad 0 \quad -\bar{\beta} B_k^f D_k \;], \\
\tilde{Q}_2 &= [\; \bar{\beta}^* B_k^f C_k \quad 0 \quad -\bar{\beta}^* B_k^f \quad 0 \quad \bar{\beta}^* B_k^f D_k \;].
\end{aligned}$$

Take $\tilde{A}_k^f = P_{2,k+1} A_k^f$ and $\tilde{B}_k^f = P_{2,k+1} B_k^f$. By Schur complement Lemma and (6.17), one has $\Xi^{(i)} \le 0$ and then the proof of this theorem follows immediately from Theorem 6.1.

Remark 6.3 *It can be seen from Theorem 6.2 that the design problem of the $\mathcal{H}_\infty$ filter can be solved for the discrete time-varying system (1) with state saturation, RONs, and SPDs if inequalities (6.16) and (6.17) hold for all $0 \le k \le N - 1$. It should be pointed out that the inequalities are nonlinear and dependent on the previous calculation values $P_{1,k}$, $P_{2,k}$, $P_{3,k}$ and system parameters. In other words, they are recursive nonlinear matrix inequalities that are generally difficult to solve. Nevertheless, for inequalities (6.16) and (6.17) in this chapter, a simple yet practical way is to choose G_k beforehand (e.g. $G_k \equiv I$), and then the inequalities (6.17) reduce a set of recursive linear matrix inequalities (RLMIs) which can be solved by the existing semi-definite programming (SDP).*

It is clear that, if the system under consideration is time-invariant, Theorem 6.2 can be reduced to the following corollary where the filter parameters described by (6.5) are also time-invariant.

Corollary 6.1 *Let the disturbance attenuation level $\gamma > 0$ be given. The $\mathcal{H}_\infty$ filtering problem is solvable for a class of discrete time-invariant systems with the same structure as (6.1) if there exist three positive definite matrices P_1, P_2, P_3, a positive scalar λ, and three real-valued matrices G, $\tilde{A}^f$, $\tilde{B}^f$ satisfying the following recursive matrix inequalities:*

$$\|G\|_\infty \le 1, \tag{6.20}$$

$$\begin{bmatrix} \tilde{\Xi}^{(i)} & \bar{S}_i^T P_1 & Q_3^T & Q_4^T \\ * & -P_1 & 0 & 0 \\ * & * & -P_2 & 0 \\ * & * & * & -P_2 \end{bmatrix} < 0, \quad i \in [1, 2^n] \tag{6.21}$$

where

$$\tilde{\Xi}^{(i)} = \begin{bmatrix} \tilde{\Xi}^*_{11} & \tilde{\Xi}^*_{12} & 0 & \tilde{\Xi}^*_{14} & \tilde{\Xi}^*_{15} \\ * & \tilde{\Xi}^*_{22} & 0 & 0 & 0 \\ * & * & \tilde{\Xi}^*_{33} & 0 & \tilde{\Xi}^*_{35} \\ * & * & * & \tilde{\Xi}^{*(i)}_{44} & 0 \\ * & * & * & * & \tilde{\Xi}^*_{55} \end{bmatrix},$$

$$\begin{aligned}
\tilde{\Xi}^*_{11} &= E^T E + \bar{\beta} C^T P_3 C - \lambda \Delta_1 - P_1, \quad \tilde{\Xi}^*_{12} = -E^T E, \\
\tilde{\Xi}^*_{14} &= \lambda \Delta_2, \quad \tilde{\Xi}^*_{15} = \bar{\beta} C^T P_3 D, \quad \tilde{\Xi}^*_{22} = E^T E - P_2, \\
\tilde{\Xi}^*_{33} &= -\bar{\beta} P_3, \quad \tilde{\Xi}^{*(i)}_{44} = \bar{\alpha}(1-\bar{\alpha}) S_i^T P_1 S_i - \lambda I, \\
\tilde{\Xi}^*_{55} &= \bar{\beta} D^T P_3 D - \gamma^2 I, \quad \bar{\beta}^* = \sqrt{\bar{\beta}(1-\bar{\beta})}, \\
\Delta_1 &= \frac{\Phi^T \Psi + \Psi^T \Phi}{2}, \quad \Delta_2 = \frac{\Phi + \Psi}{2}, \\
\bar{S}_i &= [\ S_i A + \Theta S_i^- G \Theta^{-1} \ \ 0 \ \ 0 \ \ \bar{\alpha} S_i \ \ S_i B, \ ] \\
Q_3 &= [\ -\bar{\beta} \tilde{B}^f C \ \ \tilde{A}^f + \tilde{B}^f C \ \ -(1-\bar{\beta}) \tilde{B}^f \ \ 0 \ \ -\bar{\beta} \tilde{B}^f D \ ], \\
Q_4 &= [\ \bar{\beta}^* \tilde{B}^f C \ \ 0 \ \ -\bar{\beta}^* \tilde{B}^f \ \ 0 \ \ \bar{\beta}^* \tilde{B}^f D \ ].
\end{aligned}$$

Furthermore, the desired filter parameters can be determined by

$$A^f = P_2^{-1} \tilde{A}^f, \quad B^f = P_2^{-1} \tilde{B}^f. \tag{6.22}$$

Let Υ be the set of n-dimensional row vectors which has only one nonzero element 1. Denote the ith element of Υ as φ_i whose ith element is 1. $\mathcal{Y}_i$ denotes the set of n-dimensional row vectors whose ith element is 1 and whose other elements are 1 or -1. There are 2^{n-1} elements in $\mathcal{Y}_i$, and let its jth element be denoted as ψ_{ij}. Therefore, $\|G_k\|_\infty$ can be guaranteed by $\varphi_i G_k \psi_{ij} \leq 1$, $i \in [1, 2^n]$. Similar to [58], we introduce the following filter design algorithm via the iterative linear matrix inequality approach to solve inequalities (6.16) and (6.17).

Remark 6.4 *The iterative linear matrix inequality (ILMI) approach is proposed to solve the inequalities (6.16) and (6.17). Note that such an iterative approach is actually LMI-based and its computational complexity is therefore dependent polynomially on the variable dimensions and the number of iterations. The inequalities (6.16) and (6.17) can be solved by using standard algorithms such as the interior-point method. Nevertheless, since the ILMIs involve previously calculated values in the iterative process, there is a possibility for the iteration to diverge or fall into endless loops. One of our future research topics would be to overcome such a computational shortcoming. Fortunately, research on LMI optimization is a very active area in the applied mathematics, optimization, and the operations research community, and substantial progress can be expected in the future.*

6.3 $\mathcal{H}_\infty$ Filtering for Partial State Saturation Case

In many practical systems, it is quite common that only partial states are subjected to saturation constraints. For example, in the case of the dynamics of a car, variables such as speed and steering angle have finite physical limits and they saturate when reaching these limits, whereas variables such as yaw velocity and roll velocity are usually assumed to have no constraints (see, e.g. [74]). The aim of this section is to extend the main results of Section 6.2 to discrete time-varying systems with partial state saturations, RONs, and SDPs.

First, let us decompose the discrete time-varying systems (6.1) as follows

$$
\begin{cases}
x_{k+1}^c = A_{11,k} x_k^c + A_{12,k} x_k^s + \alpha_k f_1(k, x_k^c, x_k^s) + B_{1,k} w_k, \\
x_{k+1}^s = \hbar \left(A_{21,k} x_k^c + A_{22,k} x_k^s + \alpha_k f_2(k, x_k^c, x_k^s) + B_{2,k} w_k \right), \\
y_{ck} = C_{1,k} x_k^c + C_{2,k} x_k^s + D_k w_k, \\
z_k = E_{1,k} x_k^c + E_{2,k} x_k^s
\end{cases}
$$

where $x_k^c \in \mathbb{R}^{n-m}$ and $x_k^s \in \mathbb{R}^m$ are the state, $A_{11,k}$, $A_{12,k}$, $A_{21,k}$, $A_{22,k}$, $B_{1,k}$, $B_{2,k}$, $C_{1,k}$, $C_{2,k}$, D_k, $E_{1,k}$, $E_{2,k}$ are time-varying matrices with appropriate dimensions. The nonlinear vector-valued function $f(k, x_k) \triangleq [\ f_1^T(k, x_k^c, x_k^s) \quad f_2^T(k, x_k^c, x_k^s)\ ]^T$ satisfies (6.3). Here, the state x_k^c is free of saturation, while the state x_k^s is under saturation constraint. According to Lemma 5.2, one has

$$
\begin{aligned}
x_{k+1}^s = \sum_{i=1}^{2^m} \delta_{i,k} \big(&\Theta S_i \Theta^{-1} (A_{21,k} x_k^c + A_{22,k} x_k^s \\
&+ \alpha_k f_2(k, x_k^n, x_k^s) + B_{2,k} w_k) + \Theta S_i^- G_k \Theta^{-1} x_k^s \big)
\end{aligned}
\tag{6.23}
$$

where $\delta_{i,k} > 0$ and $\sum_{i=1}^{2^m} \delta_{i,k} = 1$.

Denoting $x_k = [x_k^{cT} \ \ x_k^{sT}]$, $\eta_k = [\ x_k^T \ \ \hat{x}_k^T \ \ y_{k-1}^T\ ]^T$ and $\tilde{z}_k = z_k - \hat{z}_k$, the following augmented system is obtained

$$
\begin{cases}
\eta_{k+1} = \sum_{i=1}^{2^m} \delta_{i,k} \{ \mathcal{A}_{i,k} \eta_k + \mathcal{B}_{i,k} w_k + \bar{\alpha} \mathcal{F}_i f(k, x_k) \\
\qquad\qquad + (\alpha_k - \bar{\alpha}) \mathcal{F}_i f(k, x_k) + (\beta_k - \bar{\beta}) \mathcal{H}_k \eta_k + (\beta_k - \bar{\beta}) \mathcal{G}_k w_k \}, \\
\tilde{z}_k = \mathcal{E}_k \eta_k
\end{cases}
\tag{6.24}
$$

where

$$\tilde{A}_{i,k} = \begin{bmatrix} A_{11,k} & A_{12,k} \\ S_i A_{21,k} & S_i A_{22,k} + \Theta S_i^- G_k \Theta^{-1} \end{bmatrix}, \quad C_k = \begin{bmatrix} C_{1,k}^T \\ C_{2,k}^T \end{bmatrix}^T,$$

$$\tilde{B}_{i,k} = \begin{bmatrix} B_{1,k} \\ S_i B_{2,k} \end{bmatrix}, \quad F_i = \begin{bmatrix} I & 0 \\ 0 & S_i \end{bmatrix}, \quad E_k = \begin{bmatrix} E_{1,k} \\ E_{2,k} \end{bmatrix},$$

$$\mathcal{A}_{i,k} = \begin{bmatrix} \tilde{A}_{i,k} & 0 & 0 \\ -\bar{\beta} B_k^f C_k & A_k^f + B_k^f C_k & -(1-\bar{\beta}) B_k^f \\ \bar{\beta} C_k & 0 & (1-\bar{\beta}) I \end{bmatrix},$$

$$\mathcal{H}_k = \begin{bmatrix} 0 & 0 & 0 \\ -B_k^f C_k & 0 & B_k^f \\ C_k & 0 & -I \end{bmatrix}, \quad \mathcal{B}_{i,k} = \begin{bmatrix} \tilde{B}_{i,k} \\ -\bar{\beta} B_k^f D_k \\ \bar{\beta} D_k \end{bmatrix},$$

$$\mathcal{E}_k = \begin{bmatrix} E_k^T \\ -E_k^T \\ 0 \end{bmatrix}^T, \quad \mathcal{G}_k = \begin{bmatrix} 0 \\ -B_k^f D_k \\ D_k \end{bmatrix}, \quad \mathcal{F}_i = \begin{bmatrix} F_i \\ 0 \\ 0 \end{bmatrix}.$$

Following similar lines as in Theorems 6.1-6.2 and Corollary 6.1, we have the following three results where only the sketches of the proofs are given to ensure conciseness.

Theorem 6.3 *Let the filter parameters A_k^f, B_k^f in (6.5) and the positive scalar $\gamma > 0$ be given. Then, under the zero initial condition, the augmented system (6.24) satisfies the $\mathcal{H}_\infty$ performance constraint (6.7) for all nonzero w_k if there exist three families of positive definite matrices $\{P_{1,k}\}_{0 \le k \le N}$, $\{P_{2,k}\}_{0 \le k \le N}$, $\{P_{3,k}\}_{0 \le k \le N}$, a set of real-valued matrices $\{G_k\}_{0 \le k \le N}$, and a family of positive scalars $\{\lambda_k\}_{0 \le k \le N}$ satisfying the following recursive matrix inequalities:*

$$\|G_k\|_\infty \le 1, \tag{6.25}$$

$$\begin{bmatrix} \Pi_{11}^{(i)} & \Pi_{12} & 0 & \Pi_{14}^{(i)} & \Pi_{15}^{(i)} \\ * & \Pi_{22} & \Pi_{23} & 0 & \Pi_{25} \\ * & * & \Pi_{33} & 0 & 0 \\ * & * & * & \Pi_{44}^{(i)} & \Pi_{45}^{(i)} \\ * & * & * & * & \Pi_{55}^{(i)} \end{bmatrix} < 0, \quad i \in [1, 2^m] \tag{6.26}$$

where

$$\Delta_1 = \frac{\Phi_k^T \Psi_k + \Psi_k^T \Phi_k}{2}, \quad \Delta_2 = \frac{\Phi_k + \Psi_k}{2},$$

$$\Pi_{11}^{(i)} = \tilde{A}_{i,k}^T P_{1,k+1} \tilde{A}_{i,k} + \bar{\beta} C_k^T B_k^{fT} P_{2,k+1} B_k^f C_k$$
$$+ \bar{\beta} C_k^T P_{3,k+1} C_k + E_k^T E_k - \lambda_k \Delta_1 - P_{1,k},$$

$$\Pi_{12} = -\bar{\beta} C_k^T B_k^{fT} P_{2,k+1} (A_k^f + B_k^f C_k) - E_k^T E_k,$$

$$\Pi_{14}^{(i)} = \bar{\alpha} \tilde{A}_{i,k}^T P_{1,k+1} F_i + \lambda_k \Delta_2,$$

$$\Pi_{15}^{(i)} = \tilde{A}_{i,k}^T P_{1,k+1} \tilde{B}_{i,k} + \bar{\beta} C_k^T B_k^{fT} P_{2,k+1} B_k^f D_k + \bar{\beta} C_k^T P_{3,k+1} D_k,$$

$$\Pi_{22} = (A_k^f + B_k^f C_k)^T P_{2,k+1} (A_k^f + B_k^f C_k) + E_k^T E_k - P_{2,k},$$

$$\Pi_{23} = -(1-\bar{\beta})(A_k^f + B_k^f C_k)^T P_{2,k+1} B_k^f,$$

$$\Pi_{25} = -\bar{\beta}(A_k^f + B_k^f C_k)^T P_{2,k+1} B_k^f D_k,$$

$$\Pi_{33} = (1-\bar{\beta}) B_k^{fT} P_{2,k+1} B_k^f + (1-\bar{\beta}) P_{3,k+1} - P_{3,k},$$

$$\Pi_{44}^{(i)} = \bar{\alpha} F_i^T P_{1,k+1} F_i - \lambda_k I, \quad \Pi_{45}^{(i)} = \bar{\alpha} F_i^T P_{1,k+1} \tilde{B}_{i,k},$$

$$\Pi_{55}^{(i)} = \tilde{B}_{i,k}^T P_{1,k+1} \tilde{B}_{i,k} + \bar{\beta} D_k^T B_k^{fT} P_{2,k+1} B_k^f D_k + \bar{\beta} D_k^T P_{3,k+1} D_k - \gamma^2 I.$$

Proof *Choosing J_k and P_k similar to ones in the proof of Theorem 6.1, we can calculate that*

$$\mathbb{E}\{J_k\} = \mathbb{E}\{\eta_{k+1}^T P_{k+1} \eta_{k+1} - \eta_k^T P_k \eta_k\}$$

$$= \mathbb{E}\Big\{ \Big[\sum_{i=1}^{2^m} \delta_{i,k} \big(\mathcal{A}_{i,k} \eta_k + \mathcal{B}_{i,k} w_k + \bar{\alpha} \mathcal{F}_i f(k, x_k) + (\alpha_k - \bar{\alpha}) \mathcal{F}_i f(k, x_k) \Big.$$

$$\Big. + (\beta_k - \bar{\beta}) \mathcal{H}_k \eta_k + (\beta_k - \bar{\beta}) \mathcal{G}_k w_k \big) \Big]^T P_{k+1} \Big[\sum_{i=1}^{2^m} \delta_{i,k} \big(\mathcal{A}_{i,k} \eta_k$$

$$+ \mathcal{B}_{i,k} w_k + \bar{\alpha} \mathcal{F}_i f(k, x_k) + (\alpha_k - \bar{\alpha}) \mathcal{F}_i f(k, x_k) + (\beta_k - \bar{\beta}) \mathcal{H}_k \eta_k$$

$$+ (\beta_k - \bar{\beta}) \mathcal{G}_k w_k \big) \Big] - \eta_k^T P_k \eta_k \Big\}$$

$$\leq \max_{i \in [1, 2^m]} \mathbb{E}\Big\{ \Big[\mathcal{A}_{i,k} \eta_k + \mathcal{B}_{i,k} w_k + \bar{\alpha} \mathcal{F}_i f(k, x_k) + (\alpha_k - \bar{\alpha}) \mathcal{F}_i f(k, x_k) \Big.$$

$$\Big. + (\beta_k - \bar{\beta}) \mathcal{H}_k \eta_k + (\beta_k - \bar{\beta}) \mathcal{G}_k w_k \Big]^T P_{k+1} \Big[\mathcal{A}_{i,k} \eta_k + \mathcal{B}_{i,k} w_k \tag{6.27}$$

$$+ \bar{\alpha} \mathcal{F}_i f(k, x_k) + (\alpha_k - \bar{\alpha}) \mathcal{F}_i f(k, x_k) + (\beta_k - \bar{\beta}) \mathcal{H}_k \eta_k$$

$$+ (\beta_k - \bar{\beta}) \mathcal{G}_k w_k \Big] - \eta_k^T P_k \eta_k \Big\}$$

$$\leq \max_{i \in [1, 2^m]} \mathbb{E}\Big\{ \eta_k^T \mathcal{A}_{i,k}^T P_{k+1} \mathcal{A}_{i,k} \eta_k + w_k^T \mathcal{B}_{i,k}^T P_{k+1} \mathcal{B}_{i,k} w_k$$

$$+ \bar{\alpha} f^T(k, x_k) \mathcal{F}_i^T P_{k+1} \mathcal{F}_i f(k, x_k) + \bar{\beta}(1-\bar{\beta}) \eta_k^T \mathcal{H}_k^T P_{k+1} \mathcal{H}_k \eta_k$$

$$+ \bar{\beta}(1-\bar{\beta}) w_k^T \mathcal{G}_k^T P_{k+1} \mathcal{G}_k w_k + 2 \eta_k^T \mathcal{A}_{i,k}^T P_{k+1} \mathcal{B}_{i,k} w_k$$

$$+ 2\bar{\alpha} \eta_k^T \mathcal{A}_{i,k}^T P_{k+1} \mathcal{F}_i f(k, x_k) + 2\bar{\alpha} f^T(k, x_k) \mathcal{F}_i^T P_{k+1} \mathcal{B}_{i,k} w_k$$

$$+ 2\bar{\beta}(1-\bar{\beta}) \eta_k^T \mathcal{H}_k^T P_{k+1} \mathcal{G}_k w_k - \eta_k^T P_k \eta_k \Big\}.$$

Then, along the same line as the proof of Theorem 6.1, this result can be easily obtained.

Theorem 6.4 *Let the disturbance attenuation level $\gamma > 0$ be given. The finite-horizon $\mathcal{H}_\infty$ filtering problem is solvable for the discrete time-varying systems (6.23) with partial state saturation, RONs, and SPDs if there* *exist three families of positive definite matrices* $\{P_{1,k}\}_{0\leq k\leq N}$, $\{P_{2,k}\}_{0\leq k\leq N}$, $\{P_{3,k}\}_{0\leq k\leq N}$, *a family of positive scalars* $\{\lambda_k\}_{0\leq k\leq N}$, *and families of real-valued matrices* $\{G_k\}_{0\leq k\leq N-1}$, $\{\tilde{A}_k^f\}_{0\leq k\leq N-1}$, $\{\tilde{B}_k^f\}_{0\leq k\leq N-1}$ *satisfying the following recursive matrix inequalities:*

$$\|G_k\|_\infty \leq 1, \tag{6.28}$$

$$\begin{bmatrix} \bar{\Pi}^{(i)} & \vec{S}_i^T P_{1,k+1} & Q_1^T & Q_2^T \\ * & -P_{1,k+1} & 0 & 0 \\ * & * & -P_{2,k+1} & 0 \\ * & * & * & -P_{2,k+1} \end{bmatrix} < 0, \quad i \in [1, 2^n] \tag{6.29}$$

where

$$\vec{S}_i = \begin{bmatrix} \tilde{A}_{i,k} & 0 & 0 & \bar{\alpha}F_i & \tilde{B}_{i,k} \end{bmatrix},$$

$$\Pi_{44}^{*(i)} = \bar{\alpha}(1-\bar{\alpha})F_i^T P_{1,k+1}F_i - \lambda_k I,$$

$$\bar{\Pi}^{(i)} = \begin{bmatrix} \Xi_{11}^* & \Xi_{12}^* & 0 & \Xi_{14}^* & \Xi_{15}^* \\ * & \Xi_{22}^* & 0 & 0 & 0 \\ * & * & \Xi_{33}^* & 0 & 0 \\ * & * & * & \Pi_{44}^{*(i)} & 0 \\ * & * & * & * & \Xi_{55}^* \end{bmatrix},$$

and other parameters are defined as in Theorem 6.2. Furthermore, the desired filter parameters can be determined by

$$A_k^f = P_{2,k+1}^{-1}\tilde{A}_k^f, \quad B_k^f = P_{2,k+1}^{-1}\tilde{B}_k^f, \tag{6.30}$$

for all $0 \leq k \leq N-1$.

Similar to Corollary 6.1, if the systems with partial state saturations are time-invariant, we have the following corollary readily.

Corollary 6.2 *Let the disturbance attenuation level* $\gamma > 0$ *be given. The* $\mathcal{H}_\infty$ *filtering problem is solvable for a class of discrete time-invariant systems with the structure as in (6.23) if there exist three positive definite matrices* P_1, P_2, P_3, *a positive scalar* λ, *and three real-valued matrices* G, $\tilde{A}^f$, $\tilde{B}^f$ *satisfying the following recursive matrix inequalities:*

$$\|G\|_\infty \leq 1, \tag{6.31}$$

$$\begin{bmatrix} \bar{\Xi}^{(i)} & \vec{S}_i^T P_1 & Q_3^T & Q_4^T \\ * & -P_1 & 0 & 0 \\ * & * & -P_2 & 0 \\ * & * & * & -P_2 \end{bmatrix} < 0, \quad i \in [1, 2^n], \tag{6.32}$$

where

$$\tilde{\Xi}^{(i)} = \begin{bmatrix} \tilde{\Xi}_{11}^* & \tilde{\Xi}_{12}^* & 0 & \tilde{\Xi}_{14}^* & \tilde{\Xi}_{15}^* \\ * & \tilde{\Xi}_{22}^* & 0 & 0 & 0 \\ * & * & \tilde{\Xi}_{33}^* & 0 & 0 \\ * & * & * & \tilde{\Pi}_{44}^{*(i)} & 0 \\ * & * & * & * & \tilde{\Xi}_{55}^* \end{bmatrix},$$

$$\vec{S}_i = \begin{bmatrix} \tilde{A}_i & 0 & 0 & \bar{\alpha}F_i & B_i \end{bmatrix}, \quad \tilde{\Pi}_{44}^{*(i)} = \bar{\alpha}(1-\bar{\alpha})F_i^T P_1 F_i - \lambda I,$$

$$\tilde{A}_i = \begin{bmatrix} A_{11} & A_{12} \\ S_i A_{21} & S_i A_{22} + \Theta S_i^- G \Theta^{-1} \end{bmatrix}, \quad B_i = \begin{bmatrix} B_1 \\ S_i B_2 \end{bmatrix},$$

and other parameters are defined as in Corollary 6.1. Furthermore, the desired filter parameters can be determined by

$$A^f = P_2^{-1} \tilde{A}^f, \quad B^f = P_2^{-1} \tilde{B}^f. \tag{6.33}$$

Remark 6.5 *In this chapter, the performance requirement on the filtering error dynamics is expressed in terms of the $\mathcal{H}_\infty$ disturbance rejection attenuation level over a given finite horizon. In the case of partial state saturations, it can be seen from Theorem 6.3 and Theorem 6.4 that the saturation level, quantified by the diagonal matrix Θ for the saturated states x_k^s, does come to play an important role in determining the feasibility of (6.25)-(6.26) (for Theorem 6.3) and (6.28)-(6.29) (for Theorem 6.4). On the other hand, although it is always desired to have an $\mathcal{H}_\infty$ attenuation level γ that is as small as possible, there does exist a lowest bound for γ especially when certain phenomena (e.g. saturations, randomly occurring nonlinearities and packet dropouts) are simultaneously present. Obviously, a bigger index for γ gives rise to more flexibility for the filter design. Therefore, for the proposed Algorithm 1, the value of the parameter γ indeed influences the feasibility of the iterative process. Moreover, it is easy to see that the saturations, if not adequately handled, may pose significant side effects in the filter design. Specifically, a) saturated nonlinear functions are likely to cause undesirable oscillatory behaviors especially at the switching between saturations and unsaturations; b) the saturated and unsaturated states may affect each other to make the system dynamics even more complicated; c) and the filter performance may be deteriorated or even impaired by the saturation constraints especially during the initial stages.*

6.4   Simulation Examples

To illustrate the effectiveness of the proposed methods, two numerical examples are given in this section.

Example 6.1 *The first example concerns the discrete time-varying systems (6.1) with state saturation whose parameters are as follows:*

$$A_k = \begin{bmatrix} -0.75 + 0.2\sin(2k) & 0.4 \\ -1.3 & 0.9 + 0.2\cos(k+2) \end{bmatrix}, \quad C_k = \begin{bmatrix} 1.1 \\ 0.85 \end{bmatrix}^T,$$

$$B_k = \begin{bmatrix} 0.11 & 0.09 + 0.5e^{-5k} \end{bmatrix}, \quad D_k = 0.11, \quad E_k = \begin{bmatrix} 0.75 & 0.85 \end{bmatrix},$$

$$f(k, x_k) = \begin{cases} \begin{bmatrix} -0.6x_{k,1} + 0.3x_{k,2} + \tanh(0.3x_{k,1}) \\ 0.6x_{k,2} - \tanh(0.2x_{k,2}) \end{bmatrix}, & 0 \le k < 11 \\[2mm] \begin{bmatrix} 0.65x_{k,1} - \tanh(0.25x_{k,1}) \\ 0.5x_{k,2} \end{bmatrix}, & 11 \le k < 21 \end{cases}$$

where the time horizon is taken as $N = 21$. Then, it is easy to see that the constraint (6.3) can be met with

$$\Phi_k = \begin{cases} \begin{bmatrix} -0.6 & 0.3 \\ 0 & 0.4 \end{bmatrix}, & 0 \le k < 11, \\[2mm] \begin{bmatrix} 0.4 & 0 \\ 0 & 0.5 \end{bmatrix}, & 11 \le k < 21, \end{cases}$$

$$\Psi_k = \begin{cases} \begin{bmatrix} -0.3 & 0.3 \\ 0 & 0.6 \end{bmatrix}, & 0 \le k < 11, \\[2mm] \begin{bmatrix} 0.65 & 0 \\ 0 & 0.5 \end{bmatrix}, & 11 \le k < 21. \end{cases}$$

    Considering both the RONs and the SPDs, the probabilities are taken as $\bar{\alpha} = 0.2$ and $\bar{\beta} = 0.75$, respectively. Moreover, the disturbance attenuation level is chosen as $\gamma = 0.96$ and the saturation level is $\Theta = \mathrm{diag}\{1.6, 1.4\}$. Using the given algorithm and MATLAB® (with YALMIP 3.0), the set of solutions to recursive matrix inequalities (6.16) and (6.17) in Theorem 6.2 are obtained and shown in Table 6.1.

Example 6.2 *The second example considers the design problem of the $\mathcal{H}_\infty$ filter for discrete time-varying systems with partial state saturation, where the second state $x_{2,k}$ is under saturation constraint with the saturation level $\Theta = 1.4$. Consider the system with*

$$A_{11,k} = -0.23 + 0.35\tanh(k+2), \quad A_{12,k} = 0, \quad A_{21,k} = 0.3,$$

$$A_{22,k} = 0.35 + \frac{0.2(k^2 - 8)}{k^2}, \quad B_{1,k} = 0.05, \quad B_{2,k} = 0.65e^{-5k},$$

$$C_{1,k} = 1.1, \quad C_{2,k} = 0.85, \quad D_k = 0.11, \quad E_{1,k} = 0.75, \quad E_{2,k} = 0.85,$$

and other parameters are the same as in the above example. Using the given algorithm and MATLAB (with YALMIP 3.0), the set of solutions to recursive matrix inequalities (6.28) and (6.29) in Theorem 6.4 are shown in Table 6.2.

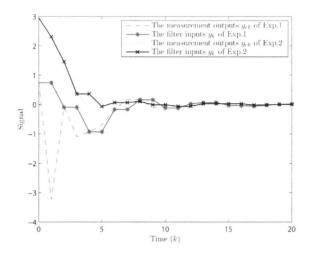

FIGURE 6.1
The measurement outputs and the filter inputs.

In the simulations, set the system initial values of two examples as $[1.5 - 1.35]^T$ and $[3.5 - 1.35]^T$, respectively. Let the initial value of $\hat{x}_0$ be $[0\ 0]^T$ and the filter input signal be $y_{-1} = 0$. The exogenous disturbance input is selected as $w_k = 5 \exp(-0.2k)\sin(k)$. The simulation results for Example 6.1 and Example 6.2 (denoted as Exp. 6.1 and Exp. 6.2) are shown in Fig. 6.1 and Fig. 6.2, where Fig. 6.1 plots the measurement outputs and the filter inputs, and Fig. 6.2 depicts the filtering errors. As shown in Fig. 6.2, there appears to be a big jump in errors in the initial stage, which is mainly due to obvious "efforts" from the filter to adjust its gains in order to compensate the big estimation errors caused by the state initialization. This kind of compensation is well reflected by the strong fluctuations in the initial transient filtering process. The simulation results have confirmed that the designed $\mathcal{H}_\infty$ filtering performs very well.

6.5   Summary

In this chapter, we have investigated the $\mathcal{H}_\infty$ filtering problem for a class of discrete time-varying systems with state saturations, RONs, and SPDs over a finite-horizon. Two sequences of mutually independent random variables that obey Bernoulli distribution are introduced to describe the addressed randomness. With a free matrix whose infinity norm is less than or equal

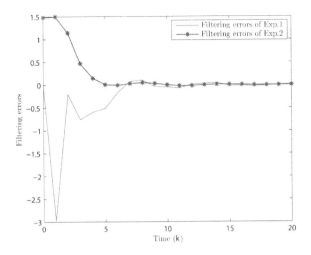

FIGURE 6.2
Filtering errors $\tilde{z}_k$.

to 1, some sufficient conditions have been derived for the finite horizon filter to satisfy the prescribed $\mathcal{H}_\infty$ performance requirement. These conditions are expressed in terms of recursive nonlinear matrix inequalities that can be checked by the presented RLMI approach. Two numerical simulation examples have been provided to demonstrate the effectiveness and applicability of the proposed design approach. As illustrated in the introduction, the obtained result could be applied to mechanical systems, digital filters, electric power systems with hard limits on power inputs, field voltages and currents, digitally controlled systems with sampled-data and digital redesign, and so on. Further research topics include the extension of the main results to the dissipative control problems for discrete-time systems with state saturations, and also to the network control systems with both state saturations and missing measurements.

The filter design algorithm

Step 1. Given the $\mathcal{H}_\infty$ performance index $\gamma > 0$, and set $k = 0$.

Step 2. Given $S_i = I$, $\forall i \in [1, 2^n]$. Solve the linear matrix inequality
(6.17) for $P_{1,k+1}$, $P_{2,k+1}$, $P_{3,k+1}$, $\tilde{A}_k^f$, $\tilde{B}_k^f$, λ_k when $k \neq 0$, or
for $P_{1,1}$, $P_{2,1}$, $P_{3,1}$, $P_{1,0}$, $P_{2,0}$, $P_{3,0}$, $\tilde{A}_0^f$, $\tilde{B}_0^f$, λ_0 when $k = 0$. If
successful in finding a feasible solution, set $\tau = 0$ and go to the
next step, else set $\mu = 1$ and go to Step 5.

Step 3. Using the $P_{1,k+1}$ obtained previously, solve the following
optimization problem for $U_k = \{ P_{2,k+1}, P_{3,k+1}, G_k, \tilde{A}_k^f, \tilde{B}_k^f, \lambda_k, \mu \}$ when $k \neq 0$, or for $U_0 = \{ P_{2,1}, P_{3,1}, P_{1,0}, P_{2,0}, P_{3,0}, G_0, \tilde{A}_0^f, \tilde{B}_0^f, \lambda_0, \mu \}$ when $k = 0$:

$$\inf_{U_k} \mu$$

$$\text{s.t.} \begin{cases} \Xi^* < \mu I, \\ \varphi_i G_k \psi_{ij} \leq 1, \quad i \in [1, 2^n], \quad j \in [1, 2^{n-1}]. \end{cases}$$

- If $\mu < 0$, then go to *Step 5*.

- If $\tau = 0$ and $\mu > 0$, then set $\tau = \tau + 1$, $\mu_\tau = \mu$ and go to the next step.

- If $\tau > 0$ and $\mu_\tau > \mu > 0$, then set $\tau = \tau + 1$, $\mu_\tau = \mu$ and go to the next step.

- If $\tau > 0$ and $\mu > \mu_\tau > 0$, then go to *Step 5*.

Step 4. Using the G_k obtained previously, solve the following LMI's
optimization problem for $U_k^* = \{ P_{1,k+1}, P_{2,k+1}, P_{3,k+1}, \tilde{A}_k^f, \tilde{B}_k^f, \lambda_k, \mu \}$ when $k \neq 0$, or for $U_0^* = \{ P_{1,1}, P_{2,1}, P_{3,1}, P_{1,0}, P_{2,0}, P_{3,0}, \tilde{A}_0^f, \tilde{B}_0^f, \lambda_0, \mu \}$ when $k = 0$:

$$\inf_{U_k^*} \mu$$

$$\text{s.t.} \quad \Xi^* < \mu I, \quad i \in [1, 2^n].$$

- If $\mu < 0$, then go to the next step.

- If $\mu_\tau > \mu > 0$, then set $\tau = \tau + 1$, $\mu_\tau = \mu$ and go to *Step 2*.

- Others go to the next step.

Step 5. If $\mu < 0$, then $A_k^f = P_{2,k+1}^{-1} \tilde{A}_k^f$, $B_k^f = P_{2,k+1}^{-1} \tilde{B}_k^f$ are suitable filter
parameters, and go to the next step, else no condition can be
drawn, and go to *Step 7*.

Step 6. If $k \leq N - 1$, then go to *Step 2*, else go to the next step.

Step 7. Stop.

TABLE 6.1
Variables and filter parameters.

k	$P_{1,k}$	$P_{2,k}$	$P_{3,k}$	A_k^f	B_k^f
0	$\begin{bmatrix} 91.9921 & -9.1519 \\ -9.1519 & 89.5407 \end{bmatrix}$	$\begin{bmatrix} 47.4558 & 1.5531 \\ 1.5531 & 46.8795 \end{bmatrix}$	39.7137	$\begin{bmatrix} -0.1048 & -0.0792 \\ 0.2632 & 0.2062 \end{bmatrix}$	$\begin{bmatrix} -0.0254 \\ -0.0135 \end{bmatrix}$
1	$\begin{bmatrix} 25.2054 & -9.6740 \\ -9.6740 & 16.6853 \end{bmatrix}$	$\begin{bmatrix} 95.1981 & 8.9999 \\ 8.9999 & 83.1322 \end{bmatrix}$	0.0392	$\begin{bmatrix} -0.3119 & -0.1971 \\ 0.7455 & 0.4896 \end{bmatrix}$	$\begin{bmatrix} -0.0278 \\ -0.0130 \end{bmatrix}$
2	$\begin{bmatrix} 25.0660 & -14.2308 \\ -14.2308 & 14.9675 \end{bmatrix}$	$\begin{bmatrix} 90.2235 & 20.8675 \\ 20.8675 & 55.4596 \end{bmatrix}$	0.0343	$\begin{bmatrix} -0.3198 & -0.1964 \\ 0.8178 & 0.5227 \end{bmatrix}$	$\begin{bmatrix} -0.0272 \\ -0.0118 \end{bmatrix}$
3	$\begin{bmatrix} 23.0777 & -11.5001 \\ -11.5001 & 12.6758 \end{bmatrix}$	$\begin{bmatrix} 90.5466 & 21.7528 \\ 21.7528 & 49.9456 \end{bmatrix}$	0.0356	$\begin{bmatrix} -0.3379 & -0.1706 \\ 0.8420 & 0.4520 \end{bmatrix}$	$\begin{bmatrix} -0.0252 \\ -0.0111 \end{bmatrix}$
4	$\begin{bmatrix} 25.3657 & -9.3007 \\ -9.3007 & 9.1380 \end{bmatrix}$	$\begin{bmatrix} 90.2014 & 22.2358 \\ 22.2358 & 49.5409 \end{bmatrix}$	0.0395	$\begin{bmatrix} -0.3249 & -0.1628 \\ 0.8775 & 0.4710 \end{bmatrix}$	$\begin{bmatrix} -0.0276 \\ -0.0113 \end{bmatrix}$
$\dots$	$\dots$	$\dots$	$\dots$	$\dots$	$\dots$
20	$\begin{bmatrix} 24.2509 & -11.5938 \\ -11.5938 & 11.7296 \end{bmatrix}$	$\begin{bmatrix} 86.9948 & 24.6079 \\ 24.6079 & 53.4379 \end{bmatrix}$	0.0387	$\begin{bmatrix} -0.3584 & -0.2300 \\ 0.8619 & 0.5711 \end{bmatrix}$	$\begin{bmatrix} -0.0280 \\ -0.0129 \end{bmatrix}$
21	$\begin{bmatrix} 20.0290 & -13.7456 \\ -13.7456 & 16.0879 \end{bmatrix}$	$\begin{bmatrix} 89.7664 & 22.2881 \\ 22.2881 & 51.4581 \end{bmatrix}$	0.0339		

TABLE 6.2
Variables and filter parameters.

k	$P_{1,k}$	$P_{2,k}$	$P_{3,k}$	A_k^f	B_k^f
0	$\begin{bmatrix} 448.5640 & -78.6224 \\ -78.6224 & 379.8219 \end{bmatrix}$	$\begin{bmatrix} 61.0930 & 2.2671 \\ 2.2671 & 60.4877 \end{bmatrix}$	18.8229	$\begin{bmatrix} -0.3166 & -0.2761 \\ 0.2555 & 0.2227 \end{bmatrix}$	$\begin{bmatrix} -0.0217 \\ -0.0340 \end{bmatrix}$
1	$\begin{bmatrix} 67.7091 & -0.6721 \\ -0.6721 & 59.2365 \end{bmatrix}$	$\begin{bmatrix} 71.0064 & 5.7322 \\ 5.7322 & 76.3465 \end{bmatrix}$	0.1090	$\begin{bmatrix} -0.1749 & -0.1708 \\ 0.1856 & 0.1804 \end{bmatrix}$	$\begin{bmatrix} -0.0035 \\ -0.0034 \end{bmatrix}$
2	$\begin{bmatrix} 93.5089 & -8.2684 \\ -8.2684 & 113.6558 \end{bmatrix}$	$\begin{bmatrix} 352.9001 & 149.7004 \\ 149.7004 & 347.6532 \end{bmatrix}$	0.1097	$\begin{bmatrix} -0.3705 & -0.3415 \\ 0.4390 & 0.4041 \end{bmatrix}$	$\begin{bmatrix} -0.0035 \\ -0.0030 \end{bmatrix}$
3	$\begin{bmatrix} 112.0331 & -7.1965 \\ -7.1965 & 134.2761 \end{bmatrix}$	$\begin{bmatrix} 381.9950 & 137.3744 \\ 137.3744 & 340.0770 \end{bmatrix}$	0.1098	$\begin{bmatrix} -0.3611 & -0.3051 \\ 0.4492 & 0.3797 \end{bmatrix}$	$\begin{bmatrix} -0.0035 \\ -0.0028 \end{bmatrix}$
4	$\begin{bmatrix} 114.8840 & -6.3026 \\ -6.3026 & 134.6109 \end{bmatrix}$	$\begin{bmatrix} 391.8702 & 132.2112 \\ 132.2112 & 338.3444 \end{bmatrix}$	0.1099	$\begin{bmatrix} -0.3889 & -0.3865 \\ 0.4943 & 0.4902 \end{bmatrix}$	$\begin{bmatrix} -0.0035 \\ -0.0028 \end{bmatrix}$
...	...	...	...	...	...
20	$\begin{bmatrix} 120.5605 & -0.6206 \\ -0.6206 & 127.2991 \end{bmatrix}$	$\begin{bmatrix} 401.2425 & 128.4070 \\ 128.4070 & 333.0420 \end{bmatrix}$	0.1099	$\begin{bmatrix} -0.4484 & -0.2502 \\ 0.5909 & 0.3322 \end{bmatrix}$	$\begin{bmatrix} -0.0035 \\ -0.0027 \end{bmatrix}$
21	$\begin{bmatrix} 120.5458 & -0.6203 \\ -0.6203 & 127.2348 \end{bmatrix}$	$\begin{bmatrix} 401.4538 & 128.1628 \\ 128.1628 & 333.3196 \end{bmatrix}$	0.1099		

7

Finite-Horizon Envelope-Constrained $\mathcal{H}_\infty$ Filtering with Fading Measurements

In signal processing and communications, the envelope-constrained index has been utilized to solve a wide range of practical engineering, such as the radar and sonar detection. In the past few years, some representative results have been reported for linear time-invariant systems. It should be pointed out that the resultant ever-increasing popularity of communication networks will lead to some new research challenges for envelope-constrained filtering with phenomenon as discussed in Chapter 2. The essential difficulties can be identified as follows: 1) how can we define the criterion for the envelope-constrained filtering of a class of time-varying systems with network-induced phenomena? and 2) how can we examine the impact from the statistical information of both fading measurements and RONs on the filtering performance?

In this chapter, the envelope-constrained $\mathcal{H}_\infty$ filtering problem is investigated for a class of discrete time-varying stochastic systems over a finite horizon. The system under consideration involves fading measurements, randomly occurring nonlinearities (RONs), and mixed (multiplicative and additive) noises. A novel envelope-constrained performance criterion is proposed to better quantify the transient dynamics of the filtering error process over the finite horizon. The purpose of the problem addressed is to design a time-varying filter such that both the $\mathcal{H}_\infty$ performance and the desired envelope constraints are achieved at each time step. By utilizing stochastic analysis techniques combined with the ellipsoid description of the estimation errors, sufficient conditions are established in the form of recursive matrix inequalities (RMIs) reflecting both the envelope information and the desired $\mathcal{H}_\infty$ performance index. The filter gain matrix is characterized by means of the solvability of the deduced RMIs. Finally, a simulation example is provided to show the effectiveness of the proposed filtering design scheme.

7.1 Modeling and Problem Formulation

Consider the following discrete time-varying stochastic system defined on $k \in [0, N]$

$$\begin{cases} x(k+1) = \big(A(k) + \xi_k D(k)\big)x(k) + \alpha_k h(k, x(k)) + E_1(k)v(k), \\ y(k) = C(k)x(k) + E_2(k)v(k), \\ z(k) = L(k)x(k) \end{cases} \quad (7.1)$$

where $x(k) \in \mathbb{R}^{n_x}$, $y(k) \in \mathbb{R}^{n_y}$ and $z(k) \in \mathbb{R}^{n_z}$ are, respectively, the state vector, the measurement output (without fading), and the signal to be estimated. $v(k) \in l([0, N]; \mathbb{R}^q)$ is the disturbance input, $A(k)$, $C(k)$, $D(k)$, $E_1(k)$, $E_2(k)$ and $L(k)$ are constant matrices with appropriate dimensions, and $\xi_k \in \mathbb{R}$ is a zero-mean random sequence with $\mathbb{E}\{\xi_k^2\} = 1$. The random variable $\alpha_k \in \mathbb{R}$, which is unrelated to ξ_k, is a Bernoulli distributed white sequence obeying the following probability distribution law

$$\text{Prob}\{\alpha_k = 0\} = 1 - \bar{\alpha}, \quad \text{Prob}\{\alpha_k = 1\} = \bar{\alpha}.$$

The nonlinear vector-valued function $h : [0, N] \times \mathbb{R}^{n_x} \to \mathbb{R}^{n_x}$ is continuous, and satisfies $h(k, 0) = 0$ and the sector-bounded condition

$$\big[h(k, x) - h(k, y) - \Phi(k)(x - y)\big]^T \big[h(k, x) - h(k, y) - \Psi(k)(x - y)\big] \le 0 \quad (7.2)$$

for all x, $y \in \mathbb{R}^{n_x}$, where $\Phi(k)$ and $\Psi(k)$ are real matrices with appropriate dimensions.

Letting the number of paths denoted by ℓ be given, the *actually* received signal by the filter is of the following form

$$\tilde{y}(k) = \sum_{s=0}^{\ell_k} \vartheta_k^s y(k - s) + E_3(k)w(k) \quad (7.3)$$

where $\ell_k = \min\{\ell, k\}$, ϑ_k^s $(s = 0, 1, \cdots, \ell_k)$ are the mutually independent channel coefficients having probability density functions $f(\vartheta_k^s)$ on the interval $[0, 1]$ with mathematical expectations $\bar{\vartheta}^s$ and variances $\tilde{\vartheta}^s$, and $w(k) \in l([0, N]; \mathbb{R})$ is an external disturbance.

For the purpose of simplicity, for $-\ell \le i \le -1$, we assume that $C(i) = 0$, $y(i) = 0$ and $[v^T(i)\ w^T(i)] = 0$. Based on the *actually* received signal $\tilde{y}_k$, the following filter is constructed

$$\begin{cases} \hat{x}(k+1) = A(k)\hat{x}(k) + \bar{\alpha}h(k, \hat{x}(k)) \\ \qquad\qquad + K(k)\Big(\tilde{y}(k) - \sum_{s=0}^{\ell} \bar{\vartheta}^s C(k - s)\hat{x}(k - s)\Big), \\ \hat{z}(k) = L(k)\hat{x}(k) \end{cases} \quad (7.4)$$

where $\hat{x}(k) \in \mathbb{R}^{n_x}$ is the estimated state, $\hat{z}(k) \in \mathbb{R}^{n_z}$ represents the estimated output, and $K(k)$ is the time-varying filter gain matrix to be designed.

Let the state estimation error be $e_k = x(k) - \hat{x}(k)$ and the output estimation error be $\tilde{z}_k = z(k) - \hat{z}(k)$. Then, the dynamics of the filtering errors can be obtained from (7.1) and (7.4) as follows

$$
\begin{cases}
e_{k+1} = \bar{A}(k)e_k + \xi_k D(k)x(k) + \bar{\alpha}(h(k, x_k) - h(k, \hat{x}_k)) \\
\qquad + (\alpha_k - \bar{\alpha})h(k, x_k) - \sum_{s=0}^{\ell} \bar{\vartheta}^s K(k)C(k-s)e_{k-s} \\
\qquad - \sum_{s=0}^{\ell} (\vartheta_k^s - \bar{\vartheta}^s)K(k)C(k-s)x(k-s) + E_1(k)v(k) \qquad (7.5) \\
\qquad - \sum_{s=0}^{\ell} \vartheta_k^s K(k)E_2(k-s)v(k-s) - K(k)E_3(k)w(k), \\
\tilde{z}_k = L(k)e_k.
\end{cases}
$$

Furthermore, denoting

$$
\eta_k = [x^T(k) \quad e_k^T]^T, \quad \zeta_k = [v^T(k) \quad w^T(k)]^T,
$$
$$
h_k = \left[h^T(k, x(k)) \quad h^T(k, x(k)) - h^T(k, \hat{x}(k)) \right]^T,
$$

we have the following augmented system

$$
\begin{cases}
\eta_{k+1} = \mathcal{A}_k \eta_k + \xi_k \mathcal{D}_k \mathcal{S}_1 \eta_k + \bar{\alpha} h_k + (\alpha_k - \bar{\alpha})\mathcal{S}_1 h_k \\
\qquad - \sum_{s=1}^{\ell} \bar{\vartheta}^s \mathcal{K}_k \mathcal{C}_{k-s} \mathcal{S}_2 \eta_{k-s} - \sum_{s=0}^{\ell} (\vartheta_k^s - \bar{\vartheta}^s)\mathcal{K}_k \mathcal{C}_{k-s} \mathcal{S}_3 \eta_{k-s} \\
\qquad - \sum_{s=0}^{\ell} \bar{\vartheta}^s \mathcal{K}_k \mathcal{E}_{k-s} \mathcal{S}_3 \zeta_{k-s} - \sum_{s=0}^{\ell} (\vartheta_k^s - \bar{\vartheta}^s)\mathcal{K}_k \mathcal{E}_{k-s} \mathcal{S}_3 \zeta_{k-s} + \mathcal{F}_k \zeta_k, \\
\tilde{z}_k = \mathcal{L}_k \eta_k
\end{cases} \qquad (7.6)
$$

where

$$
\begin{aligned}
\mathcal{A}_k &= \operatorname{diag}\{A(k), A(k) - \bar{\vartheta}^0 K(k)C(k)\}, \\
\mathcal{C}_k &= \operatorname{diag}\{C(k), C(k)\}, \quad \mathcal{D}_k = \operatorname{diag}\{D(k), D(k)\}, \\
\mathcal{K}_k &= \operatorname{diag}\{K(k), K(k)\}, \quad \mathcal{S}_2 = \operatorname{diag}\{0, I\}, \\
\mathcal{S}_1 &= \begin{bmatrix} I & 0 \\ I & 0 \end{bmatrix}, \ \mathcal{S}_3 = \begin{bmatrix} 0 & 0 \\ I & 0 \end{bmatrix}, \ \mathcal{L}_k = \begin{bmatrix} 0 \\ L^T(k) \end{bmatrix}^T, \\
\mathcal{E}_k &= \begin{bmatrix} 0 & 0 \\ E_2(k) & 0 \end{bmatrix}, \ \mathcal{F}_k = \begin{bmatrix} E_1(k) & 0 \\ E_1(k) & -K(k)E_3(k) \end{bmatrix}.
\end{aligned}
$$

Our aim in this chapter is to design an envelope-constrained $\mathcal{H}_\infty$ filter of the form (7.4) such that the following requirements are met simultaneously:

a) ($\mathcal{H}_\infty$ requirement) for any nonzero ζ_k, the output $\tilde{z}_k$ of the augmented system (7.6) satisfies

$$\sum_{k=0}^{N} \mathbb{E}\{||\tilde{z}_k||^2\} \le \gamma^2 \sum_{k=0}^{N} ||\zeta_k||^2 + \gamma^2 \sum_{i=-\ell}^{0} \mathbb{E}\{\eta_i^T \mathcal{W}_i \eta_i\} \tag{7.7}$$

where γ is a prescribed positive scalar and $\mathcal{W}_i > 0$ ($-\ell \le i \le 0$) are some weighting matrices;

b) (envelope constraints) under the zero-initial condition, for the given input signal

$$\zeta_k^* = \begin{cases} 1, & k = 0, \\ 0, & 1 \le k \le N, \end{cases}$$

the corresponding output $\tilde{z}_k^*$ of the augmented system (7.6) satisfies

$$d_k^i - \varepsilon_k^i \le \mathbb{E}\{\mathbf{I}_i \tilde{z}_k^*\} \le d_k^i + \varepsilon_k^i, \quad k \in [1, N], \quad i \in [1, n_z] \tag{7.8}$$

where $\mathbf{I}_i := \big[\underbrace{0, \cdots, 0}_{i-1}, 1, \underbrace{0, \cdots, 0}_{n_z - i}\big]$, and $\{d_k^i\}_{k \in [0,\ N]}$ and $\{\varepsilon_k^i\}_{k \in [0,\ N]}$ are the sequences of the desired output and the tolerance band, respectively. Obviously, ε_k^i ($k \in [1, N]$, $i \in [1, n_z]$) are positive scalars.

Remark 7.1 *(7.8) gives the envelope constraint on the individual estimation error in the mean square, which can be understood as the stochastic version of the envelope definition in [124] over a finite horizon. Similar to [124], an envelope-constrained $\mathcal{H}_\infty$ filtering system is shown in Fig. 7.1. The signal $y(k)$ is the measurement of the plant system with an energy-bounded disturbance input $v(k)$ and the filter input signal $\tilde{y}(k)$ is the output of transmission channel corrupted by an energy-bounded noise $w(k)$. $\hat{z}(k)$ is the estimate of $z(k)$. The aim of the envelope-constrained filtering problem is twofold: 1) a filter gain is designed to reconstruct $z(k)$ by using the distorted signal $\tilde{y}(k)$; and 2) for the given input signal $w(k)$ and $v(k)$, $\tilde{z}_k$ is guaranteed to lie within the specified envelope at each time step. It should be pointed out that the $\mathcal{H}_\infty$ criterion is concerned with the performance requirement as a whole over the finite-horizon $[0, N]$. In contrast, the envelope-constrained requirement (7.8) can be used to describe the transient dynamics of the filtering error process at each time step, that is, the outputs $\{\tilde{z}_k^*\}_{k \in [0,N]}$ stimulated by the given input $\{\zeta_k^*\}_{k \in [0,N]}$ are included in a desired envelope in the mean square sense at each step.*

In the following section, by resorting to stochastic analysis combined with the ellipsoid description of the estimation errors, some sufficient conditions are proposed to guarantee the $\mathcal{H}_\infty$ performance and achieve the desired envelope constraints for the given $\mathcal{H}_\infty$ filter over the given finite horizon.

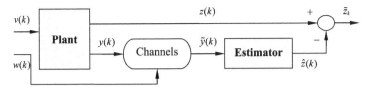

FIGURE 7.1
The envelope-constrained $\mathcal{H}_\infty$ filtering system model.

7.2 $\mathcal{H}_\infty$ Performance Analysis

Denote

$$\eta_k^\ell = [\eta_{k-1}^T \;\; \cdots \;\; \eta_{k-\ell}^T]^T, \quad \zeta_k^\ell = [\zeta_k^T \;\; \cdots \;\; \zeta_{k-\ell}^T]^T, \quad \tilde{\eta}_k = [\eta_k^T \;\; h_k^T \;\; (\eta_k^\ell)^T \;\; (\zeta_k^\ell)^T]^T.$$

The following lemma will be used in deriving our main results.

Lemma 7.1 *Let the external disturbances ζ_k and the initial values $\{\eta_k\}_{k \in [-\ell, \, 0]}$ be given. For the function*

$$V_k = \eta_k^T \mathcal{P}_k \eta_k + \sum_{j=1}^{\ell} \sum_{i=k-j}^{k-1} \eta_i^T \mathcal{R}_{i,j} \eta_i \tag{7.9}$$

where $\mathcal{P}_k$ and $\mathcal{R}_{i,j}$ are symmetric positive definite matrices with appropriate dimensions, the following relationship

$$\mathbb{E}\{\Delta V_k\} := \mathbb{E}\{V_{k+1} - V_k\} = \mathbb{E}\{\tilde{\eta}_k^T \Pi_1^k \tilde{\eta}_k\} \tag{7.10}$$

is true, where

$$\Pi_1^k = \begin{bmatrix} \Pi_{11}^k & \Pi_{12}^k & \Pi_{13}^k & \Pi_{14}^k \\ * & \Pi_{22}^k & \Pi_{23}^k & \Pi_{24}^k \\ * & * & \Pi_{33}^k & \Pi_{34}^k \\ * & * & * & \Pi_{44}^k \end{bmatrix},$$

$$\Lambda_0 = [\bar{\vartheta}^0 I \;\; \bar{\vartheta}^1 I \;\; \cdots \;\; \bar{\vartheta}^\ell I], \quad \Lambda_1 = [\bar{\vartheta}^1 I \;\; \bar{\vartheta}^2 I \;\; \cdots \;\; \bar{\vartheta}^\ell I],$$

$$\mathcal{I}_{0v} = diag\{\sqrt{\tilde{\vartheta}^0} I, \; \sqrt{\tilde{\vartheta}^1} I, \; \cdots, \; \sqrt{\tilde{\vartheta}^\ell} I\},$$

$$\mathcal{I}_{1v} = diag\{\sqrt{\tilde{\vartheta}^1} I, \; \sqrt{\tilde{\vartheta}^2} I, \; \cdots, \; \sqrt{\tilde{\vartheta}^\ell} I\},$$

$$\bar{\mathcal{C}}_{0k} = diag\{\mathcal{C}_k, \mathcal{C}_{k-1}, \cdots, \mathcal{C}_{k-\ell}\}, \quad \bar{\mathcal{S}}_{31} = diag_\ell\{\mathcal{S}_3\},$$

$$\bar{\mathcal{C}}_{1k} = diag\{\mathcal{C}_{k-1}, \mathcal{C}_{k-2}, \cdots, \mathcal{C}_{k-\ell}\}, \quad \bar{\mathcal{S}}_{21} = diag_\ell\{\mathcal{S}_2\},$$

$$\bar{\mathcal{E}}_{0k} = diag\{\mathcal{E}_k, \mathcal{E}_{k-1}, \cdots, \mathcal{E}_{k-\ell}\}, \quad \mathcal{I}_D = [\, I \;\; 0 \;\; \cdots \;\; 0\,],$$

$$\bar{\mathcal{E}}_{1k} = diag\{\mathcal{E}_{k-1}, \mathcal{E}_{k-2}, \cdots, \mathcal{E}_{k-\ell}\}, \quad \tilde{\alpha} = \bar{\alpha}(1 - \bar{\alpha}),$$

$$\Pi_{11}^k = \mathcal{A}_k^T \mathcal{P}_{k+1} \mathcal{A}_k - \mathcal{P}_k + \mathcal{S}_1^T \mathcal{D}_k^T \mathcal{P}_{k+1} \mathcal{D}_k \mathcal{S}_1$$
$$+ \tilde{\vartheta}^0 \mathcal{S}_3^T \mathcal{C}_k^T \mathcal{K}_k^T \mathcal{P}_{k+1} \mathcal{K}_k \mathcal{C}_k \mathcal{S}_3 + \sum_{j=1}^{\ell} \mathcal{R}_{k,j},$$

$$\Pi_{12}^k = \bar{\alpha} \mathcal{A}_k^T \mathcal{P}_{k+1}, \quad \Pi_{13}^k = -\mathcal{A}_k^T \mathcal{P}_{k+1} \mathcal{K}_k \Lambda_1 \bar{\mathcal{C}}_{1k} \bar{\mathcal{S}}_{21},$$

$$\Pi_{14}^k = -\mathcal{A}_k^T \mathcal{P}_{k+1} \mathcal{K}_k \Lambda_0 \bar{\mathcal{E}}_{0k} + \mathcal{A}_k^T \mathcal{P}_{k+1} \mathcal{F}_k \mathcal{I}_D + \tilde{\vartheta}^0 \left(\mathcal{K}_k \mathcal{C}_k \mathcal{S}_3 \right)^T \mathcal{P}_{k+1} \mathcal{K}_k \mathcal{E}_k \mathcal{I}_D,$$

$$\Pi_{22}^k = \bar{\alpha}^2 \mathcal{P}_{k+1} + \tilde{\alpha} \mathcal{S}_1^T \mathcal{P}_{k+1} \mathcal{S}_1, \quad \Pi_{23}^k = -\bar{\alpha} \mathcal{P}_{k+1} \mathcal{K}_k \Lambda_1 \bar{\mathcal{C}}_{1k} \bar{\mathcal{S}}_{21},$$

$$\Pi_{24}^k = -\bar{\alpha} \mathcal{P}_{k+1} \mathcal{K}_k \Lambda_0 \bar{\mathcal{E}}_{0k} + \bar{\alpha} \mathcal{P}_{k+1} \mathcal{F}_k \mathcal{I}_D,$$

$$\Pi_{33}^k = \bar{\mathcal{S}}_{21}^T \bar{\mathcal{C}}_{1k}^T \Lambda_1^T \mathcal{K}_k^T \mathcal{P}_{k+1} \mathcal{K}_k \Lambda_1 \bar{\mathcal{C}}_{1k} \bar{\mathcal{S}}_{21} + (\bar{\mathcal{C}}_{1k} \bar{\mathcal{S}}_{31} \mathcal{I}_{1v})^T (I \otimes (\mathcal{K}_k^T \mathcal{P}_{k+1} \mathcal{K}_k))$$
$$\times (\bar{\mathcal{C}}_{1k} \bar{\mathcal{S}}_{31} \mathcal{I}_{1v}) - diag\{\mathcal{R}_{k-1,1}, \mathcal{R}_{k-2,2}, \cdots, \mathcal{R}_{k-\ell,\ell}\},$$

$$\Pi_{34}^k = \bar{\mathcal{S}}_{21}^T \bar{\mathcal{C}}_{1k}^T \Lambda_1^T \mathcal{K}_k^T \mathcal{P}_{k+1} \mathcal{K}_k \Lambda_0 \bar{\mathcal{E}}_{0k} - \bar{\mathcal{S}}_{21}^T \bar{\mathcal{C}}_{1k}^T \Lambda_1^T \mathcal{K}_k^T \mathcal{P}_{k+1} \mathcal{F}_k \mathcal{I}_D$$
$$+ [0 \quad (\bar{\mathcal{C}}_{1k} \bar{\mathcal{S}}_{31} \mathcal{I}_{1v})^T (I \otimes (\mathcal{K}_k^T \mathcal{P}_{k+1} \mathcal{K}_k))(\bar{\mathcal{E}}_{1k} \mathcal{I}_{1v})],$$

$$\Pi_{44}^k = (\mathcal{K}_k \Lambda_0 \bar{\mathcal{E}}_{0k} - \mathcal{F}_k \mathcal{I}_D)^T \mathcal{P}_{k+1} (\mathcal{K}_k \Lambda_0 \bar{\mathcal{E}}_{0k} - \mathcal{F}_k \mathcal{I}_D)$$
$$+ \mathcal{I}_{0v}^T \bar{\mathcal{E}}_{0k}^T (I \otimes (\mathcal{K}_k^T \mathcal{P}_{k+1} \mathcal{K}_k)) \bar{\mathcal{E}}_{0k} \mathcal{I}_{0v}.$$

Proof *By calculating the difference of the first term in V_k along the trajectory of the system (7.6) and taking the mathematical expectation, one has*

$$\mathbb{E}\{\eta_{k+1}^T \mathcal{P}_{k+1} \eta_{k+1} - \eta_k^T \mathcal{P}_k \eta_k\}$$
$$= \mathbb{E}\Big\{ \eta_k^T \mathcal{A}_k^T \mathcal{P}_{k+1} \mathcal{A}_k \eta_k - \eta_k^T \mathcal{P}_k \eta_k + 2\bar{\alpha} \eta_k^T \mathcal{A}_k^T \mathcal{P}_{k+1} h_k$$
$$- 2\eta_k^T \mathcal{A}_k^T \mathcal{P}_{k+1} \mathcal{K}_k \Lambda_1 \bar{\mathcal{C}}_{1k} \bar{\mathcal{S}}_{21} \eta_k^\ell - 2\eta_k^T \mathcal{A}_k^T \mathcal{P}_{k+1} \mathcal{K}_k \Lambda_0 \bar{\mathcal{E}}_{0k} \zeta_k^\ell$$
$$+ 2\eta_k^T \mathcal{A}_k^T \mathcal{P}_{k+1} \mathcal{F}_k \mathcal{I}_D \zeta_k^\ell + \eta_k^T \mathcal{S}_1^T \mathcal{D}_k^T \mathcal{P}_{k+1} \mathcal{D}_k \mathcal{S}_1 \eta_k$$
$$+ \bar{\alpha}^2 h_k^T \mathcal{P}_{k+1} h_k - 2\bar{\alpha} h_k^T \mathcal{P}_{k+1} \mathcal{K}_k \Lambda_1 \bar{\mathcal{C}}_{1k} \bar{\mathcal{S}}_{21} \eta_k^\ell$$
$$- 2\bar{\alpha} h_k^T \mathcal{P}_{k+1} \mathcal{K}_k \Lambda_0 \bar{\mathcal{E}}_{0k} \zeta_k^\ell + 2\bar{\alpha} h_k^T \mathcal{P}_{k+1} \mathcal{F}_k \mathcal{I}_D \zeta_k^\ell$$
$$+ \bar{\alpha}(1-\bar{\alpha}) h_k^T \mathcal{S}_1^T \mathcal{P}_{k+1} \mathcal{S}_1 h_k + (\eta_k^\ell)^T \bar{\mathcal{S}}_{21}^T \bar{\mathcal{C}}_{1k}^T \Lambda_1^T \mathcal{K}_k^T \mathcal{P}_{k+1} \mathcal{K}_k \Lambda_1 \bar{\mathcal{C}}_{1k} \bar{\mathcal{S}}_{21} \eta_k^\ell$$
$$+ 2(\eta_k^\ell)^T \bar{\mathcal{S}}_{21}^T \bar{\mathcal{C}}_{1k}^T \Lambda_1^T \mathcal{K}_k^T \mathcal{P}_{k+1} \mathcal{K}_k \Lambda_0 \bar{\mathcal{E}}_{0k} \zeta_k^\ell - 2(\eta_k^\ell)^T \bar{\mathcal{S}}_{21}^T \bar{\mathcal{C}}_{1k}^T \Lambda_1^T \mathcal{K}_k^T \mathcal{P}_{k+1} \mathcal{F}_k \mathcal{I}_D \zeta_k^\ell$$
$$+ \tilde{\vartheta}^0 \eta_k^T \left(\mathcal{K}_k \mathcal{C}_k \mathcal{S}_3 \right)^T \mathcal{P}_{k+1} \mathcal{K}_k \mathcal{C}_k \mathcal{S}_3 \eta_k + 2\tilde{\vartheta}^0 \eta_k^T \mathcal{S}_3^T \mathcal{C}_k^T \mathcal{K}_k^T \mathcal{P}_{k+1} \mathcal{K}_k \mathcal{E}_k \mathcal{I}_D \zeta_k^\ell$$
$$+ (\eta_k^\ell)^T (\bar{\mathcal{C}}_{1k} \bar{\mathcal{S}}_{31} \mathcal{I}_{1v})^T (I \otimes (\mathcal{K}_k^T \mathcal{P}_{k+1} \mathcal{K}_k))(\bar{\mathcal{C}}_{1k} \bar{\mathcal{S}}_{31} \mathcal{I}_{1v}) \eta_k^\ell$$
$$+ 2\eta_k^{\ell\,T} [0 \quad (\bar{\mathcal{C}}_{1k} \bar{\mathcal{S}}_{31} \mathcal{I}_{1v})^T (I \otimes (\mathcal{K}_k^T \mathcal{P}_{k+1} \mathcal{K}_k))(\bar{\mathcal{E}}_{1k} \mathcal{I}_{1v})] \zeta_k^\ell$$
$$+ (\zeta_k^\ell)^T \bar{\mathcal{E}}_{0k}^T \Lambda_0^T \mathcal{K}_k^T \mathcal{P}_{k+1} \mathcal{K}_k \Lambda_0 \bar{\mathcal{E}}_{0k} \zeta_k^\ell - 2(\zeta_k^\ell)^T \bar{\mathcal{E}}_{0k}^T \Lambda_0^T \mathcal{K}_k^T \mathcal{P}_{k+1} \mathcal{F}_k \mathcal{I}_D \zeta_k^\ell$$
$$+ (\zeta_k^\ell)^T (\bar{\mathcal{E}}_{0k} \mathcal{I}_{0v})^T (I \otimes (\mathcal{K}_k^T \mathcal{P}_{k+1} \mathcal{K}_k))(\bar{\mathcal{E}}_{0k} \mathcal{I}_{0v}) \zeta_k^\ell + (\zeta_k^\ell)^T \mathcal{I}_D^T \mathcal{F}_k^T \mathcal{P}_{k+1} \mathcal{F}_k \mathcal{I}_D \zeta_k^\ell \Big\}.$$

On the other hand, it is not difficult to show that

$$\mathbb{E}\Big\{ \sum_{j=1}^{\ell} \Big(\sum_{i=k-j+1}^{k} \eta_i^T \mathcal{R}_{i,j} \eta_i - \sum_{i=k-j}^{k-1} \eta_i^T \mathcal{R}_{i,j} \eta_i \Big) \Big\}$$

$$= \sum_{j=1}^{\ell} \mathbb{E}\left\{ \eta_k^T \mathcal{R}_{k,j} \eta_k - \eta_{k-j}^T \mathcal{R}_{k-j,j} \eta_{k-j} \right\} \tag{7.11}$$

$$= \mathbb{E}\left\{ \sum_{j=1}^{\ell} \eta_k^T \mathcal{R}_{k,j} \eta_k - \eta_k^{\ell\,T} diag\{\mathcal{R}_{k-1,1}, \mathcal{R}_{k-2,2}, \cdots, \mathcal{R}_{k-\ell,\ell}\} \eta_k^{\ell} \right\}.$$

Therefore, it follows from the above two equations that the equality (7.10) holds, which completes the proof.

Theorem 7.1 *Let the positive scalar $\gamma > 0$, the positive definite matrices $\mathcal{W}_i > 0$ ($-\ell \leq i \leq 0$) and the filter gain matrices $\{K(k)\}_{k \in [0,\,N]}$ be given. For the augmented system (7.6), the finite-horizon $\mathcal{H}_\infty$ performance requirement defined in (7.7) is guaranteed for all nonzero ζ_k if there exist families of positive scalars $\{\lambda_k\}_{k \in [0,\,N]}$, positive definite matrices $\{\mathcal{P}_k\}_{k \in [0,\,N+1]}$ and $\{\mathcal{R}_{i,j}\}_{i \in [-\ell,\,N],\, j \in [1,\,\ell]}$ satisfying the following recursive matrix inequalities*

$$\Pi_2^k = \begin{bmatrix} \Pi_{11}^k - \lambda_k \mathcal{U}_{2k} + \mathcal{L}_k^T \mathcal{L}_k & \Pi_{12}^k + \lambda_k \mathcal{U}_{1k}^T & \Pi_{13}^k & \Pi_{14}^k \\ * & \Pi_{22}^k - \lambda_k I & \Pi_{23}^k & \Pi_{24}^k \\ * & * & \Pi_{33}^k & \Pi_{34}^k \\ * & * & * & \Pi_{44}^k - \frac{\gamma^2}{\ell+1} I \end{bmatrix} < 0 \tag{7.12}$$

and

$$\mathcal{P}_0 \leq \gamma^2 \mathcal{W}_0, \quad \sum_{j=i}^{\ell} \mathcal{R}_{-i,j} \leq \gamma^2 \mathcal{W}_{-i}, \quad i = 1, 2, \cdots, \ell \tag{7.13}$$

where

$$\mathcal{U}_{1k} = I \otimes (\Phi(k) + \Psi(k))/2, \ \mathcal{U}_{2k} = I \otimes (\Phi^T(k)\Psi(k) + \Psi^T(k)\Phi(k))/2$$

and the other corresponding matrices are defined in Lemma 7.1.

Proof *In order to analyze the $\mathcal{H}_\infty$ performance of the augmented system (7.6), we introduce the following function*

$$\mathcal{J}(k) = \eta_{k+1}^T \mathcal{P}_{k+1} \eta_{k+1} - \eta_k^T \mathcal{P}_k \eta_k$$
$$+ \sum_{j=1}^{\ell} \left(\sum_{i=k-j+1}^{k} \eta_i^T \mathcal{R}_{i,j} \eta_i - \sum_{i=k-j}^{k-1} \eta_i^T \mathcal{R}_{i,j} \eta_i \right). \tag{7.14}$$

It is easy to see from (7.2) that

$$\left[h_k - (I \otimes \Phi(k))\eta_k \right]^T \left[h_k - (I \otimes \Psi(k))\eta_k \right] \leq 0. \tag{7.15}$$

Then, substituting (7.10) and (7.15) into (7.14) results in

$$\mathbb{E}\{\mathcal{J}(k)\} \leq \mathbb{E}\left\{ \tilde{\eta}_k^T \Pi_1^k \tilde{\eta}_k \right.$$
$$\left. - \lambda_k \left[h_k - (I \otimes \Phi(k))\eta_k \right]^T \left[h_k - (I \otimes \Psi(k))\eta_k \right] \right\}. \tag{7.16}$$

Due to $\{\zeta_k\}_{k\in[-\ell,\,-1]} = 0$, adding the zero term

$$\tilde{z}_k^T \tilde{z}_k - \gamma^2 \zeta_k^T \zeta_k - (\tilde{z}_k^T \tilde{z}_k - \gamma^2 \zeta_k^T \zeta_k)$$

to $\mathbb{E}\{\mathcal{J}(k)\}$ yields

$$
\begin{aligned}
\mathbb{E}\{\mathcal{J}(k)\} \leq \mathbb{E}\Big\{ &\tilde{\eta}_k^T \Pi_1 \tilde{\eta}_k + ||\tilde{z}_k||^2 - \frac{\gamma^2}{\ell+1} \sum_{s=0}^{\ell} ||\zeta_{k-s}||^2 \\
&- \lambda_k \big[h_k - (I \otimes \Phi(k))\eta_k\big]^T \big[h_k - (I \otimes \Psi(k))\eta_k\big] \\
&+ \frac{\gamma^2}{\ell+1} \sum_{s=0}^{\ell} ||\zeta_{k-s}||^2 - \gamma^2 ||\zeta_k||^2 \Big\} \\
&- \mathbb{E}\Big\{ ||\tilde{z}_k||^2 - \gamma^2 ||\zeta_k||^2 \Big\} \\
\leq \mathbb{E}\Big\{ &\tilde{\eta}_k^T \Pi_2^k \tilde{\eta}_k \Big\} - \mathbb{E}\Big\{ ||\tilde{z}_k||^2 - \gamma^2 ||\zeta_k||^2 \Big\} \\
&+ \mathbb{E}\Big\{ \frac{\gamma^2}{\ell+1} \sum_{s=0}^{\ell} ||\zeta_{k-s}||^2 - \gamma^2 ||\zeta_k||^2 \Big\}.
\end{aligned}
\tag{7.17}
$$

Summing up (7.17) on both sides from 0 to N with respect to k, one has

$$
\begin{aligned}
\sum_{k=0}^{N} \mathbb{E}\{\mathcal{J}(k)\} = \ &\mathbb{E}\{V_{N+1}\} - \mathbb{E}\{V_0\} \\
\leq \ &\mathbb{E}\Big\{ \sum_{k=0}^{N} \tilde{\eta}_k^T \Pi_2^k \tilde{\eta}_k \Big\} - \mathbb{E}\Big\{ \sum_{k=0}^{N} (||\tilde{z}_k||^2 - \gamma^2 ||\zeta_k||^2) \Big\} \\
&+ \mathbb{E}\Big\{ \frac{\gamma^2}{\ell+1} \sum_{s=0}^{\ell} \sum_{k=0}^{N} (||\zeta_{k-s}||^2 - ||\zeta_k||^2) \Big\}.
\end{aligned}
\tag{7.18}
$$

Finally, it can be easily concluded from (7.12), (7.13) and (7.18) that

$$
\begin{aligned}
0 \leq \ &\mathbb{E}\{V_{N+1}\} + \mathbb{E}\Big\{ \gamma^2 \sum_{i=-\ell}^{0} \eta_i^T \mathcal{W}_i \eta_i - V_0 \Big\} \\
\leq \ &\mathbb{E}\Big\{ \sum_{k=0}^{N} (\gamma^2 ||\zeta_k||^2 - ||\tilde{z}_k||^2) + \gamma^2 \sum_{i=-\ell}^{0} \eta_i^T \mathcal{W}_i \eta_i \Big\},
\end{aligned}
\tag{7.19}
$$

which means that the $\mathcal{H}_\infty$ performance index (7.7) holds, and the proof is now complete.

7.3 Envelope Constraint Analysis

Let us now deal with the analysis issue on the envelope constraints for the addressed discrete time-varying stochastic systems by employing the idea of ellipsoid description borrowed from the set-membership filtering method.

Theorem 7.2 *Let the filter gain matrices* $\{K(k)\}_{k\in[0,N]}$ *as well as the sequence* $\{d_k^i, \varepsilon_k^i\}_{k\in[1,N]}$ *of desired outputs and tolerance bands be given. For the augmented system (7.6) with* $\eta_i = 0$ *($-\ell \le i \le 0$), the envelope constraints defined in (7.8) are guaranteed for the given input* $\{\zeta_k^*\}_{k\in[0,N]}$ *if there exist families of positive scalars* $\{\mu_{k+1}, \pi_k, \tau_0^k, \tau_1^k, \cdots, \tau_\ell^k\}_{k\in[0,\,N]}$ *and positive definite matrices* $\{\mathcal{Q}_k\}_{k\in[1,\,N+1]}$ *satisfying the following recursive matrix inequalities*

$$
\begin{cases}
\Xi_k = \begin{bmatrix}
-(\varepsilon_{k+1}^i)^2 + \mu_{k+1} & 0 & \bar{\mathcal{M}}_k^T \\
* & -\mu_{k+1}\mathcal{Q}_{k+1}^{-1} & \mathcal{Q}_{k+1}^{-1}\mathcal{L}_{k+1}^T\mathbf{I}_i^T \\
* & * & -I
\end{bmatrix} < 0, \quad (7.20\text{a}) \\[4mm]
\Omega_k = \begin{bmatrix}
\Xi_{0k} & \Xi_{1k}^T & \Xi_{2k}^T & \Xi_{3k}^T & \Xi_{4k}^T \\
* & -\mathcal{Q}_{k+1}^{-1} & 0 & 0 & 0 \\
* & * & -I \otimes \mathcal{Q}_{k+1}^{-1} & 0 & 0 \\
* & * & * & -\mathcal{Q}_{k+1}^{-1} & 0 \\
* & * & * & * & -I \otimes \mathcal{Q}_{k+1}^{-1}
\end{bmatrix} < 0 \quad (7.20\text{b})
\end{cases}
$$

where $i \in [1,\, n_z]$ *and*

$$
\Xi_{0k} = \begin{bmatrix}
\Xi_{5k} & -\pi_k \eta_k^{*T}\mathcal{U}_{2k}\Theta_k & 0 & \pi_k \eta_k^{*T}\mathcal{U}_{1k}^T \\
* & -\pi_k \Theta_k^T\mathcal{U}_{2k}\Theta_k - \tau_0^k I & 0 & \pi_k\Theta_k^T\mathcal{U}_{1k}^T \\
* & * & -I \otimes \Gamma_k & 0 \\
* & * & * & -\pi_k I
\end{bmatrix},
$$

$$
\Xi_{1k} = \begin{bmatrix} -\bar{\alpha}\bar{h}_k & \mathcal{A}_k\Theta_k & -\mathcal{K}_k\Lambda_1\bar{\mathcal{C}}_{1k}\bar{\mathcal{S}}_{21}\Theta_k & \bar{\alpha}I \end{bmatrix},
$$

$$
\Xi_{2k} = \begin{bmatrix}
\mathcal{D}_k\mathcal{S}_1\bar{\eta}_k^* & \mathcal{D}_k\mathcal{S}_1\Theta_k & 0 & 0 \\
0 & 0 & 0 & \sqrt{\bar{\alpha}}I
\end{bmatrix},
$$

$$
\Xi_{3k} = \begin{bmatrix} \sqrt{\bar{\vartheta}^0}\mathcal{K}_k(\mathcal{C}_k\mathcal{S}_3\bar{\eta}_k^* + \mathcal{E}_k\zeta_k^*) & \sqrt{\bar{\vartheta}^0}\mathcal{K}_k\mathcal{C}_k\mathcal{S}_3\Theta_k & 0 & 0 \end{bmatrix},
$$

$$
\Xi_{4k} = \begin{bmatrix} (I \otimes \mathcal{K}_k)(\bar{\mathcal{C}}_{1k}\bar{\mathcal{S}}_{31}\mathcal{I}_{1v}\bar{\eta}_k^{\ell*} + \bar{\mathcal{E}}_{1k}\mathcal{I}_{1v}\zeta_k^{\ell*}) & 0 & (I \otimes \mathcal{K}_k)\bar{\mathcal{C}}_{1k}\bar{\mathcal{S}}_{31}\Theta_k\mathcal{I}_{1v} & 0 \end{bmatrix},
$$

$$
\Xi_{5k} = -\pi_k \eta_k^{*T}\mathcal{U}_{2k}\eta_k^* + \sum_{i=0}^{\ell}\tau_i^k - 1, \quad \Gamma_k = diag\{\tau_1^k, \tau_2^k, \cdots, \tau_\ell^k\},
$$

$$
\bar{\mathcal{M}}_k = \mathbf{I}_i\mathcal{L}_{k+1}\mathcal{M}_k - d_{k+1}^i, \quad \bar{\Theta}_k = diag\{\Theta_{k-1}, \cdots, \Theta_{k-\ell}\},
$$

$$
\mathcal{M}_k = \mathcal{A}_k\bar{\eta}_k^* + \bar{\alpha}\bar{h}_k^* - \mathcal{K}_k\Lambda_1\bar{\mathcal{C}}_{1k}\bar{\mathcal{S}}_{21}\bar{\eta}_k^{\ell*} - \mathcal{K}_k\Lambda_1\bar{\mathcal{E}}_{1k}\zeta_k^{\ell*} - \bar{\vartheta}^0\mathcal{K}_k\mathcal{E}_k\zeta_k^* + \mathcal{F}_k\zeta_k^*,
$$

$$
\Theta_k = I \ (k = -\ell, -\ell+1, \cdots, 0), \quad \Theta_k\Theta_k^T = \mathcal{Q}_k^{-1} \ (k > 0),
$$

$$
\bar{\eta}_k^{\ell*} = \begin{bmatrix} \bar{\eta}_{k-1}^{*T}, & \bar{\eta}_{k-2}^{*T}, & \cdots, & \bar{\eta}_{k-\ell}^{*T} \end{bmatrix}^T, \quad \bar{\eta}_0^* = 0, \quad \bar{\eta}_0^{\ell*} = 0,
$$

$$
\zeta_k^{\ell*} = \begin{bmatrix} \zeta_{k-1}^{*T}, & \zeta_{k-2}^{*T}, & \cdots, & \zeta_{k-\ell}^{*T} \end{bmatrix}^T, \quad \zeta_0^{\ell*} = 0.
$$

Here, $\bar{\eta}_k^*$ satisfies the following recursive dynamics

$$\bar{\eta}_{k+1}^* = \mathcal{A}_k \bar{\eta}_k^* + \bar{\alpha} \bar{h}_k^* - \sum_{s=1}^{\ell} \bar{\vartheta}^s \mathcal{K}_k \mathcal{C}_{k-s} \mathcal{S}_2 \bar{\eta}_{k-s}^* - \sum_{s=0}^{\ell} \bar{\vartheta}^s \mathcal{K}_k \mathcal{E}_{k-s} \zeta_{k-s}^* + \mathcal{F}_k \zeta_k^* \quad (7.21)$$

with $\bar{h}_k^* = \left[h^T \left(k, [I\ 0] \bar{\eta}_k^* \right) \quad h^T \left(k, [I\ 0] \bar{\eta}_k^* \right) - h^T \left(k, [I\ 0] \bar{\eta}_k^* - [0\ I] \bar{\eta}_k^* \right) \right]^T.$

Proof *To start with, let us first propose an ellipsoid description of the errors between the states of the augmented system (7.6) (with the given input ζ_k^*) and the recursive dynamics (7.21). Then, based on the obtained ellipsoid description, the envelope constraints given in (7.8) can be transformed into a set of recursive matrix inequalities that are easy to handle. For this purpose, define an ellipsoid $\Omega(\mathcal{Q}, \eta_s, 1)$ in the mean-square sense as follows*

$$\Omega(\mathcal{Q}, \eta_s, 1) = \left\{ \eta \in \mathbb{R}^{2n_x} : \eta_s \in \mathbb{R}^{2n_x},\ \mathbb{E}\{(\eta - \eta_s)^T \mathcal{Q}(\eta - \eta_s)\} \leq 1 \right\} \quad (7.22)$$

where $\mathcal{Q} \in \mathbb{R}^{2n_x \times 2n_x}$ is a positive definite matrix and η_s is the center of the ellipsoid $\Omega(\mathcal{Q}, \eta_s, 1)$.

For $-\ell \leq i \leq 0$, because of $\bar{\eta}_i^* = \eta_i = 0$, one has that η_i belongs to the ellipsoid $\Omega(I, \bar{\eta}_i^*, 1)$. Similar to [43], for $-\ell \leq s \leq 0$, there exists a set of random vectors ϖ_s $(-\ell \leq s \leq 0)$ with $\mathbb{E}\{\varpi_s^T \varpi_s\} \leq 1$ satisfying

$$\eta_s = \bar{\eta}_s^* + \Theta_s \varpi_s \quad (7.23)$$

where $\Theta_s = I$.

In what follows, by using the mathematical induction method, we will first prove the following assertion.

Assertion: *The solution $\mathcal{Q}_k$ of (7.20b) satisfies*

$$\mathbb{E}\{(\eta_k - \bar{\eta}_k^*)^T \mathcal{Q}_k (\eta_k - \bar{\eta}_k^*)\} \leq 1, \quad 1 \leq k \leq N, \quad (7.24)$$

that is, $\eta_k \in \Omega(\mathcal{Q}_k, \bar{\eta}_k^*, 1)$, where $\bar{\eta}_k^*$ is determined by the dynamics (7.21).

The proof of the above assertion is divided into two steps, namely, the initial step and the inductive step.

Initial step. For $i = 1$, it follows from (7.6) and (7.23) that

$$\eta_1 - \bar{\eta}_1^*$$
$$= \xi_0 \mathcal{D}_0 \mathcal{S}_1 \bar{\eta}_0^* + \xi_0 \mathcal{D}_0 \mathcal{S}_1 \Theta_0 \varpi_0 + \mathcal{A}_0 \Theta_0 \varpi_0 + \bar{\alpha} h_0 + (\alpha_0 - \bar{\alpha}) \mathcal{S}_1 h_0$$

$$- \sum_{s=0}^{\ell} (\vartheta_0^s - \bar{\vartheta}^s) \mathcal{K}_0 \mathcal{C}_{-s} \mathcal{S}_3 \bar{\eta}_{-s}^* - \sum_{s=1}^{\ell} \bar{\vartheta}^s \mathcal{K}_0 \mathcal{C}_{-s} \mathcal{S}_2 \Theta_{-s} \varpi_{-s} \quad (7.25)$$

$$- \sum_{s=0}^{\ell} (\vartheta_0^s - \bar{\vartheta}^s) \mathcal{K}_0 \mathcal{C}_{-s} \mathcal{S}_3 \Theta_{-s} \varpi_{-s} - \sum_{s=0}^{\ell} (\vartheta_0^s - \bar{\vartheta}^s) \mathcal{K}_0 \mathcal{E}_{-s} \zeta_{-s}^* - \bar{\alpha} \bar{h}_0^*$$

and then

$$\mathbb{E}\big\{(\eta_1 - \bar{\eta}_1^*)^T \mathcal{Q}_1 (\eta_1 - \bar{\eta}_1^*)\big\}$$

$$= \mathbb{E}\Big\{ \bar{\eta}_0^{*T} \mathcal{S}_1^T \mathcal{D}_0^T \mathcal{Q}_1 \mathcal{D}_0 \mathcal{S}_1 \bar{\eta}_0^* + 2\bar{\eta}_0^{*T} \mathcal{S}_1^T \mathcal{D}_0^T \mathcal{Q}_1 \mathcal{D}_0 \mathcal{S}_1 \Theta_0 \varpi_0$$

$$+ \varpi_0^T \Theta_0^T \mathcal{S}_1^T \mathcal{D}_0^T \mathcal{Q}_1 \mathcal{D}_0 \mathcal{S}_1 \Theta_0 \varpi_0 + \varpi_0^T \Theta_0^T \mathcal{A}_0^T \mathcal{Q}_1 \mathcal{A}_0 \Theta_0 \varpi_0$$

$$+ 2\bar{\alpha} \varpi_0^T \Theta_0^T \mathcal{A}_0^T \mathcal{Q}_1 h_0 - 2\varpi_0^T \Theta_0^T \mathcal{A}_0^T \mathcal{Q}_1 \mathcal{K}_0 \Lambda_1 \bar{\mathcal{C}}_{10} \bar{\mathcal{S}}_{21} \bar{\Theta}_{10} \varpi_0^\ell$$

$$- 2\bar{\alpha} \varpi_0^T \Theta_0^T \mathcal{A}_0^T \mathcal{Q}_1 \bar{h}_0^* + \bar{\alpha}^2 h_0^T \mathcal{Q}_1 h_0 - 2\bar{\alpha} h_0^T \mathcal{Q}_1 \mathcal{K}_0 \Lambda_1 \bar{\mathcal{C}}_{10} \bar{\mathcal{S}}_{21} \bar{\Theta}_{10} \varpi_0^\ell$$

$$- 2\bar{\alpha} h_0^T \mathcal{Q}_1 \bar{h}_0^* + \tilde{\alpha} h_0^T \mathcal{S}_1^T \mathcal{Q}_1 \mathcal{S}_1 h_0 + (\bar{\eta}_0^*)^T \tilde{\vartheta}^0 \mathcal{S}_3^T \mathcal{C}_0^T \mathcal{K}_0^T \mathcal{Q}_1 \mathcal{K}_0 \mathcal{C}_0 \mathcal{S}_3 \bar{\eta}_0^*$$

$$+ (\bar{\eta}_0^{\ell*})^T \mathcal{I}_{1v}^T \bar{\mathcal{S}}_{31}^T \bar{\mathcal{C}}_{10}^T (I \otimes (\mathcal{K}_0^T \mathcal{Q}_1 \mathcal{K}_0)) \bar{\mathcal{C}}_{10} \bar{\mathcal{S}}_{31} \mathcal{I}_{1v} \bar{\eta}_0^{\ell*}$$

$$+ 2(\bar{\eta}_0^*)^T \tilde{\vartheta}^0 \mathcal{S}_3^T \mathcal{C}_0^T \mathcal{K}_0^T \mathcal{Q}_1 \mathcal{K}_0 \mathcal{C}_0 \mathcal{S}_3 \Theta_0 \varpi_0 + 2(\bar{\eta}_0^*)^T \tilde{\vartheta}^0 \mathcal{S}_3^T \mathcal{C}_0^T \mathcal{K}_0^T \mathcal{Q}_1 \mathcal{K}_0 \mathcal{E}_0 \zeta_0^*$$

$$+ (\bar{\eta}_0^{\ell*})^T \mathcal{I}_{1v}^T \bar{\mathcal{S}}_{31}^T \bar{\mathcal{C}}_{10}^T (I \otimes (\mathcal{K}_0^T \mathcal{Q}_1 \mathcal{K}_0)) \bar{\mathcal{C}}_{10} \bar{\mathcal{S}}_{31} \bar{\Theta}_0 \mathcal{I}_{1v} \varpi_0^\ell$$

$$+ 2(\bar{\eta}_0^{\ell*})^T \mathcal{I}_{1v}^T \bar{\mathcal{S}}_{31}^T \bar{\mathcal{C}}_{10}^T (I \otimes (\mathcal{K}_0^T \mathcal{Q}_1 \mathcal{K}_0)) \bar{\mathcal{E}}_{10} \mathcal{I}_{1v} \zeta_0^{*}$$

$$+ (\varpi_0^\ell)^T \bar{\Theta}_0^T \bar{\mathcal{S}}_{21}^T \bar{\mathcal{C}}_{10}^T \Lambda_1^T \mathcal{K}_0^T \mathcal{Q}_1 \mathcal{K}_0 \Lambda_1 \bar{\mathcal{C}}_{10} \bar{\mathcal{S}}_{21} \bar{\Theta}_0 \varpi_0^\ell$$

$$+ 2\bar{\alpha} (\varpi_0^\ell)^T \bar{\Theta}_0^T \bar{\mathcal{S}}_{21}^T \bar{\mathcal{C}}_{10}^T \Lambda_1^T \mathcal{K}_0^T \mathcal{Q}_1 \bar{h}_0^* + \varpi_0^T \tilde{\vartheta}^0 \Theta_0^T \mathcal{S}_3^T \mathcal{C}_0^T \mathcal{K}_0^T \mathcal{Q}_1 \mathcal{K}_0 \mathcal{C}_0 \mathcal{S}_3 \Theta_0 \varpi_0$$

$$+ 2(\varpi_0^\ell)^T \mathcal{I}_{1v}^T \bar{\Theta}_0^T \bar{\mathcal{S}}_{31}^T \bar{\mathcal{C}}_{10}^T (I \otimes (\mathcal{K}_0^T \mathcal{Q}_1 \mathcal{K}_0)) \bar{\mathcal{E}}_{10} \mathcal{I}_{1v} \zeta_0^{\ell*}$$

$$+ 2\varpi_0^T \tilde{\vartheta}^0 \Theta_0^T \mathcal{S}_3^T \mathcal{C}_0^T \mathcal{K}_0^T \mathcal{Q}_1 \mathcal{K}_0 \mathcal{E}_0 \zeta_0^* + \tilde{\vartheta}^0 \zeta_0^T \mathcal{E}_0^T \mathcal{K}_0^T \mathcal{Q}_1 \mathcal{K}_0 \mathcal{E}_0 \zeta_0$$

$$+ (\varpi_0^\ell)^T \mathcal{I}_{1v}^T \bar{\Theta}_0^T \bar{\mathcal{S}}_{31}^T \bar{\mathcal{C}}_{10}^T (I \otimes (\mathcal{K}_0^T \mathcal{Q}_1 \mathcal{K}_0)) \bar{\mathcal{C}}_{10} \bar{\mathcal{S}}_{31} \bar{\Theta}_0 \mathcal{I}_{1v} \varpi_0^\ell$$

$$+ (\zeta_0^{\ell*})^T \mathcal{I}_{1v}^T \bar{\mathcal{E}}_{10}^T (I \otimes (\mathcal{K}_0^T \mathcal{Q}_1 \mathcal{K}_0)) \bar{\mathcal{E}}_{10} \mathcal{I}_{1v} \zeta_0^{\ell*} + \bar{\alpha}^2 \bar{h}_0^{*T} \mathcal{Q}_1 \bar{h}_0^* \Big\}$$

$$= \mathbb{E}\Big\{ \varrho_0^T \Big(\Xi_{10}^T \mathcal{Q}_1 \Xi_{10} + \Xi_{20}^T (I \otimes \mathcal{Q}_1) \Xi_{20} + \Xi_{30}^T \mathcal{Q}_1 \Xi_{30} + \Xi_{40}^T (I \otimes \mathcal{Q}_1) \Xi_{40} \Big) \varrho_0 \Big\}$$

where $\varpi_0^\ell = [\, \varpi_{-1}^T, \cdots, \varpi_{-\ell}^T \,]^T$ and $\varrho_0 = [\, 1, \, \varpi_0^T, \, (\varpi_0^\ell)^T, \, h_0^T \,]^T$.
Now, considering $\mathbb{E}\{\varpi_s^T \varpi_s\} < 1$ $(s = -\ell, \cdots, 0)$ and

$$\big[h_0 - (I \otimes \Phi(0))(\bar{\eta}_0^* + \Theta_0 \varpi_0) \big]^T \big[h_0 - (I \otimes \Psi(0))(\bar{\eta}_0^* + \Theta_0 \varpi_0) \big] \leq 0,$$

one has

$$\mathbb{E}\big\{(\eta_1 - \bar{\eta}_1^*)^T \mathcal{Q}_1 (\eta_1 - \bar{\eta}_1^*)\big\} - 1$$

$$\leq \mathbb{E}\Big\{ \varrho_0^T \Big(\Xi_{10}^T \mathcal{Q}_1 \Xi_{10} + \Xi_{20}^T (I \otimes \mathcal{Q}_1) \Xi_{20} + \Xi_{30}^T \mathcal{Q}_1 \Xi_{30}$$

$$+ \Xi_{40}^T (I \otimes \mathcal{Q}_1) \Xi_{40} \Big) \varrho_0 - 1 - \sum_{s=-\ell}^{0} \tau_s^0 (\varpi_s^T \varpi_s - 1)$$

$$- \pi_0 \big[h_0 - (I \otimes \Phi(0))(\bar{\eta}_0^* + \Theta_0 \varpi_0) \big]^T \tag{7.26}$$

$$\times \big[h_0 - (I \otimes \Psi(0))(\bar{\eta}_0^* + \Theta_0 \varpi_0) \big] \Big\}$$

$$= \mathbb{E}\Big\{ \varrho_0^T \Big(\bar{\Xi}_{00}^T + \Xi_{10}^T \mathcal{Q}_1 \Xi_{10} + \Xi_{20}^T (I \otimes \mathcal{Q}_1) \bar{\Xi}_{20}$$

$$+ \bar{\Xi}_{30}^T \mathcal{Q}_1 \bar{\Xi}_{30} + \bar{\Xi}_{40}^T (I \otimes \mathcal{Q}_1) \bar{\Xi}_{40} \Big) \varrho_0 \Big\}.$$

Therefore, by using the Schur complement lemma, it can be verified from (7.26) that the solution $\mathcal{Q}_1$ of (7.20b) satisfies

$$\mathbb{E}\{(\eta_1 - \bar{\eta}_1^*)^T \mathcal{Q}_1 (\eta_1 - \bar{\eta}_1^*)\} \leq 1,$$

which means that η_1 belongs to the ellipsoid $\Omega(\mathcal{Q}_1, \bar{\eta}_1^*, 1)$.

Inductive step. So far, we have proved that the assertion is true of $i = 1$. Next, given that the assertion is true for $i = k$, we aim to show that the same assertion is true for $i = k + 1$.

Since the assertion is true for $i = k$, it follows again from [43] that there exists a set of random vectors ϖ_i (with $\mathbb{E}\{\varpi_i^T \varpi_i\} \leq 1$) satisfying $\eta_i = \bar{\eta}_i^* + \Theta_i \varpi_i$ where Θ_i is a factorization of $\mathcal{Q}_i^{-1} = \Theta_i \Theta_i^T$. It remains to show that, for $i = k + 1$, the solution $\mathcal{Q}_{k+1}$ of the recursive matrix inequalities (7.20b) guarantees $\eta_{k+1} \in \Omega(\mathcal{Q}_{k+1}, \bar{\eta}_{k+1}^*, 1)$.

For notational simplicity, denote

$$\varpi_k^\ell = [\, \varpi_{k-1}^T, \cdots, \varpi_{k-\ell}^T \,]^T, \quad \varrho_k = [\, 1, \, \varpi_k^T, \, (\varpi_k^\ell)^T, \, h_k^T \,]^T.$$

Similar to the initial step for $i = 1$, it can be derived that

$$
\begin{aligned}
&\mathbb{E}\big\{(\eta_{k+1} - \bar{\eta}_{k+1}^*)^T \mathcal{Q}_{k+1}(\eta_{k+1} - \bar{\eta}_{k+1}^*)\big\} - 1 \\
&\leq \mathbb{E}\Big\{\varrho_k^T \Big(\Xi_{1k}^T \mathcal{Q}_{k+1}\Xi_{1k} + \Xi_{2k}^T(I \otimes \mathcal{Q}_{k+1})\Xi_{2k} + \Xi_{3k}^T \mathcal{Q}_{k+1}\Xi_{3k} \\
&\quad + \Xi_{4k}^T(I \otimes \mathcal{Q}_{k+1})\Xi_{4k}\Big)\varrho_k - 1 - \sum_{s=k-\ell}^{k} \tau_s^k(\varpi_s^T \varpi_s - 1) \qquad\qquad (7.27) \\
&\quad - \pi_k \big[h_k - (I \otimes \Phi(k))(\bar{\eta}_k^* + \Theta_k \varpi_k)\big]^T \big[h_k - (I \otimes \Psi(k))(\bar{\eta}_k^* + \Theta_k \varpi_k)\big]\Big\} \\
&= \mathbb{E}\Big\{\varrho_k^T \Big(\Xi_{0k}^T + \Xi_{1k}^T \mathcal{Q}_{k+1}\Xi_{1k} + \Xi_{2k}^T(I \otimes \mathcal{Q}_{k+1})\Xi_{2k} \\
&\quad + \Xi_{3k}^T \mathcal{Q}_{k+1}\Xi_{3k} + \Xi_{4k}^T(I \otimes \mathcal{Q}_{k+1})\Xi_{4k}\Big)\varrho_k\Big\},
\end{aligned}
$$

which implies that η_{k+1} belongs to $\Omega(\mathcal{Q}_{k+1}, \bar{\eta}_{k+1}^*, 1)$ if (7.20b) holds. Therefore, by the induction, it can be concluded that the solution $\mathcal{Q}_k$ of (7.20b) satisfies (7.24).

Having proved the assertion, let us now consider the envelope constraints (7.8) which are equivalent to the following inequalities

$$\big[\mathbb{E}\{(\mathbf{I}_i \tilde{z}_{k+1} - d_{k+1}^i)\}\big]^2 \leq (\varepsilon_{k+1}^i)^2, \quad k \in [0, N-1], \quad i \in [1, n_z]. \qquad (7.28)$$

Noticing that there exists a random vector ϖ_{k+1} satisfying

$$\eta_{k+1} = \bar{\eta}_{k+1}^* + \Theta_{k+1}\varpi_{k+1}, \quad \mathbb{E}\{\varpi_{k+1}^T \varpi_{k+1}\} \leq 1,$$

one has

$$\left[\mathbb{E}\{(\mathbf{I}_i\tilde{z}_{k+1} - d^i_{k+1})\}\right]^2 - (\varepsilon^i_{k+1})^2$$

$$= \left[\mathbb{E}\{(\mathbf{I}_i\mathcal{L}_{k+1}(\bar{\eta}^*_{k+1} + \Theta_{k+1}\varpi_{k+1}) - d^i_{k+1})\}\right]^2 - (\varepsilon^i_{k+1})^2$$

$$= \left[(\mathbf{I}_i\mathcal{L}_{k+1}\mathcal{M}_k - d^i_{k+1}) + \mathbf{I}_i\mathcal{L}_{k+1}\Theta_{k+1}\mathbb{E}\{\varpi_{k+1}\}\right]^2 - (\varepsilon^i_{k+1})^2 \qquad (7.29)$$

$$= \chi_k^T \left\{ \begin{bmatrix} -(\varepsilon^i_{k+1})^2 & 0 \\ 0 & 0 \end{bmatrix} + \begin{bmatrix} \bar{\mathcal{M}}_k^T \\ \Theta_{k+1}^T\mathcal{L}_{k+1}^T\mathbf{I}_i^T \end{bmatrix} \begin{bmatrix} \bar{\mathcal{M}}_k^T \\ \Theta_{k+1}^T\mathcal{L}_{k+1}^T\mathbf{I}_i^T \end{bmatrix}^T \right\} \chi_k$$

where $\chi_k := [\,1 \quad \mathbb{E}\{\varpi^T_{k+1}\}\,]^T$.

In light of $\mathbb{E}\{\varpi^T_{k+1}\}\mathbb{E}\{\varpi_{k+1}\} \le \mathbb{E}\{\varpi^T_{k+1}\varpi_{k+1}\} < 1$, it follows from (7.29) that

$$\left[\mathbb{E}\{(\mathbf{I}_i\tilde{z}_{k+1} - d^i_{k+1})\}\right]^2 - (\varepsilon^i_{k+1})^2$$

$$< \left[\mathbb{E}\{(\mathbf{I}_i\tilde{z}_{k+1} - d^i_{k+1})\}\right]^2 - (\varepsilon^i_{k+1})^2 + \mu_{k+1} - \mu_{k+1}\mathbb{E}\{\varpi^T_{k+1}\}\mathbb{E}\{\varpi_{k+1}\}$$

$$= \chi_k^T \left\{ \begin{bmatrix} -(\varepsilon^i_{k+1})^2 + \mu_{k+1} & 0 \\ 0 & -\mu_{k+1}I \end{bmatrix} \right. \qquad (7.30)$$

$$\left. + \begin{bmatrix} \bar{\mathcal{M}}_k^T \\ \Theta_{k+1}^T\mathcal{L}_{k+1}^T\mathbf{I}_i^T \end{bmatrix} \begin{bmatrix} \bar{\mathcal{M}}_k^T \\ \Theta_{k+1}^T\mathcal{L}_{k+1}^T\mathbf{I}_i^T \end{bmatrix}^T \right\} \chi_k$$

$$= \chi_k^T \tilde{\Upsilon}_k \chi_k.$$

Obviously, it follows from (7.30) that (7.28) (or (7.8)) holds if $\tilde{\Upsilon}_k < 0$ *which is, according to the Schur complement lemma, equivalent to*

$$\begin{bmatrix} -(\varepsilon^i_{k+1})^2 + \mu_{k+1} & 0 & \bar{\mathcal{M}}_k^T \\ * & -\mu_{k+1}I & \Theta_{k+1}^T\mathcal{L}_{k+1}^T\mathbf{I}_i^T \\ * & * & -I \end{bmatrix} < 0. \qquad (7.31)$$

By performing the congruence transformation $\mathrm{diag}\{I,\ \Theta_{k+1},\ I\}$ to (7.31), it is not difficult to obtain the inequality (7.20a) and therefore (7.31) is true. It can now be concluded that the envelope constraints (7.8) are achieved, which completes the proof.

Remark 7.2 *It is worth mentioning that the idea of the set-membership filtering is to construct an ellipsoidal state estimation set of all system states consistent with the measured outputs and the given disturbance information (i.e. a specified ellipsoid description) [104, 143]. Different from the traditional point estimation approaches (e.g. the $\mathcal{H}_\infty$ state estimation, the Bayes' estimation and the method of moments), the set-membership filtering approach can be utilized to obtain a certain region encompassing the system states rather than the estimation vector. In this chapter, borrowed from the set-membership filtering method, the idea of employing the ellipsoid description of the estimation errors is used to convert the envelope constraints (7.8) into a set of matrix inequalities (7.20a) which can be easily handled via standard software package.*

7.4   Envelope-Constrained $\mathcal{H}_\infty$ Filter Design

Having established the analysis results, we are in a position to deal with the filter design problem. For this purpose, denote

$$\bar{\Pi}_{11} = \sum_{j=1}^{\ell} \mathcal{R}_{k,j} - \mathcal{P}_k - \lambda_k \mathcal{U}_{2k} + \mathcal{L}_k^T \mathcal{L}_k,$$

$$\bar{\Pi}_{33} = \text{diag}\{\mathcal{R}_{k-1,1}, \mathcal{R}_{k-2,2}, \cdots, \mathcal{R}_{k-\ell,\ell}\},$$

$$\Upsilon_{0k} = \begin{bmatrix} \bar{\Pi}_{11} & \lambda_k \mathcal{U}_{1k} & 0 & 0 \\ * & -\lambda_k I & 0 & 0 \\ * & * & -\bar{\Pi}_{33} & 0 \\ * & * & * & -\frac{\gamma^2}{\ell+1} I \end{bmatrix},$$

$$\Upsilon_{1k} = \begin{bmatrix} \mathcal{A}_k & \bar{\alpha} I & -\mathcal{K}_k \Lambda_1 \bar{\mathcal{C}}_{1k} \bar{\mathcal{S}}_{21} & \mathcal{F}_k \mathcal{I}_D - \mathcal{K}_k \Lambda_0 \bar{\mathcal{E}}_{0k} \end{bmatrix},$$

$$\Upsilon_{2k} = \begin{bmatrix} \sqrt{\tilde{\vartheta}^0} \mathcal{K}_k \mathcal{C}_k \mathcal{S}_3 & 0 & 0 & \sqrt{\tilde{\vartheta}^0} \mathcal{K}_k \mathcal{E}_k \mathcal{I}_D \end{bmatrix},$$

$$\Upsilon_{3k} = \begin{bmatrix} 0 & 0 & (I \otimes \mathcal{K}_k) \bar{\mathcal{C}}_{1k} \bar{\mathcal{S}}_{31} \mathcal{I}_{1v} & \begin{bmatrix} 0 & (I \otimes \mathcal{K}_k) \bar{\mathcal{E}}_{1k} \mathcal{I}_{1v} \end{bmatrix} \end{bmatrix},$$

$$\Upsilon_{4k} = \begin{bmatrix} \mathcal{D}_k \mathcal{S}_1 & 0 & 0 & 0 \\ 0 & \sqrt{\bar{\alpha}} \mathcal{S}_1 & 0 & 0 \end{bmatrix},$$

$$\tilde{\mathcal{P}}_{k+1} = \mathcal{P}_{k+1}^{-1}, \quad \tilde{\mathcal{Q}}_{k+1} = \mathcal{Q}_{k+1}^{-1}.$$

It is not difficult to see that the inequalities (7.12), (7.20a) and (7.20b) in Theorem 7.1 and 7.2 are, respectively, equivalent to

$$\begin{cases} \Pi_k = \begin{bmatrix} \Upsilon_{0k} & \Upsilon_{1k}^T & \Upsilon_{2k}^T & \Upsilon_{3k}^T & \Upsilon_{4k}^T \\ * & -\tilde{\mathcal{P}}_{k+1} & 0 & 0 & 0 \\ * & * & -\tilde{\mathcal{P}}_{k+1} & 0 & 0 \\ * & * & * & -I \otimes \tilde{\mathcal{P}}_{k+1} & 0 \\ * & * & * & * & -I \otimes \tilde{\mathcal{P}}_{k+1} \end{bmatrix} < 0, \quad (7.32a) \\[2em] \tilde{\bar{\Xi}}_k = \begin{bmatrix} -(\varepsilon_{k+1}^i)^2 + \mu_{k+1} & 0 & \mathcal{M}_k^T \\ * & -\mu_{k+1} I & \tilde{\mathcal{Q}}_{k+1} \mathcal{L}_{k+1}^T \mathbf{I}_i^T \\ * & * & -I \end{bmatrix} \\ \quad + \text{diag}\Big\{0, \mu_{k+1}(I - \tilde{\mathcal{Q}}_{k+1}), 0\Big\} < 0, \quad i \in [1, n_z], \quad (7.32b) \\[2em] \tilde{\Omega}_k = \begin{bmatrix} \Xi_{0k} & \Xi_{1k}^T & \Xi_{2k}^T & \Xi_{3k}^T & \Xi_{4k}^T \\ * & -\tilde{\mathcal{Q}}_{k+1} & 0 & 0 & 0 \\ * & * & -I \otimes \tilde{\mathcal{Q}}_{k+1} & 0 & 0 \\ * & * & * & -\tilde{\mathcal{Q}}_{k+1} & 0 \\ * & * & * & * & -I \otimes \tilde{\mathcal{Q}}_{k+1} \end{bmatrix} < 0. \quad (7.32c) \end{cases}$$

Furthermore, it is apparent that $\tilde{\bar{\Xi}}_k < 0$ if both

$$\tilde{\mathcal{Q}}_{k+1} \geq I \qquad\qquad (7.33)$$

and

$$\begin{bmatrix} -(\varepsilon_{k+1}^i)^2 + \mu_{k+1} & 0 & \bar{\mathcal{M}}_k^T \\ * & -\mu_{k+1}I & \tilde{\mathcal{Q}}_{k+1}\mathcal{L}_{k+1}^T \mathbf{I}_i^T \\ * & * & -I \end{bmatrix} < 0, \quad i \in [1, n_z] \quad (7.34)$$

hold. Note that the introduction of (7.33) and (7.34) is for computational convenience.

According to Theorem 7.1 and Theorem 7.2, we have the following filter design scheme.

Theorem 7.3 *Let the disturbance attenuation level $\gamma > 0$, the positive definite matrices $\mathcal{W}_i$ ($i = -\ell, \cdots, 0$), and the sequence of desired output and tolerance band $\{d_k^i, \varepsilon_k^i\}_{k \in [1, N]}$ be given. For the discrete time-varying stochastic system (7.1) with the envelope-constrained $\mathcal{H}_\infty$ filter (7.4), if there exist*

- *a positive scalar sequence $\{\lambda_k, \mu_{k+1}, \pi_k, \tau_0^k, \tau_1^k, \cdots, \tau_\ell^k\}_{k \in [0, N]}$,*

- *a positive definite matrix sequence $\{\mathcal{P}_0, \tilde{\mathcal{P}}_k, \tilde{\mathcal{Q}}_k\}_{k \in [1, N+1]}$ with $\tilde{\mathcal{Q}}_{k+1} \geq I$,*

- *a positive definite matrix sequence $\{\mathcal{R}_{i,j}\}_{i \in [-\ell, N], j \in [1, \ell]}$, and*

- *a real-valued matrix sequence $\{K(k)\}_{k \in [0, N]}$*

to guarantee (7.13), (7.32a), (7.32c), and (7.34) with the parameters updated by $\mathcal{P}_{k+1} = \tilde{\mathcal{P}}_{k+1}^{-1}$ and $\mathcal{Q}_{k+1} = \tilde{\mathcal{Q}}_{k+1}^{-1}$, then the output estimation errors $\{\tilde{z}_k\}_{k \in [0,N]}$ satisfy both the desired $\mathcal{H}_\infty$ performance (7.7) for any nonzero inputs and the envelope constraint requirement (7.8) for the certain inputs $\{\zeta_k^*\}_{k \in [0,N]}.$

By means of Theorem 7.3, the algorithm for designing the $\mathcal{H}_\infty$ filter gains with envelope constraints can be outlined as follows.

Remark 7.3 *In this chapter, the envelope-constrained $\mathcal{H}_\infty$ filtering problem is investigated for a class of discrete time-varying stochastic systems with fading measurements, RONs, and mixed noises. The main result established in Theorem 7.3 contains all the information about the $\mathcal{H}_\infty$ index, the envelope constraints, the occurring probability of RONs, and the statistical information of channel coefficients. The main novelty is twofold: 1) a new envelope-constrained performance criterion is proposed to describe the transient dynamics of the filtering error process; and 2) by employing the ellipsoid description on the estimation errors, the envelope constraints (7.8) are transformed into a set of matrix inequalities and the filter gain matrix is obtained by solving these matrix inequalities.*

Filter design algorithm

Step 1.	Give the $\mathcal{H}_\infty$ performance index γ, the desired output and tolerance band $\{d_k^i,\ \varepsilon_k^i\}_{k\in[1,\ N]}$, and the positive definite matrices $\mathcal{W}_i\ (i = -\ell, \cdots, 0)$.
Step 2.	Set $k = 0$ and solve the matrix inequalities (7.13), $\check{\mathcal{Q}}_1 \geq I$, and the recursive matrix inequalities (7.32a), (7.32c) and (7.34) to obtain the values of matrices $\mathcal{P}_0$, $\tilde{\mathcal{P}}_1$, $\check{\mathcal{Q}}_1$ and the filter gain matrix $K(0)$.
Step 3.	Update the matrices $\mathcal{P}_1 = \tilde{\mathcal{P}}_1^{-1}$, $\mathcal{Q}_1 = \check{\mathcal{Q}}_1^{-1}$, and then obtain the matrix Θ_1 by using the matrix decomposition method.
Step 4.	For the instant k, solve the matrix inequalities $\check{\mathcal{Q}}_{k+1} \geq I$, and the recursive matrix inequalities (7.32a), (7.32c) and (7.34) to obtain the values of matrices $\tilde{\mathcal{P}}_{k+1}$, $\check{\mathcal{Q}}_{k+1}$ and the filter gain matrix $K(k)$.
Step 5.	Update the matrices $\mathcal{P}_{k+1}$, $\mathcal{Q}_{k+1}$, obtain Θ_{k+1}, and set $k = k+1$.
Step 6.	If $k < N$, then go to Step 4, else go to Step 7.
Step 7.	Stop.

7.5   Simulation Examples

In this section, a numerical example is presented to illustrate the effectiveness of the proposed envelope-constrained $\mathcal{H}_\infty$ filter design scheme for discrete time-varying stochastic systems (7.1) with fading measurements (7.3). The corresponding parameters are given as follows

$$A = \begin{bmatrix} 0.48 + 0.2\sin(1.5k) & 0.3 \\ -0.32 & 0.44 \end{bmatrix}, \quad D = \begin{bmatrix} 0.05 & -0.1 \\ 0 & 0.02 \end{bmatrix},$$

$$C = \begin{bmatrix} -0.4 & 0.5 \end{bmatrix}, \quad L = \begin{bmatrix} -0.08 & 0.06 \end{bmatrix},$$

$$E_1 = \begin{bmatrix} 0.2 & -0.08 \end{bmatrix}^T, \quad E_2 = 0.03, \quad E_3 = 0.01.$$

The probability of RONs is taken as $\bar{\alpha} = 0.4$ and the nonlinear vector-valued function $h(k, x_k)$ is chosen as

$$h(k, x(k)) = \begin{cases} \begin{bmatrix} -0.06x_1(k) + 0.03x_2(k) + \tanh(0.03x_1(k)) \\ 0.06x_2(k) - \tanh(0.02x_2(k)) \end{bmatrix}, & 0 \leq k < 6, \\ \begin{bmatrix} 0.04x_1(k) - \tanh(0.02x_1(k)) \\ 0.05x_2(k) \end{bmatrix}, & 6 \leq k \leq 20 \end{cases}$$

where $x_i(k)$ $(i = 1, 2)$ denotes the i-th element of the system state $x(k)$. It is easy to see that constraint (7.2) is met with

$$\Phi(k) = \begin{cases} \begin{bmatrix} -0.03 & 0.03 \\ 0 & 0.06 \end{bmatrix}, & 0 \le k < 6, \\ \begin{bmatrix} 0.02 & 0 \\ 0 & 0.05 \end{bmatrix}, & 6 \le k \le 20, \end{cases}$$

$$\Psi(k) = \begin{cases} \begin{bmatrix} -0.06 & 0.03 \\ 0 & 0.04 \end{bmatrix}, & 0 \le k < 6, \\ \begin{bmatrix} 0.04 & 0 \\ 0 & 0.05 \end{bmatrix}, & 6 \le k \le 20. \end{cases}$$

The order of the fading model is $\ell = 1$ and the probability density functions of channel coefficients are as follows:

$$\begin{cases} f(\vartheta^0) = 0.0005(e^{9.89\vartheta^0} - 1), & 0 \le \vartheta^0 \le 1, \\ f(\vartheta^1) = 8.5017e^{-8.5\vartheta^1}, & 0 \le \vartheta^2 \le 1. \end{cases}$$

It can be obtained that the mathematical expectation $\bar{\vartheta}_s$ and variance $\tilde{\vartheta}_s$ $(s = 0, 1)$ are 0.8991, 0.1174, 0.0133 and 0.01364, respectively. Furthermore, the given input ζ_k^*, the desired output d_k and the tolerance band ε_k are, respectively, selected as

$$d_k = 0, \quad 0 \le k \le 20; \quad \varepsilon_k = \begin{cases} 0.013, & 0 \le k < 4, \\ 0.005, & 4 \le k < 8, \\ 0.003, & 8 \le k \le 20. \end{cases}$$

The desired envelope is shown in Fig. 7.2. Let the positive scalar γ and the corresponding matrices be taken as $\gamma = 0.96$, $W_0 = 21I$ and $W_i = I$ $(i = -\ell, \cdots, -1)$. By applying Theorem 7.3, the desired filter parameters are obtained and shown in Tab. 7.1. Other matrices are omitted to save space.

TABLE 7.1
The filter parameter $K(k)$.

k	0	1	2	3	$\cdots$
$K(k)$	-0.0982	0.0272	0.6190	0.5049	$\cdots$
	0.0747	0.3120	-0.0859	-0.0297	

In the simulation, the exogenous disturbance inputs are selected as

$$w_k = 0.5e^{-0.2k}\sin(k), \quad v_k = \frac{4}{k+1}\cos(k).$$

The initial values $x(k)$ are randomly generated and obey uniform distribution

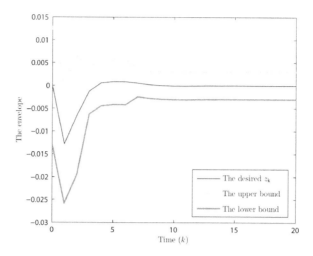

FIGURE 7.2
The desired envelope.

over $[-0.5, \ 0.5]$. The simulation results are shown in Figs. 7.3-7.5, where Fig. 7.3 plots the measurement outputs by sensors and the received signal by filter, and Figs. 7.4 and 7.5 depict the outputs and the filtering errors, respectively. The simulation results have confirmed that the filter performs very well.

7.6   Summary

In this chapter, we have addressed the envelope-constrained $\mathcal{H}_\infty$ filtering problem for a class of discrete time-varying stochastic systems with fading measurements, randomly occurring nonlinearities (RONs), and mixed noises. Some unrelated random variables have been introduced, respectively, to govern the phenomena of RONs and fading measurements. A novel envelope-constrained performance criterion has been proposed to describe the transient dynamics of the filtering error process. By employing the stochastic analysis approach combined with the ellipsoid description of the estimation errors, some sufficient conditions have been established in the form of recursive matrix inequalities and the desired filter gain matrices have been obtained in terms of the solution to these matrix inequalities. A numerical example was provided at the end of the chapter to demonstrate the effectiveness of the proposed design schemes.

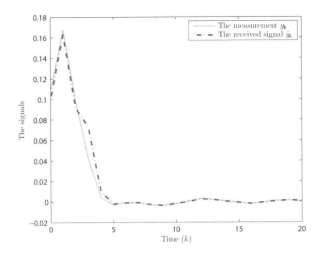

FIGURE 7.3
The measurement and the received signal.

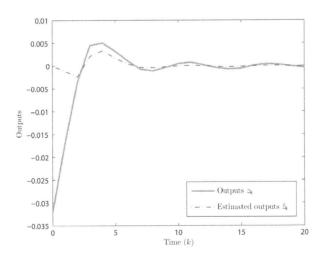

FIGURE 7.4
The outputs.

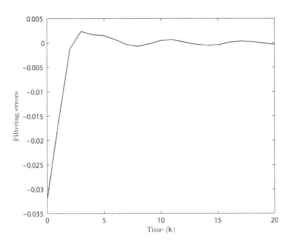

FIGURE 7.5
The filtering errors.

8

Distributed Filtering under Uniform Quantizations and Deception Attacks through Sensor Networks

As the scale of sensor networks increases, the complexities are inevitably enhanced, which gives rise to the urgent necessity for developing new distributed state estimation technologies in order to meet the need of practical engineering. It is not surprising that, in the past few years, the distributed state estimation problem with network-enhanced complexities has become an interesting and imperative yet challenging topic. Different from the traditional central state estimation techniques, the difficulty in designing distributed state estimation algorithms stems from both the complicated coupling between the sensor nodes according to a given topology and the effects from network-enhanced complexities. In addition, due to physical constraints or technological limitations, data among sensors are usually transmitted over common networks without proper security protections. Specifically, the interconnection of low-cost sensor nodes makes it complicated to protect against inherent physical vulnerabilities therein. Typical attacks include denial of service (DoS) attacks, replay attacks and deception attacks. It is worth mentioning that deception attacks in different scenarios can also be called false data-injection attacks and malicious attacks, to just name a few. Therefore, it is of great importance to determine the impact of distributed filtering.

This chapter is concerned with the distributed recursive filtering problem for a class of discrete time-delayed stochastic systems subject to both uniform quantization and deception attack effects on the measurement outputs. As deception attacks cannot be regarded as an energy-bounded input, the method developed in Chapter 4 is invalid. Therefore, we have to develop a new design and analysis method to handle this challenge. In addition, we endeavor to answer the following questions: 1) How can we design a distributed filter that effectively fuses the unreliable data corrupted by noises, quantization errors, and possible deception attacks? 2) How can we develop an efficient filtering algorithm that would help reduce the computation burden resulting from time-delays and the large number of sensor nodes? 3) How can we cope with the complicated coupling issues between the filtering errors and observed states in the performance analysis? With the help of covariance analysis, an upper bound for the filtering error covariance is derived and subsequently minimized by properly designing the filter parameters via a gradient-based method at

143

each sampling instant. Furthermore, by utilizing mathematical induction, a sufficient condition is established to ensure the asymptotic boundedness of the sequence of the error covariance in the simultaneous presence of time-delays, deception attacks, and uniform quantization effects.

8.1   Modeling and Problem Formulation

In this chapter the underlying sensor network has n sensor nodes which are distributed in space according to a fixed network topology represented by a directed graph $\mathscr{G} = (\mathscr{V}, \mathscr{E}, \mathscr{H})$ of order n with the set of nodes $\mathscr{V} = \{1, 2, \cdots, n\}$, the set of edges $\mathscr{E} \in \mathscr{V} \times \mathscr{V}$, and the weighted adjacency matrix $\mathscr{H} = [h_{ij}]$ with nonnegative adjacency element h_{ij}. An edge of $\mathscr{G}$ is denoted by the ordered pair (i, j). The adjacency elements associated with the edges of the graph are positive, i.e., $h_{ji} > 0 \iff (j, i) \in \mathscr{E}$, which means that sensor i can obtain information from sensor j. The set of neighbors of node $i \in \mathscr{V}$ is denoted by $\mathcal{N}_i = \{j \in \mathscr{V} : (j, i) \in \mathscr{E}\}$. The in-degree of node i is defined as $\hbar_{\mathrm{in}}^i = \sum_{j \in \mathcal{N}_i} h_{ji}$. In this chapter assume $\hbar_{\mathrm{in}}^i \neq 0$ for all nodes.

Let the target plant be described by the following discrete-time stochastic system with multiplicative noises:

$$
\begin{aligned}
x_{k+1} = &\Big(A_{0,k} + \sum_{s=1}^{r} \omega_{s,k} A_{s,k}\Big) x_k \\
&+ \Big(A_{0,k}^d + \sum_{s=1}^{r} \omega_{s,k} A_{s,k}^d\Big) x_{k-\tau} + B_k w_k
\end{aligned}
\tag{8.1}
$$

with n sensors modeled by

$$
\tilde{y}_{i,k} = \big(C_{0,k} + \varpi_{i,k} C_{i,k}\big) x_k + D_k v_{i,k}, \quad i = 1, 2, \cdots, n
\tag{8.2}
$$

where $x_k \in \mathbb{R}^{n_x}$ is the state of the target plant that cannot be observed directly, $\tilde{y}_{i,k} \in \mathbb{R}^{n_y}$ is the ideal measurement output (without quantization) from sensor i. $w_k \in \mathbb{R}^s$ and $v_{i,k} \in \mathbb{R}^p$ ($i = 1, 2, \cdots, n$) are the white noises with zero-mean and unity covariance, and are mutually uncorrelated in k and i. $\omega_{s,k} \in \mathbb{R}$ ($s = 1, 2, \cdots, r$) and $\varpi_{i,k} \in \mathbb{R}$ ($i = 1, 2, \cdots, n$) are multiplicative noises with zero-mean and unity variances, and are mutually uncorrelated in k. r and τ are two known positive integers. $A_{s,k}$, $A_{s,k}^d$ ($s = 0, 1, \cdots, r$) and $C_{i,k}$ ($i = 0, 1, \cdots, n$) are known time-varying matrices with compatible dimensions.

In this chapter, a uniform quantizer is taken into account. Assume that the overall quantizer range is $[-M, M]$ with $M > 0$. The length of the quantizer level u is defined as $u = 2M/(2^b - 1)$ where b is the number of bits for digital sensors. The uniform quantizer [102] is denoted by

$$\bar{y}_{i,k} = Q(\tilde{y}_{i,k}) = \tilde{y}_{i,k} + q_{i,k} \tag{8.3}$$

where $\bar{y}_{i,k}$ is the quantized signal and $q_{i,k} \in \mathbb{R}^{n_y}$ is the quantization error process. Here, $q_{i,k}$ is an additive white uniform distributed noise with each element being uniformly distributed in $[-0.5u, 0.5u]$. Obviously, the variance of such a quantization error process is $\frac{u^2}{12}I$.

As discussed in the introduction, from the defenders' perspectives, the successful cyber-attacks could be intermittent or random in the implementation. In this case, the signals during the network transmission are subject to deception attacks modeled as follows:

$$y_{i,k} = Q(\tilde{y}_{i,k}) + \alpha_{i,k}\zeta_{i,k} \tag{8.4}$$

where $y_{i,k}$ is the *received signal* by neighboring nodes, $\zeta_{i,k} \in \mathbb{R}^{n_y}$ stands for the signal sent by attackers that is described as $\zeta_{i,k} = -Q(\tilde{y}_{i,k}) + \xi_k$. Here, the non-zero ξ_k satisfying $\|\xi_k\| \le \delta$ is an arbitrary limited-magnitude signal and the bound δ is a known positive scalar that can be estimated through statistical tests or specified by security requirements. The variable $\alpha_{i,k}$ is a Bernoulli distributed white sequence taking values of 0 or 1 with the following probabilities

$$\text{Prob}\{\alpha_{i,k} = 0\} = 1 - \bar{\alpha}, \quad \text{Prob}\{\alpha_{i,k} = 1\} = \bar{\alpha}$$

where $\bar{\alpha} \in [0,1]$ is a known constant.

In this chapter, the Kalman-type recursive filter on node i is of the following form:

$$\begin{cases} \hat{x}_{i,k+1|k} = A_{0,k}\hat{x}_{i,k|k} + A_{0,k}^d\hat{x}_{i,k-\tau|k-\tau}, & (8.5a) \\ \hat{x}_{i,k+1|k+1} = \hat{x}_{i,k+1|k} + \hbar_{in}^i K_{i,k+1}^1 \\ \quad \times (\bar{y}_{i,k+1} - C_{0,k+1}\hat{x}_{i,k+1|k}) \\ \quad + K_{i,k+1}^2 \sum_{j \in \mathcal{N}_i} h_{ji}(y_{j,k+1} - (1-\bar{\alpha})C_{0,k+1}\hat{x}_{i,k+1|k}) & (8.5b) \end{cases}$$

where $\hat{x}_{i,k|k}$ and $\hat{x}_{i,k+1|k}$ are, respectively, the state estimate and the one-step prediction at time k, and $K_{i,k+1}^1$ and $K_{i,k+1}^2$ are the filter parameters to be determined.

Remark 8.1 *For distributed filtering problems, the information available on each node is not only from itself but also from its neighbors according to the given topology. From the problem addressed in this chapter, the data received from any neighboring sensors might be unreliable due to the possible deception attacks. As such, the innovation in (8.5b) is divided into two parts, that is, $\hbar_{in}^i(\bar{y}_{i,k+1} - C_{0,k+1}\hat{x}_{i,k+1|k})$ regarding the data from the node itself and $\sum_{j \in \mathcal{N}_i} h_{ji}(y_{j,k+1} - (1-\bar{\alpha})C_{0,k+1}\hat{x}_{i,k+1|k})$ accounting for the data from the neighboring nodes. Therefore, the proposed filter model (8.5b) can be utilized to effectively fuse the data from two different sources, thereby improving the filtering performance.*

Let us denote the one-step prediction error and the filtering error on node i as $e_{i,k+1|k} := x_{k+1} - \hat{x}_{i,k+1|k}$ and $e_{i,k|k} := x_k - \hat{x}_{i,k|k}$, respectively. The objective of this chapter is twofold:

R1) Design a Kalman-type filter of the form (8.5a) and (8.5b) such that, in the presence of deception attacks, an upper bound for the filtering error covariance is guaranteed, i.e., there exists a sequence of positive-definite matrices $\Pi_{i,k|k}$ satisfying

$$\mathbb{E}\{e_{i,k|k}e_{i,k|k}^T\} \leq \Pi_{i,k|k}, \quad \forall k > 0. \tag{8.6}$$

Furthermore, the sequence of upper bounds $\Pi_{i,k|k}$ is minimized by the designed filter parameters $K_{i,k+1}^1$ and $K_{i,k+1}^2$ through a recursive scheme.

R2) For designed filter parameters $K_{i,k+1}^1$ and $K_{i,k+1}^2$, find a condition under which the sequence $\Pi_{i,k|k}$ is asymptotically bounded as time tends to infinity.

8.2 Distributed Filter Design

In this section, by resorting to the stochastic analysis combined with some special matrix inequalities, a sufficient condition on the filter design is proposed by solving two Riccati-like difference equations in order to guarantee an upper bound of the filtering error covariance. Moreover, such an upper bound is minimized based on the designed filter. Before proceeding further, we introduce the following lemmas which will be needed for the derivation of our main results. In addition, all proofs of lemmas and theorems are moved to the appendixes for clarity of presentation.

Lemma 8.1 *[131] (Matrix Inverse Lemma) Let X, Y, U and V be given matrices with appropriate dimensions. If X, Y and $Y^{-1} + VX^{-1}U$ are invertible, then the following holds*

$$(X + UYV)^{-1} = X^{-1} - X^{-1}U(Y^{-1} + VX^{-1}U)^{-1}VX^{-1}.$$

Lemma 8.2 *[126] Suppose that $X = X^T > 0$, $\Phi_k(X) = \Phi_k^T(X) \in \mathbb{R}^{n_x \times n_x}$ and $\Psi_k(X) = \Psi_k^T(X) \in \mathbb{R}^{n_x \times n_x}$. If there exists $Y = Y^T > X$ such that*

$$\Phi_k(Y) \geq \Phi_k(X), \quad \Psi_k(Y) \geq \Phi_k(Y), \tag{8.7}$$

then the solutions M_k and N_k to the following difference equations

$$M_{k+1} = \Phi_k(M_k), \ N_{k+1} = \Psi_k(N_k), \ M_0 = N_0 > 0 \tag{8.8}$$

satisfy $M_k \leq N_k$.

Lemma 8.3 *For the addressed system (8.1), the state covariance matrix* $X_{k+1} = \mathbb{E}\{x_{k+1}x_{k+1}^T\}$ *obeys the following inequality:*

$$
X_{k+1} \leq (1+\varepsilon_k)\sum_{s=0}^{r} A_{s,k}X_kA_{s,k}^T + B_kB_k^T
$$
$$
+ (1+\varepsilon_k^{-1})\sum_{s=0}^{r} A_{s,k}^d X_{k-\tau}(A_{s,k}^d)^T. \tag{8.9}
$$

Proof *Along the trajectory of system (8.1), it can be derived that*

$$
X_{k+1} = B_kB_k^T + \sum_{s=0}^{r}\Big\{A_{s,k}\mathbb{E}\{x_kx_k^T\}A_{s,k}^T
$$
$$
+ A_{s,k}^d\mathbb{E}\{x_{k-\tau}x_{k-\tau}^T\}(A_{s,k}^d)^T
$$
$$
+ A_{s,k}\mathbb{E}\{x_kx_{k-\tau}^T\}(A_{s,k}^d)^T \tag{8.10}
$$
$$
+ A_{s,k}^d\mathbb{E}\{x_{k-\tau}x_k^T\}A_{s,k}^T\Big\}.
$$

By utilizing the element inequality $xy^T + yx^T \leq \varepsilon xx^T + \varepsilon^{-1}yy^T$ for $\forall x,y \in \mathbb{R}^n$, one has

$$
X_{k+1} \leq B_kB_k^T + \sum_{s=0}^{r}\Big\{(1+\varepsilon_k)A_{s,k}\mathbb{E}\{x_kx_k^T\}A_{s,k}^T
$$
$$
+ (1+\varepsilon_k^{-1})A_{s,k}^d\mathbb{E}\{x_{k-\tau}x_{k-\tau}^T\}(A_{s,k}^d)^T\Big\}
$$
$$
= (1+\varepsilon_k)\sum_{s=0}^{r} A_{s,k}X_kA_{s,k}^T + B_kB_k^T \tag{8.11}
$$
$$
+ (1+\varepsilon_k^{-1})\sum_{s=0}^{r} A_{s,k}^d X_{k-\tau}(A_{s,k}^d)^T,
$$

which completes the proof.

In light of Lemma 8.3, the upper bounds of both the covariance matrix $P_{i,k+1|k}$ of one-step prediction error and the filtering error covariance $P_{i,k+1|k+1}$ are presented in the following theorem.

Theorem 8.1 *For the addressed system (8.1) with measurements (8.3) suffering from attacks (8.4), the covariance matrix $P_{i,k+1|k}$ of one-step prediction errors and the filtering error covariance $P_{i,k+1|k+1}$ satisfy*

$$
P_{i,k+1|k} \leq \Pi_{i,k+1|k}, \quad P_{i,k+1|k+1} \leq \Pi_{i,k+1|k+1} \tag{8.12}
$$

where

$$
\begin{aligned}
\Pi_{i,k+1|k} &= (1+\varepsilon_k)A_{0,k}\Pi_{i,k|k}A_{0,k}^T + B_kB_k^T \\
&\quad + (1+\varepsilon_k^{-1})A_{0,k}^d\Pi_{i,k-\tau|k-\tau}(A_{0,k}^d)^T \\
&\quad + (1+\varepsilon_k)\sum_{i=1}^{r} A_{i,k}X_kA_{i,k}^T \\
&\quad + (1+\varepsilon_k^{-1})\sum_{i=1}^{r} A_{i,k}^dX_{k-\tau}(A_{i,k}^d)^T, \\
\Pi_{i,k+1|k+1} &= \Psi_{i,k+1}^0\Pi_{i,k+1|k}\Psi_{i,k+1}^{0T} + K_{i,k+1}^1\Psi_{i,k+1}^1 \\
&\quad \times K_{i,k+1}^{1T} + K_{i,k+1}^2\Psi_{i,k+1}^{23}K_{i,k+1}^{2T}
\end{aligned}
\tag{8.13}
$$

with

$$
\tilde{s}_i = \sum_{l=1}^{n}\sum_{j=1}^{n}\bar{\alpha}^2 h_{li}h_{ji} + \sum_{l=1}^{n}(\bar{\alpha}-\bar{\alpha}^2)(h_{li})^2,
$$

$$
\Psi_{i,k+1}^0 = I - \hbar_{in}^i K_{i,k+1}^1 C_{0,k+1} - \hbar_{in}^i(1-\bar{\alpha})K_{i,k+1}^2 C_{0,k+1},
$$

$$
\Psi_{i,k+1}^1 = (\hbar_{in}^i)^2\Big(C_{i,k+1}X_{k+1}C_{i,k+1}^T + D_{k+1}D_{k+1}^T + \frac{u^2 I}{12}\Big),
$$

$$
\Psi_{i,k+1}^2 = \sum_{j=1}^{n}(1-\bar{\alpha})(h_{ji})^2\Big(\bar{\alpha}(1+\varepsilon_k)C_{0,k+1}X_{k+1}C_{0,k+1}^T
$$

$$
\qquad\qquad + C_{j,k+1}X_{k+1}C_{j,k+1}^T\Big),
$$

$$
\Psi_{i,k+1}^3 = (1+2\varepsilon_k^{-1})\tilde{s}_i\delta^2 I
$$

$$
\qquad\qquad + \sum_{j=1}^{n}(1-\bar{\alpha})(h_{ji})^2\Big(D_{k+1}D_{k+1}^T + \frac{u^2 I}{12}\Big),
$$

$$
\Psi_{i,k+1}^{23} = \Psi_{i,k+1}^2 + \Psi_{i,k+1}^3, \quad \Pi_{i,j|j} = P_{i,j|j}\ (-\tau \le j \le 0).
$$

Proof *First, the covariance matrix $P_{i,k+1|k}$ of one-step prediction error is given by*

$$
\begin{aligned}
&P_{i,k+1|k} \\
&= \mathbb{E}\big\{(x_{k+1}-\hat{x}_{i,k+1|k})(x_{k+1}-\hat{x}_{i,k+1|k})^T\big\} \\
&= \mathbb{E}\bigg\{\Big[A_{0,k}e_{i,k|k} + A_{0,k}^d e_{i,k-\tau|k-\tau} + \sum_{i=1}^{r}\omega_{i,k}A_{i,k}x_k \\
&\qquad + \sum_{i=1}^{r}\omega_{i,k}A_{i,k}^d x_{k-\tau} + B_k w_k\Big] \\
&\qquad \times \Big[A_{0,k}e_{i,k|k} + A_{0,k}^d e_{i,k-\tau|k-\tau} + \sum_{i=1}^{r}\omega_{i,k}A_{i,k}x_k \\
&\qquad + \sum_{i=1}^{r}\omega_{i,k}A_{i,k}^d x_{k-\tau} + B_k w_k\Big]^T\bigg\}
\end{aligned}
$$

$$
\begin{aligned}
&= A_{0,k}\mathbb{E}\{e_{i,k|k}e_{i,k|k}^T\}A_{0,k}^T + A_{0,k}\mathbb{E}\{e_{i,k|k}e_{i,k-\tau|k-\tau}^T\} \\
&\quad \times (A_{0,k}^d)^T + A_{0,k}^d\mathbb{E}\{e_{i,k-\tau|k-\tau}e_{i,k|k}^T\}A_{0,k}^T \\
&\quad + A_{0,k}^d\mathbb{E}\{e_{i,k-\tau|k-\tau}e_{i,k-\tau|k-\tau}^T\}(A_{0,k}^d)^T \\
&\quad + \sum_{i=1}^{r} A_{i,k}\mathbb{E}\{x_k x_k^T\}A_{i,k}^T + \sum_{i=1}^{r} A_{i,k}^d\mathbb{E}\{x_{k-\tau}x_{k-\tau}^T\} \\
&\quad \times (A_{i,k}^d)^T + \sum_{i=1}^{r} A_{i,k}\mathbb{E}\{x_k x_{k-\tau}^T\}(A_{i,k}^d)^T \\
&\quad + \sum_{i=1}^{r} A_{i,k}^d\mathbb{E}\{x_{k-\tau}x_k^T\}A_{i,k}^T + B_k B_k^T \\
&\le (1+\varepsilon_k)A_{0,k}P_{i,k|k}A_{0,k}^T + B_k B_k^T + (1+\varepsilon_k^{-1})A_{0,k}^d \\
&\quad \times P_{i,k-\tau|k-\tau}(A_{0,k}^d)^T + (1+\varepsilon_k)\sum_{i=1}^{r} A_{i,k}X_k A_{i,k}^T \\
&\quad + (1+\varepsilon_k^{-1})\sum_{i=1}^{r} A_{i,k}^d X_{k-\tau}(A_{i,k}^d)^T.
\end{aligned}
\tag{8.14}
$$

Using Lemma 8.2, one has $P_{i,k+1|k} \le \Pi_{i,k+1|k}$. In what follows, for the purpose of simplicity, we introduce the notations:

$$
\begin{aligned}
\mathcal{C}_{k+1} &= [\ \mathcal{C}_{1,k+1}^T \quad \cdots \quad \mathcal{C}_{n,k+1}^T\ ]^T, \\
\mathcal{H}_i &= [\ \underbrace{0 \quad \cdots \quad 0}_{i-1} \quad \hbar_{in}^i \quad \underbrace{0 \quad \cdots \quad 0}_{n-i}\ ], \\
W_{k+1} &= diag\{\varpi_{1,k+1}, \cdots, \varpi_{n,k+1}\}, \\
\Lambda_{k+1}^1 &= [\ \alpha_{1,k+1} \quad \cdots \quad \alpha_{n,k+1}\ ]^T, \\
\Lambda_{k+1}^2 &= [\ \bar{\alpha}-\alpha_{1,k+1} \quad \cdots \quad \bar{\alpha}-\alpha_{n,k+1}\ ]^T, \\
\Lambda_{k+1}^3 &= diag\{1-\alpha_{1,k+1}, \cdots, 1-\alpha_{n,k+1}\}.
\end{aligned}
$$

Subtracting (8.5b) from (8.1) leads to

$$
\begin{aligned}
&e_{i,k+1|k+1} \\
&= x_{k+1} - \hat{x}_{i,k+1|k+1} \\
&= \Big(I - \hbar_{in}^i K_{i,k+1}^1 C_{0,k+1} - \sum_{j\in\mathcal{N}_i}(1-\bar{\alpha})h_{ji}K_{i,k+1}^2 \\
&\quad \times C_{0,k+1}\Big)e_{i,k+1|k} - \hbar_{in}^i K_{i,k+1}^1\big(\varpi_{i,k+1}C_{i,k+1}x_{k+1} \\
&\quad + D_{k+1}v_{i,k+1} + q_{i,k+1}\big) - K_{i,k+1}^2\sum_{j\in\mathcal{N}_i}h_{ji}\Big((\bar{\alpha} \\
&\quad - \alpha_{j,k+1})C_{0,k+1} + (1-\alpha_{j,k+1})\varpi_{j,k+1}C_{j,k+1}\Big)x_{k+1}
\end{aligned}
$$

$$- K_{i,k+1}^2 \sum_{j \in \mathcal{N}_i} h_{ji}(\alpha_{j,k+1}\xi_{k+1} + (1 - \alpha_{j,k+1})q_{j,k+1})$$

$$- K_{i,k+1}^2 \sum_{j \in \mathcal{N}_i} (1 - \alpha_{j,k+1})h_{ji}D_{k+1}v_{j,k+1} \qquad (8.15)$$

$$= \Psi_{i,k+1}^0 e_{i,k+1|k} - K_{i,k+1}^1(\mathcal{H}_i \otimes I)(W_{k+1} \otimes I)\mathcal{C}_{k+1}x_{k+1}$$
$$- K_{i,k+1}^1(\mathcal{H}_i \otimes I)(I \otimes D_{k+1})v_{k+1} - K_{i,k+1}^1(\mathcal{H}_i$$
$$\otimes I)q_{k+1} - K_{i,k+1}^2(\bar{\mathcal{H}}_i \otimes I)(\Lambda_{k+1}^2 \otimes I)C_{0,k+1}x_{k+1}$$
$$- K_{i,k+1}^2(\bar{\mathcal{H}}_i \otimes I)(\Lambda_{k+1}^3 W_{k+1} \otimes I)\mathcal{C}_{k+1}x_{k+1}$$
$$- K_{i,k+1}^2(\bar{\mathcal{H}}_i \otimes I)(\Lambda_{k+1}^1 \otimes I)\xi_{k+1}$$
$$- K_{i,k+1}^2(\bar{\mathcal{H}}_i \otimes I)(\Lambda_{k+1}^3 \otimes I)q_{k+1}$$
$$- K_{i,k+1}^2(\bar{\mathcal{H}}_i \otimes I)(\Lambda_{k+1}^3 \otimes I)(I \otimes D_{k+1})v_{k+1}$$

with

$$\bar{\mathcal{H}}_i = [\hbar_j^i]_{1 \times n} = \begin{cases} \hbar_j^i = h_{ji}, & j \in \mathcal{N}_i, \\ \hbar_j^i = 0, & j \notin \mathcal{N}_i. \end{cases}$$

Now, let us calculate the filtering error covariance $P_{i,k+1|k+1}$ as follows:

$$P_{i,k+1|k+1}$$
$$= \mathbb{E}\{(x_{k+1} - \hat{x}_{i,k+1|k+1})(x_{k+1} - \hat{x}_{i,k+1|k+1})^T\}$$
$$= \Psi_{i,k+1}^0 \mathbb{E}\{e_{i,k+1|k}e_{i,k+1|k}^T\}\Psi_{i,k+1}^{0T}$$
$$- \Psi_{i,k+1}^0 \mathbb{E}\{e_{i,k+1|k}\xi_{k+1}^T(\Lambda_{k+1}^1 \otimes I)^T\}(\bar{\mathcal{H}}_i \otimes I)^T K_{i,k+1}^{2T}$$
$$- K_{i,k+1}^2(\mathcal{S}_i \otimes I)\mathbb{E}\{(\Lambda_{k+1}^1 \otimes I)\xi_{k+1}e_{i,k+1|k}^T\}\Psi_{i,k+1}^{0T}$$
$$+ K_{i,k+1}^1(\mathcal{H}_i \otimes I)\mathbb{E}\{(W_{k+1} \otimes I)\mathcal{C}_{k+1}x_{k+1}$$
$$\times x_{k+1}^T \mathcal{C}_{k+1}^T(W_{k+1} \otimes I)^T\}(\mathcal{H}_i \otimes I)^T K_{i,k+1}^{1T}$$
$$+ K_{i,k+1}^1(\mathcal{H}_i \otimes I)(I \otimes D_{k+1})\mathbb{E}\{v_{k+1}$$
$$\times v_{k+1}^T\}(I \otimes D_{k+1})^T(\mathcal{H}_i \otimes I)^T K_{i,k+1}^{1T}$$
$$+ K_{i,k+1}^1(\mathcal{H}_i \otimes I)\mathbb{E}\{q_{k+1}q_{k+1}^T\}(\mathcal{H}_i \otimes I)^T K_{i,k+1}^{1T}$$
$$+ K_{i,k+1}^2(\bar{\mathcal{H}}_i \otimes I)\mathbb{E}\{(\Lambda_{k+1}^2 \otimes I)C_{0,k+1}x_{k+1}$$
$$\times x_{k+1}^T C_{0,k+1}^T(\Lambda_{k+1}^2 \otimes I)^T\}(\bar{\mathcal{H}}_i \otimes I)^T K_{i,k+1}^{2T}$$
$$- K_{i,k+1}^2(\bar{\mathcal{H}}_i \otimes I)\mathbb{E}\{(\Lambda_{k+1}^2 \otimes I)C_{0,k+1}x_{k+1}$$
$$\times \xi_{k+1}^T(\Lambda_{k+1}^1 \otimes I)^T\}(\bar{\mathcal{H}}_i \otimes I)^T K_{i,k+1}^{2T}$$
$$- K_{i,k+1}^2(\bar{\mathcal{H}}_i \otimes I)\mathbb{E}\{(\Lambda_{k+1}^1 \otimes I)\xi_{k+1}$$
$$\times x_{k+1}^T C_{0,k+1}^T(\Lambda_{k+1}^2 \otimes I)^T\}(\bar{\mathcal{H}}_i \otimes I)^T K_{i,k+1}^{2T}$$
$$+ K_{i,k+1}^2(\bar{\mathcal{H}}_i \otimes I)\mathbb{E}\{(\Lambda_{k+1}^3 W_{k+1} \otimes I)\mathcal{C}_{k+1}x_{k+1}$$
$$\times x_{k+1}^T \mathcal{C}_{k+1}^T(\Lambda_{k+1}^3 W_{k+1} \otimes I)^T\}(\bar{\mathcal{H}}_i \otimes I)^T K_{i,k+1}^{2T}$$
$$+ K_{i,k+1}^2(\bar{\mathcal{H}}_i \otimes I)\mathbb{E}\{(\Lambda_{k+1}^1 \otimes I)\xi_{k+1}$$

$$\times \xi_{k+1}^T(\Lambda_{k+1}^1 \otimes I)^T\}(\bar{\mathcal{H}}_i \otimes I)^T K_{i,k+1}^{2T}$$
$$+ K_{i,k+1}^2(\bar{\mathcal{H}}_i \otimes I)\mathbb{E}\{(\Lambda_{k+1}^3 \otimes I)(I \otimes D_{k+1})v_{k+1}$$
$$\times v_{k+1}^T(I \otimes D_{k+1})^T(\Lambda_{k+1}^3 \otimes I)^T\}(\bar{\mathcal{H}}_i \otimes I)^T K_{i,k+1}^{2T}$$
$$+ K_{i,k+1}^2(\bar{\mathcal{H}}_i \otimes I)\mathbb{E}\{(\Lambda_{k+1}^3 \otimes I)q_{k+1}$$
$$\times q_{k+1}^T(\Lambda_{k+1}^3 \otimes I)^T\}(\bar{\mathcal{H}}_i \otimes I)^T K_{i,k+1}^{2T}$$
$$\leq (1 + \varepsilon_k)\Psi_{i,k+1}^0 P_{i,k+1|k}\Psi_{i,k+1}^{0T}$$
$$+ K_{i,k+1}^1(\mathcal{H}_i \otimes I)X_{k+1}^C(\mathcal{H}_i \otimes I)^T K_{i,k+1}^{1T}$$
$$+ K_{i,k+1}^1(\mathcal{H}_i\mathcal{H}_i^T \otimes D_{k+1}D_{k+1}^T)K_{i,k+1}^{1T}$$
$$+ \frac{u^2}{12}K_{i,k+1}^1(\mathcal{H}_i\mathcal{H}_i^T \otimes I)K_{i,k+1}^{1T}$$
$$+ (1 + \varepsilon_k)\bar{\alpha}(1 - \bar{\alpha})K_{i,k+1}^2(\bar{\mathcal{H}}_i\bar{\mathcal{H}}_i^T$$
$$\otimes C_{0,k+1}X_{k+1}C_{0,k+1}^T)K_{i,k+1}^{2T}$$
$$+ (1 - \bar{\alpha})K_{i,k+1}^2(\bar{\mathcal{H}}_i \otimes I)X_{k+1}^C(\bar{\mathcal{H}}_i \otimes I)^T K_{i,k+1}^{2T}$$
$$+ (1 + 2\varepsilon_k^{-1})\delta^2 K_{i,k+1}^2(\bar{\mathcal{H}}_i \otimes I)(\bar{\alpha}^2(\mathbf{1}\mathbf{1}^T) \otimes I$$
$$+ (\bar{\alpha} - \bar{\alpha}^2)I)(\bar{\mathcal{H}}_i \otimes I)^T K_{i,k+1}^{2T}$$
$$+ (1 - \bar{\alpha})K_{i,k+1}^2(\bar{\mathcal{H}}_i\bar{\mathcal{H}}_i^T \otimes D_{k+1}D_{k+1}^T)K_{i,k+1}^{2T}$$
$$+ \frac{(1 - \bar{\alpha})u^2}{12}K_{i,k+1}^2(\bar{\mathcal{H}}_i\bar{\mathcal{H}}_i^T \otimes I)K_{i,k+1}^{2T}. \tag{8.16}$$

Using the properties of the Kronecker product in the above matrix inequality results in

$$P_{i,k+1|k+1} \leq \Psi_{i,k+1}^0 P_{i,k+1|k}\Psi_{i,k+1}^{0T} \\ + K_{i,k+1}^1\Psi_{i,k+1}^1 K_{i,k+1}^{1T} + K_{i,k+1}^2\Psi_{i,k+1}^{23}K_{i,k+1}^{2T}, \tag{8.17}$$

which implies that the second inequality in (8.12) is true.

Up to now, the upper bound for the filtering error covariance has been derived. We are now in a position to consider the filter gain design problem for the addressed problem. The following result can be accessed by using the gradient-based approach.

Theorem 8.2 *For the addressed system (8.1) with measurements (8.3) suffering from attacks (8.4), the gain matrices of the recursive filter (8.5a) and (8.5b) are given as follows*

$$K_{i,k+1}^1 = \hbar_{in}^i\Pi_{i,k+1|k}C_{0,k+1}^T\big(I - (\hbar_{in}^i)^2(1 - \bar{\alpha})^2 \\ \times (\mathcal{S}_{i,k+1}^2)^{-1}\mathcal{S}_{i,k+1}^0\big)(\Omega_{i,k+1}^1)^{-1},$$
$$K_{i,k+1}^2 = (1 - \bar{\alpha})\hbar_{in}^i\Pi_{i,k+1|k}C_{0,k+1}^T\big(I - (\hbar_{in}^i)^2 \\ \times (\mathcal{S}_{i,k+1}^1)^{-1}\mathcal{S}_{i,k+1}^0\big)(\Omega_{i,k+1}^2)^{-1} \tag{8.18}$$

where

$$\mathcal{S}_{i,k+1}^0 = C_{0,k+1} \Pi_{i,k+1|k} C_{0,k+1}^T,$$

$$\mathcal{S}_{i,k+1}^1 = (\hbar_{in}^i)^2 \mathcal{S}_{i,k+1}^0 + \Psi_{i,k+1}^1,$$

$$\mathcal{S}_{i,k+1}^2 = (\hbar_{in}^i)^2 (1-\bar{\alpha})^2 \mathcal{S}_{i,k+1}^0 + \Psi_{i,k+1}^{23},$$

$$\Omega_{i,k+1}^1 = \mathcal{S}_{i,k+1}^1 - (1-\bar{\alpha})^2 (\hbar_{in}^i)^4 \mathcal{S}_{i,k+1}^0 (\mathcal{S}_{i,k+1}^2)^{-1} \mathcal{S}_{i,k+1}^0,$$

$$\Omega_{i,k+1}^2 = \mathcal{S}_{i,k+1}^2 - (1-\bar{\alpha})^2 (\hbar_{in}^i)^4 \mathcal{S}_{i,k+1}^0 (\mathcal{S}_{i,k+1}^1)^{-1} \mathcal{S}_{i,k+1}^0.$$

Furthermore, the upper bound of the filtering error covariance $\Pi_{i,k+1|k+1}$ is recursively calculated by Riccati-like difference equation (8.13).

Proof *According to Theorem 8.1, the design of gains K_{k+1}^1 and K_{k+1}^2 need to minimize $Trace(\Pi_{i,k+1|k+1})$. For this purpose, taking the partial derivative of $Trace(\Pi_{i,k+1|k+1})$ with respect to K_{k+1}^1 and K_{k+1}^2, and letting the derivative be zero, one has*

$$\frac{\partial Trace(\Pi_{i,k+1|k+1})}{\partial K_{i,k+1}^1} = 0$$

$$\Rightarrow -\hbar_{in}^i \Psi_{i,k+1}^0 \Pi_{i,k+1|k} C_{0,k+1}^T + K_{i,k+1}^1 \Psi_{i,k+1}^1 = 0,$$

$$\frac{\partial Trace(\Pi_{i,k+1|k+1})}{\partial K_{i,k+1}^2} = 0$$

$$\Rightarrow -\hbar_{in}^i (1-\bar{\alpha}) \Psi_{i,k+1}^0 \Pi_{i,k+1|k} C_{0,k+1}^T + K_{i,k+1}^2 \Psi_{i,k+1}^{23} = 0.$$

Subsequently, we have

$$-\hbar_{in}^i \big(I - \hbar_{in}^i (1-\bar{\alpha}) K_{i,k+1}^2 C_{0,k+1} \big) \Pi_{i,k+1|k} C_{0,k+1}^T$$
$$+ K_{i,k+1}^1 \big((\hbar_{in}^i)^2 C_{0,k+1} \Pi_{i,k+1|k} C_{0,k+1}^T + \Psi_{i,k+1}^1 \big) = 0,$$
$$-\hbar_{in}^i (1-\bar{\alpha}) \big(I - \hbar_{in}^i K_{i,k+1}^1 C_{0,k+1} \big) \Pi_{i,k+1|k} C_{0,k+1}^T$$
$$+ K_{i,k+1}^2 \big((\hbar_{in}^i)^2 (1-\bar{\alpha})^2 C_{0,k+1}$$
$$\times \Pi_{i,k+1|k} C_{0,k+1}^T + \Psi_{i,k+1}^{23} \big) = 0,$$

which can be further simplified as follows:

$$K_{i,k+1}^1 \mathcal{S}_{i,k+1}^1 + (1-\bar{\alpha})(\hbar_{in}^i)^2 K_{i,k+1}^2 \mathcal{S}_{i,k+1}^0$$
$$- \hbar_{in}^i \Pi_{i,k+1|k} C_{0,k+1}^T = 0,$$
$$K_{i,k+1}^2 \mathcal{S}_{i,k+1}^2 + (1-\bar{\alpha})(\hbar_{in}^i)^2 K_{i,k+1}^1 \mathcal{S}_{i,k+1}^0$$
$$- (1-\bar{\alpha}) \hbar_{in}^i \Pi_{i,k+1|k} C_{0,k+1}^T = 0.$$

$$(8.19)$$

Furthermore, the following

$$
\begin{aligned}
\Omega^1_{i,k+1} &= \mathcal{S}^1_{i,k+1} - (1-\bar{\alpha})^2 (\hbar^i_{in})^4 \mathcal{S}^0_{i,k+1} \left(\mathcal{S}^2_{i,k+1}\right)^{-1} \mathcal{S}^0_{i,k+1} \\
&= \Psi^1_{i,k+1} + (\hbar^i_{in})^2 \Big[\mathcal{S}^0_{i,k+1} - \Big((1-\bar{\alpha})^2 (\hbar^i_{in})^2 \mathcal{S}^0_{i,k+1} \\
&\quad + \Psi^{23}_{i,k+1} - \Psi^{23}_{i,k+1}\Big)\left(\mathcal{S}^2_{i,k+1}\right)^{-1} \mathcal{S}^0_{i,k+1} \Big] \\
&= \Psi^1_{i,k+1} + (\hbar^i_{in})^2 \Psi^{23}_{i,k+1} \left(\mathcal{S}^2_{i,k+1}\right)^{-1} \mathcal{S}^0_{i,k+1} \\
&= \Psi^1_{i,k+1} + (1-\bar{\alpha})^{-2} \\
&\quad \times \left(\Psi^{23}_{i,k+1} - \Psi^{23}_{i,k+1}\left(\mathcal{S}^2_{i,k+1}\right)^{-1} \Psi^{23}_{i,k+1} \right) \\
&> \Psi^1_{i,k+1}
\end{aligned}
\tag{8.20}
$$

and

$$
\begin{aligned}
\Omega^2_{i,k+1} &= \mathcal{S}^2_{i,k+1} - (1-\bar{\alpha})^2 (\hbar^i_{in})^4 \mathcal{S}^0_{i,k+1} \left(\mathcal{S}^1_{i,k+1}\right)^{-1} \mathcal{S}^0_{i,k+1} \\
&= \Psi^{23}_{i,k+1} + (1-\bar{\alpha})^2 \\
&\quad \times \left(\Psi^1_{i,k+1} - \Psi^1_{i,k+1}\left(\mathcal{S}^1_{i,k+1}\right)^{-1} \Psi^1_{i,k+1} \right) \\
&> \Psi^{23}_{i,k+1}
\end{aligned}
\tag{8.21}
$$

are true. Therefore, taking (8.19)-(8.21) into consideration, we can obtain the desired filter gain matrices. Furthermore, the upper bound for the filtering error covariance $\Pi_{i,k+1|k+1}$ is recursively calculated by Riccati-like difference equation (8.13).

Remark 8.2 *In the above theorem, the desired filter gain matrices are obtained with the aid of the solution of Riccati-like difference equations. It is not difficult to see that, for $\bar{\alpha} = 1$, the gain $K^1_{i,k+1}$ and $K^2_{i,k+1}$ will reduce to the case of a single node without neighbors ($K^2_{i,k+1} = 0$ in this case), which implies that the filter on each node will refuse to fuse the information from the neighboring nodes when the network is completely unreliable. Furthermore, it should be pointed out that these two gains are designed by minimizing the trace of an upper bound for the filtering error covariance due to the effect from time-delays, deception attacks, and uniform quantization. As such, the proposed filter is only a suboptimal one.*

Remark 8.3 *In this chapter, Theorem 8.2 is concerned with the distributed recursive filtering algorithm. Denote the number of neighbors as ϕ_i on node i. For each step of the proposed algorithm, the implementation on node i includes 6 instances of the matrix inversion operation and $4r+5\phi_i+22$ instances of the matrix multiplication operation. With the help of the dimensions of $x_k \in \mathbb{R}^{n_x}$, $\tilde{y}_{i,k} \in \mathbb{R}^{n_y}$, $w_k \in \mathbb{R}^s$ and $v_{i,k} \in \mathbb{R}^p$, it is not difficult to calculate the overall computational complexity on node i as $O((4r+4)n_x^3 + 7n_y^3 + (4\phi_i+8)n_x n_y^2 + (\phi_i+1)n_y^2 p + n_x^2 s)$, which depends linearly on the number of neighbors and* *is independent of the scale of whole sensor networks. Due to the sparseness of sensor networks, such an algorithm with low computation burden is really suitable for online application. In addition, the given algorithm does not need any auxiliary information from neighbors in calculating all matrices $\Pi_{i,k+1|k}$, $\mathcal{S}_{i,k+1}^0$, $\mathcal{S}_{i,k+1}^1$, $\mathcal{S}_{i,k+1}^2$, $\Omega_{i,k+1}^1$ and $\Omega_{i,k+1}^2$ in equation (8.18), and it is therefore a truly distributed one.*

8.3 Boundedness Analysis

So far, we have derived an upper bound for the filtering error covariance, and such an upper bound is subsequently minimized by properly designing the filter parameters via a gradient-based method at each sampling instant. In reality, we would also be interested in understanding when the sequence of the upper bounds is asymptotically bounded in order to evaluate the performance of the designed recursive filter. In this section, for obtained filter gains, we will propose a sufficient condition ensuring the boundedness of the sequence $\Pi_{i,k|k}$ with respect to the filtering error covariance. For this purpose, we firstly introduce the following crucial assumption and lemmas. Similar to Section 8.2, all proofs of lemmas and theorems are moved to the appendixes for clarity of presentation.

Assumption 8.1 *There are positive real constants $\bar{f}_i$, $\bar{f}_i^d$, $\underline{b}$, $\bar{b}$, $\underline{d}$, $\bar{d}$ and $\bar{c}_i$ $(i = 0, 1, 2, \cdots, r)$ such that the system parameter matrices are bounded:*

$$A_{i,k} A_{i,k}^T \leq \bar{f}_i I, \quad A_{i,k}^d A_{i,k}^{dT} \leq \bar{f}_i^d I, \quad C_{i,k} C_{i,k}^T \leq \bar{c}_i I,$$
$$\underline{b} I \leq B_k B_k^T \leq \bar{b} I, \quad \underline{d} I \leq D_k D_k^T \leq \bar{d} I.$$

Lemma 8.4 *[79] Let A and B be two $n \times n$ symmetric matrices with their eigenvalues listed as follows:*

$$\lambda_{\max}(A) = \lambda_1(A) \geq \lambda_2(A) \geq \cdots \geq \lambda_n(A) = \lambda_{\min}(A),$$
$$\lambda_{\max}(B) = \lambda_1(B) \geq \lambda_2(B) \geq \cdots \geq \lambda_n(B) = \lambda_{\min}(B).$$

Then, one has

$$\min_{1 \leq i \leq s} \left(\lambda_k(A) + \lambda_{s+1-i}(B) \right) \geq \lambda_s(A + B)$$

$$\geq \max_{s \leq i \leq n} \left(\lambda_i(A) + \lambda_{n+s-i}(B) \right), \quad 1 \leq s \leq n.$$

Furthermore, when either A or B is positive definite, the following are true *for any $s \leq n$:*

$$\min_{1 \leq i \leq s} \left(\lambda_i(A) \lambda_{s+1-i}(B) \right) \geq \lambda_s(AB)$$

$$\geq \max_{s \leq i \leq n} \left(\lambda_i(A) \lambda_{n+s-i}(B) \right),$$

$$\min_{1 \leq i \leq s} \left(\lambda_i(A) \lambda_{s+1-i}(BB^T) \right) \geq \lambda_s(BAB^T)$$

$$\geq \max_{s \leq i \leq n} \left(\lambda_i(A) \lambda_{n+s-i}(BB^T) \right).$$

The following lemma is easily accessible by using the mathematical induction method and its proof is therefore omitted.

Lemma 8.5 *Let $X_i \leq \bar{p}_x I$ ($-\tau \leq i \leq 0$) hold for the given positive scalar $\bar{p}_x$. For the addressed system (8.1), if there exists a positive scalar ε satisfying*

$$\left((1+\varepsilon) \sum_{i=0}^{r} \bar{f}_i + (1+\varepsilon^{-1}) \sum_{i=0}^{r} \bar{f}_i^d \right) \bar{p}_x + \bar{b} \leq \bar{p}_x,$$

then $\underline{p}_x I \leq X_k \leq \bar{p}_x I$ is true for all $k > 0$ where $\underline{p}_x$ is a known positive scalar satisfying $\underline{p}_x \leq \underline{b}$.

Now, let us give the main result, whose proof is moved to the appendix for clarity of presentation.

Theorem 8.3 *Under Assumption 8.1, let $\Pi_{i,j|j} \leq \bar{p}_{\pi_i} I$ ($-\tau \leq j \leq 0$) where $\bar{p}_{\pi_i}$ is a known positive scalar. For the addressed system (8.1) with the filtering dynamics (8.5a)-(8.5b), $\Pi_{i,k+1|k+1} \leq \bar{p}_{\pi_i} I$ holds for any $k \geq 1$ if there exists a positive scalar ε satisfying*

$$(3 + \xi_0) \varrho_1 \leq \bar{p}_{\pi_i} \tag{8.22}$$

where

$$\xi_0 = \frac{(\hbar_{in}^i \phi_{i,1}^{m\Psi})^2 \bar{c}_0 \varrho_1}{(\phi_{i,1}^{n\Psi})^3} + \frac{(\hbar_{in}^i \phi_{i,23}^{m\Psi})^2 (1-\bar{\alpha})^2 \bar{c}_0 \varrho_1}{(\phi_{i,23}^{n\Psi})^3}$$

$$+ \frac{3(\hbar_{in}^i)^4 \bar{c}_0^2 \varrho_1^2}{(\phi_{i,1}^{n\Psi})^2} + \frac{3(\hbar_{in}^i)^4 \bar{c}_0^2 \varrho_1^2}{(\phi_{i,23}^{n\Psi})^2},$$

$$\phi_{i,1}^{m\Psi} = (\hbar_{in}^i)^2 \left(\bar{C}_i \bar{p}_x + \bar{d} + u^2/12 \right), \quad \phi_{i,1}^{n\Psi} = (\hbar_{in}^i u)^2/12,$$

$$\phi_{i,23}^{m\Psi} = \sum_{j=1}^{n} (1-\bar{\alpha})(h_{ji})^2 \left(\bar{\alpha}(1+\varepsilon) \bar{c}_0 \bar{p}_x \right.$$

$$\left. + \bar{c}_j \bar{p}_x + \bar{d} + u^2/12 \right) + (1 + 2\varepsilon^{-1}) \tilde{s}_i \delta^2,$$

$$\phi_{i,23}^{n\Psi} = \sum_{j=1}^{n} (1-\bar{\alpha})(h_{ji})^2 u^2/12 + (1 + 2\varepsilon^{-1}) \tilde{s}_i \delta^2,$$

$$\varrho_1 = (1+\varepsilon) \bar{f}_0 \bar{p}_{\pi_i} + (1+\varepsilon^{-1}) \bar{f}_0^d \bar{p}_{\pi_i}$$

$$+ \sum_{i=1}^{r} \left((1+\varepsilon) \bar{f}_i + (1+\varepsilon^{-1}) \bar{f}_i^d \right) \bar{p}_x + \bar{b}.$$

Proof *In order to prove the boundedness, the mathematical induction method is utilized to deal with the difficulty coming from two filtering gains and the coupling with plant states. For this purpose, suppose that $\underline{p}_{\pi_i} I \leq \Pi_{i,k|k} \leq \bar{p}_{\pi_i} I$ holds and then show that it is also true at the inductive step (that is, at $k+1$ instants).*

For the sake of simplicity, we fix $\varepsilon_k = \varepsilon$ for any k, and can obtain from (8.13) that

$$
\begin{aligned}
&\Pi_{i,k+1|k} \\
&= (1+\varepsilon)A_{0,k}\Pi_{i,k|k}A_{0,k}^T \\
&\quad + (1+\varepsilon^{-1})A_{0,k}^d\Pi_{i,k-\tau|k-\tau}(A_{0,k}^d)^T \\
&\quad + (1+\varepsilon)\sum_{i=1}^{r} A_{i,k}X_kA_{i,k}^T \\
&\quad + (1+\varepsilon^{-1})\sum_{i=1}^{r} A_{i,k}^d X_{k-\tau}(A_{i,k}^d)^T + B_kB_k^T \\
&\leq \varrho_1 I.
\end{aligned}
\tag{8.23}
$$

In what follows, it is not difficult to see that

$$
\begin{aligned}
&\Psi_{i,k+1}^0\Pi_{i,k+1|k}\Psi_{i,k+1}^{0T} \\
&= \left(I - \hbar_{in}^i K_{i,k+1}^1 C_{0,k+1} - \hbar_{in}^i(1-\bar{\alpha})K_{i,k+1}^2 C_{0,k+1}\right) \\
&\quad \times \Pi_{i,k+1|k}\big(I - \hbar_{in}^i K_{i,k+1}^1 C_{0,k+1} \\
&\quad - \hbar_{in}^i(1-\bar{\alpha})K_{i,k+1}^2 C_{0,k+1}\big)^T \\
&\leq 3\Pi_{i,k+1|k} + 3(\hbar_{in}^i)^2 K_{i,k+1}^1 \mathcal{S}_{i,k+1}^0(K_{i,k+1}^1)^T \\
&\quad + 3(\hbar_{in}^i)^2(1-\bar{\alpha})^2 K_{i,k+1}^2 \mathcal{S}_{i,k+1}^0(K_{i,k+1}^2)^T.
\end{aligned}
\tag{8.24}
$$

Substituting (8.24) into (8.13) leads to

$$
\begin{aligned}
&\Pi_{i,k+1|k+1} \\
&\leq 3\Pi_{i,k+1|k} + K_{i,k+1}^1\Psi_{i,k+1}^1(K_{i,k+1}^1)^T \\
&\quad + K_{i,k+1}^2\Psi_{i,k+1}^{23}(K_{i,k+1}^2)^T \\
&\quad + 3(\hbar_{in}^i)^2 K_{i,k+1}^1 \mathcal{S}_{i,k+1}^0(K_{i,k+1}^1)^T \\
&\quad + 3(\hbar_{in}^i)^2(1-\bar{\alpha})^2 K_{i,k+1}^2 \mathcal{S}_{i,k+1}^0(K_{i,k+1}^2)^T
\end{aligned}
\tag{8.25}
$$

where $\Psi_{i,k+1}^{23} = \Psi_{i,k+1}^2 + \Psi_{i,k+1}^3$.

On the other hand, reviewing (8.20)-(8.21) and using the property of positive definite matrices, one has

$$
\begin{aligned}
&(\Omega_{i,k+1}^1)^{-1}\,\mathcal{S}_{i,k+1}^0(\Omega_{i,k+1}^1)^{-1} \\
&\quad \leq \lambda_{\max}(\mathcal{S}_{i,k+1}^0)(\Omega_{i,k+1}^1\Omega_{i,k+1}^1)^{-1} \\
&\quad \leq \lambda_{\max}(\mathcal{S}_{i,k+1}^0)\lambda_1^2((\Psi_{i,k+1}^1)^{-1})I,
\end{aligned}
\tag{8.26}
$$

$$(\Omega^2_{i,k+1})^{-1}\, \mathcal{S}^0_{i,k+1}(\Omega^2_{i,k+1})^{-1}$$
$$\leq \lambda_{\max}(\mathcal{S}^0_{i,k+1})(\Omega^2_{i,k+1}\Omega^2_{i,k+1})^{-1}$$
$$\leq \lambda_{\max}(\mathcal{S}^0_{i,k+1})\lambda^2_1((\Psi^{23}_{i,k+1})^{-1})I. \tag{8.27}$$

Then, denoting

$$\Lambda_{i,k+1} = (\hbar^i_{in})^2 \mathcal{S}^0_{i,k+1} - (1-\bar{\alpha})^2(\hbar^i_{in})^4 \mathcal{S}^0_{i,k+1}, (\mathcal{S}^2_{i,k+1})^{-1}\mathcal{S}^0_{i,k+1},$$

we can check

$$\Lambda_{i,k+1}$$
$$= (\hbar^i_{in})^2 \mathcal{S}^0_{i,k+1} - (\hbar^i_{in})^2(\mathcal{S}^2_{i,k+1}$$
$$\quad - \Psi^{23}_{i,k+1})(\mathcal{S}^2_{i,k+1})^{-1}\mathcal{S}^0_{i,k+1}$$
$$= (\hbar^i_{in})^2\Psi^{23}_{i,k+1}(\mathcal{S}^2_{i,k+1})^{-1}\mathcal{S}^0_{i,k+1} \tag{8.28}$$
$$= (1-\bar{\alpha})^2\Psi^{23}_{i,k+1}(\mathcal{S}^2_{i,k+1})^{-1}(\mathcal{S}^2_{i,k+1} - \Psi^{23}_{i,k+1})$$
$$= (1-\bar{\alpha})^2\big(\Psi^{23}_{i,k+1} - \Psi^{23}_{i,k+1}(\mathcal{S}^2_{i,k+1})^{-1}\Psi^{23}_{i,k+1}\big)$$
$$\geq 0.$$

According to the above inequality, it can be derived that

$$(\Omega^1_{i,k+1})^{-1}\Psi^1_{i,k+1}(\Omega^1_{i,k+1})^{-1}$$
$$= \big(\Psi^1_{i,k+1} + \Lambda_{i,k+1}\big)^{-1}\Psi^1_{i,k+1}\big(\Psi^1_{i,k+1} + \Lambda_{i,k+1}\big)^{-1}$$
$$= \big(\Psi^1_{i,k+1} + 2\Lambda_{i,k+1} + \Lambda_{i,k+1}(\Psi^1_{i,k+1})^{-1}\Lambda_{i,k+1}\big)^{-1} \tag{8.29}$$
$$\leq (\Psi^1_{i,k+1})^{-1}.$$

Furthermore, using the same line of the above inequality, we have

$$(\Omega^2_{i,k+1})^{-1}\Psi^{23}_{i,k+1}(\Omega^2_{i,k+1})^{-1} \leq (\Psi^{23}_{i,k+1})^{-1}. \tag{8.30}$$

In addition, we can calculate

$$\big(I - (\hbar^i_{in})^2(1-\bar{\alpha})^2(\mathcal{S}^2_{i,k+1})^{-1}\mathcal{S}^0_{i,k+1}\big)$$
$$\times \big(I - (\hbar^i_{in})^2(1-\bar{\alpha})^2(\mathcal{S}^2_{i,k+1})^{-1}\mathcal{S}^0_{i,k+1}\big)^T$$
$$= I - (\hbar^i_{in})^2(1-\bar{\alpha})^2(\mathcal{S}^2_{i,k+1})^{-1}\mathcal{S}^0_{i,k+1}$$
$$\quad - (\hbar^i_{in})^2(1-\bar{\alpha})^2\mathcal{S}^0_{i,k+1}(\mathcal{S}^2_{i,k+1})^{-1}$$
$$\quad + (\hbar^i_{in})^4(1-\bar{\alpha})^4(\mathcal{S}^2_{i,k+1})^{-1}\mathcal{S}^0_{i,k+1}\mathcal{S}^0_{i,k+1}(\mathcal{S}^2_{i,k+1})^{-1}$$
$$= (\mathcal{S}^2_{i,k+1})^{-1}\Psi^{23}_{i,k+1} + \Psi^{23}_{i,k+1}(\mathcal{S}^2_{i,k+1})^{-1} - I \tag{8.31}$$
$$\quad + (\hbar^i_{in})^4(1-\bar{\alpha})^4(\mathcal{S}^2_{i,k+1})^{-1}\mathcal{S}^0_{i,k+1}\mathcal{S}^0_{i,k+1}(\mathcal{S}^2_{i,k+1})^{-1}$$
$$= (\mathcal{S}^2_{i,k+1})^{-1}\Psi^{23}_{i,k+1}\Psi^{23}_{i,k+1}(\mathcal{S}^2_{i,k+1})^{-1}$$
$$\leq \frac{(\phi^{m\Psi}_{i,23})^2}{(\phi^{n\Psi}_{i,23})^2}I.$$

Similarly, it can be obtained that

$$
\begin{aligned}
&\left(I - (\hbar_{in}^i)^2 (\mathcal{S}_{i,k+1}^1)^{-1} \mathcal{S}_{i,k+1}^0\right) \\
&\times \left(I - (\hbar_{in}^i)^2 (\mathcal{S}_{i,k+1}^1)^{-1} \mathcal{S}_{i,k+1}^0\right)^T \leq \frac{(\phi_{i,1}^{m\Psi})^2}{(\phi_{i,1}^{n\Psi})^2} I.
\end{aligned}
\tag{8.32}
$$

Finally, in light of Lemma 8.4 and Lemma 8.5, substituting (8.26), (8.29), (8.30), (8.31) and (8.32) into (8.25) results in

$$
\Pi_{i,k+1|k+1} \leq (3+\xi_0)\Pi_{i,k+1|k} \leq (3+\xi_0)\varrho_1 I \leq \bar{p}_{\pi_i} I.
\tag{8.33}
$$

Therefore, by induction, it can be concluded that $\Pi_{i,k|k} < \bar{p}_{\pi_i} I$ is true for any $k \geq 1$, which completes the proof.

Remark 8.4 *It is worth noting that condition (8.22) in Theorem 8.3 includes all the information about plant dynamics, quantizations, cyber-attacks, filtering scheme, as well as the communication topology. In practice, the bound of system parameters in Assumption 8.1, the quantizer level, the communication topology, as well as the statistical characteristics of noises can all be made available to the filter design through parameter identification. Therefore, condition (8.22) is essentially an algebraic polynomial inequality on ε and $\bar{p}_{\pi_i}$. It is fairly easy, via MATLAB® software tools, to verify whether or not there exists a region of $(\varepsilon, \bar{p}_{\pi_i})$ that guarantees (8.22). In addition, it can be checked if the initial condition $\Pi_{i,j|j} = P_{i,j|j} \leq \bar{p}_{\pi_i} I \; (-\tau \leq j \leq 0)$ holds for some $\bar{p}_{\pi_i}$.*

Remark 8.5 *In this chapter, the filter parameters $K_{i,k+1}^1$ and $K_{i,k+1}^2$ can be obtained in an online manner (as described in Theorem 8.2) in order to minimize the upper bound of filtering error covariance. Furthermore, the boundedness of the obtained sequence on the filtering error covariance is discussed in Theorem 8.3, which is actually concerned with the stability issues of the recursive Kalman filtering algorithms that have received particular attention in the area. As opposed to the single filter parameter in traditional stability analysis [65, 97], two filter parameters are considered in our chapter that deal with, respectively, the reliable data from the node itself and the possibly unreliable data from the neighboring nodes. The introduction of the two filter parameters reflects the distributed nature of the filter dynamics but also brings in essential difficulties in the stability analysis as traditional methods (e.g. those in [65, 97]) are no longer applicable. In this chapter, the mathematical induction method combined with the properties of matrix analysis is utilized to overcome the difficulties and obtain the desired sufficient conditions that are related to both the quantization and the attack. Actually, it can be seen from Theorem 8.3 that the solvability of (8.22) is dependent on both δ and u.*

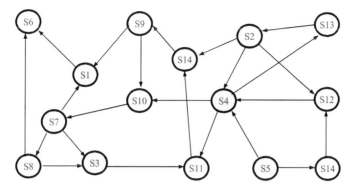

FIGURE 8.1
Topological structure of the sensor network.

8.4 Simulation Examples

In this section, we present a simulation example to illustrate the effectiveness of the proposed distributed filter design scheme for a discrete-time stochastic system with multiplicative noises through sensor networks.

The target plant considered is modeled by (8.1) with $r = \tau = 1$ and the following parameters:

$$A_{0,k} = \begin{bmatrix} 1.0 & 0.12 + 0.1\cos(0.25k) \\ -0.2 & 0.92 \end{bmatrix},$$

$$A_{1,k} = \begin{bmatrix} 0 & 0 \\ 0 & 0.1 \end{bmatrix}, \quad A_{0,k}^d = \begin{bmatrix} -0.02 & 0 \\ 0 & 0.04 \end{bmatrix},$$

$$A_{1,k}^d = \begin{bmatrix} -0.01 & 0 \\ 0 & 0 \end{bmatrix}, \quad B_k = \begin{bmatrix} 0.01 \\ -0.01 \end{bmatrix},$$

$$C_{0,k} = \begin{bmatrix} 1 & 0.25 \end{bmatrix}, \quad C_{1,k} = \begin{bmatrix} -0.01 & 0 \end{bmatrix}, \quad D_k = 0.05.$$

Sensor network is shown in Fig. 8.1, and its element in adjacency matrix $\mathscr{H}$ is $h_{ji} = 0.5$ when sensor i can obtain information from sensor j, otherwise $h_{ji} = 0$.

In this example, the probability $\bar{\alpha}$ of cyber-attacks, the length u of the quantizer level, and the bound δ are, respectively, selected as 0.15, 0.01, and 0.2. The initial conditions are set as $x_{-1} = \begin{bmatrix} 0.4 & 0.3 \end{bmatrix}^T$, $x_0 = \begin{bmatrix} 2.5 & -3 \end{bmatrix}^T$, $P_{i,-1|-1} = P_{i,0|0} = I$ $(i = 1, 2, \cdots, 5)$, $X_{-1} = 0.1I$ and $X_0 = 0.2I$. In addition, the attack signal is set to be $0.2\sin(r_{i,k})$ where $r_{i,k}$ obeys the Gaussian distribution $\mathcal{N}(0, 1^2)$.

In order to compare with the centralized Kalman filtering, we consider the node S5 without neighbors, that is, the in-degree $h_{in}^5 = 0$. On this node, the

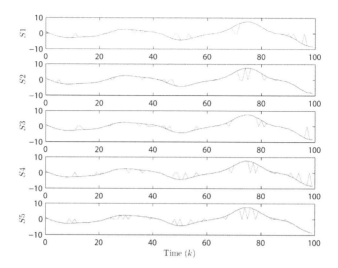

FIGURE 8.2
The measurements and the received signals.

traditional Kalman filtering algorithm is utilized to obtain the estimated state and the filtering error covariance. Simulation results are shown in Figs. 8.2-8.5, where Fig. 8.2 plots the measurements and the actually received signals where the red broken lines show the attack phenomena. Fig. 8.3 depicts the trajectories for the system states (blue lines) and their estimates (red lines). In addition, we can find from Fig. 8.4 that the innovation information from the node itself undertakes an important role to improve the filtering performance. In Fig. 8.5, we aim to examine how the success ratio of the launched attacks influences the filter performance. It can be observed that the trace of $\Pi_{2,k|k}$ at $k = 100$ increases with increased $\bar{\alpha}$. Obviously, when $\bar{\alpha} > 0.2$, the effect from attacks is quite limited, which means that the proposed distributed filter is of the expected "robustness" against unreliable measurements resulting from multiplicative noises, uniform quantizations, and intermittent deception attacks.

8.5   Summary

In this chapter, the distributed recursive filtering problem has been investigated a class of discrete time-delayed stochastic systems subject to effects of both the uniform quantization and the deception attacks. A

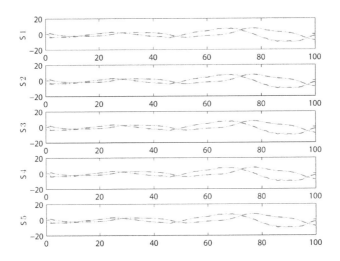

FIGURE 8.3
The system states and their estimation.

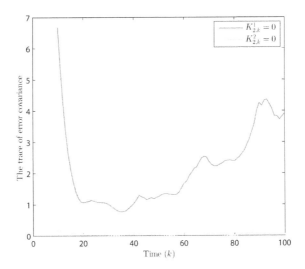

FIGURE 8.4
The trace of filtering error covariance.

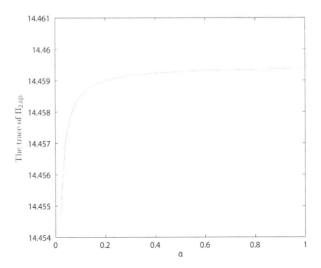

FIGURE 8.5
The trace of $\Pi_{2,k|k}$ for varying $\bar{\alpha}$ on $k = 100$.

distributed filter has been designed that fuses unreliable data corrupted by noises, quantization errors, and possible deception attacks, where the filtering algorithm is shown to be efficient in reducing the computation burden resulting from time-delays and the large number of sensor nodes. The complicated coupling issues between the filtering errors and observed states have been addressed in the performance analysis. An upper bound on the filtering error covariance has been guaranteed and then minimized by means of solving two Riccati-like difference equations. In addition, by utilizing the mathematical induction method, a sufficient condition has been proposed under which the filtering error covariance is bounded as time trends to infinity. Finally, an example has been provided to illustrate the effectiveness of the proposed filter approach.

9

Event-Triggered Distributed $\mathcal{H}_\infty$ State Estimation with Packet Dropouts through Sensor Networks

In Chapter 4 and Chapter 8, distributed filtering issues have been investigated for two classes of stochastic systems subject to network-induced phenomena or cyber-attacks. In these two topics, the data exchange is executed in a periodical paradigm, which widely neglects the behavior of communication scheduling. Recently, due to the use of scheduling, the event-triggered control and filtering issue has attracted ever-increasing research interest. Such a situation results into two challenging issues identified as follows. First, a sensor network is often subject to various network-induced phenomena (e.g., packet dropouts and stochastic nonlinearities) even if the event-triggered communication protocol is exploited. So, we need to develop a reasonable model to describe event-triggering communication mechanisms and network-induced phenomena in a unified framework. Second, the key issue in designing distributed estimators for sensor networks is how to fuse the information available for the estimator both from itself (without event-triggered mechanism) and from its neighbors (with event-triggered mechanism). In other words, it is very critical to construct a suitable distributed estimator such that the information from different sources is adequately integrated.

This chapter is concerned with the event-triggered distributed $\mathcal{H}_\infty$ state estimation problem for a class of discrete-time stochastic nonlinear systems with packet dropouts in a sensor network. An event-triggered communication mechanism is adopted over the sensor network to attempt to reduce the communication burden and energy consumption, where the measurements of each sensor are transmitted only when a certain triggering condition is violated. Furthermore, a novel distributed state estimator is designed where the available innovations are not only from the individual sensor but also from its neighboring sensors according to the given topology. By utilizing the property of the Kronecker product and the stochastic analysis approaches, sufficient conditions are established under which the addressed state estimation problem is recast as a convex optimization problem that can be easily solved via available software packages.

9.1   Modeling and Problem Formulation

In this chapter, it is assumed that the sensor network has n sensor nodes which are distributed in space according to a fixed network topology represented by a undirected graph $\mathcal{G} = (\mathcal{V}, \mathcal{E}, \mathcal{H})$ of order n with the set of nodes $\mathcal{V} = \{1, 2, \cdots, n\}$, the set of edges $\mathcal{E} \in \mathcal{V} \times \mathcal{V}$, and the weighted adjacency matrix $\mathcal{H} = [h_{ij}]$ with nonnegative adjacency element h_{ij}. An edge of $\mathcal{G}$ is denoted by the ordered pair (i, j). The adjacency elements associated with the edges of the graph are positive, i.e., $h_{ij} > 0 \Longleftrightarrow (i, j) \in \mathcal{E}$, which means that sensor i can obtain information from sensor j. The set of neighbors of node $i \in \mathcal{V}$ is denoted by $\mathcal{N}_i = \{j \in \mathcal{V} : (i, j) \in \mathcal{E}\}$.

In this chapter, a target plant is the system whose states are to be estimated through the distributed sensors. Let the target plant be described by the following discrete-time stochastic nonlinear system:

$$\begin{cases} x_{k+1} = Ax_k + A_d x_{k-\tau} + f(x_k, \vartheta_k) + Bw_k, \\ z_k = Lx_k \end{cases} \tag{9.1}$$

with n sensors modeled by

$$y_{i,k} = C_i x_k + D_i v_k, \quad i = 1, 2, \cdots, n \tag{9.2}$$

where $x_k \in \mathbb{R}^{n_x}$ is the state of the target plant that cannot be observed directly, $y_{i,k} \in \mathbb{R}^{n_y}$ is the measurement output from sensor i, $z_k \in \mathbb{R}^{n_z}$ is the output to be estimated, and $w_k, v_k \in l_2([0, \infty); \mathbb{R})$ are external disturbances. τ is a known positive scalar, and A, A_d, B, L, C_i and D_i are known constant matrices with compatible dimensions.

The function $f(x_k, \vartheta_k)$ with $f(0, \vartheta_k) = 0$ is a stochastic nonlinear function having the following first moment for all x_k:

$$\mathbb{E}\{f(x_k, \vartheta_k)|x_k\} = 0 \tag{9.3}$$

and the covariance given by

$$\mathbb{E}\{f(x_k, \vartheta_k)f^T(x_j, \vartheta_j)|x_k\} = 0, \quad k \neq j, \tag{9.4}$$

and

$$\mathbb{E}\{f(x_k, \vartheta_k)f^T(x_k, \vartheta_k)|x_k\} = \sum_{i=1}^{s} \Pi_i x_k^T \Gamma_i x_k \tag{9.5}$$

where s is a known nonnegative integer, and Π_i and Γ_i $(i = 1, 2, \cdots, s)$ are known matrices with appropriate dimensions.

For the purpose of presentation clarity, on sensor node i, denote the estimation of x_k and the innovation sequence, respectively, as $\hat{x}_{i,k}$ and

$$r_{i,k} = y_{i,k} - C_i \hat{x}_{i,k}.$$

It should be pointed out that a distributed state estimation is capable of fusing the information available for the estimator on node i from both sensor i itself and its neighbors. A further objective of this chapter is to take the event-triggered communication mechanism into consideration in order to reduce the communication burden. For this purpose, we define event generator functions $\psi_i(\cdot, \cdot) : \mathbb{R}^{n_y} \times \mathbb{R} \to \mathbb{R}$ $(i = 1, 2, \cdots, n)$ as follows:

$$\psi_i(e_{i,k}, \delta_i) = e_{i,k}^T e_{i,k} - \delta_i r_{i,k}^T r_{i,k}. \tag{9.6}$$

Here, $e_{i,k} = r_{i,k}^t - r_{i,k}$ where $r_{i,k}^t$ is the broadcast innovation at the latest event instant and δ_i is a given positive scalar. The executions are triggered as long as the condition $\psi_i(e_{i,k}, \delta_i) > 0$ is satisfied. Therefore, the sequence of event-triggered instants $0 \leq s_0^i < s_1^i < \cdots < s_l^i < \cdots$ is determined iteratively by

$$s_{l+1}^i = \inf\{k \in \mathbb{N} | k > s_l^i, \ \psi_i(e_{i,k}, \delta_i) > 0\}. \tag{9.7}$$

As is well known, due to limited network bandwidth, broadcast innovation could be subject to packet dropouts. To provide for the phenomenon of packet dropouts, the received information for neighbors of node i can be described as

$$\tilde{r}_{i,k}^t = \alpha_{i,k} r_{i,k}^t \tag{9.8}$$

where the stochastic variables $\alpha_{i,k}$ $(i = 1, 2, \cdots, n)$ are employed to govern the stochastic occurring packet dropouts. These variables are assumed to be mutually independent Bernoulli-distributed white sequences taking values of 0 or 1 with the following probabilities

$$\text{Prob}\{\alpha_{i,k} = 0\} = 1 - \bar{\alpha}, \quad \text{Prob}\{\alpha_{i,k} = 1\} = \bar{\alpha}.$$

In this chapter, the distributed state estimators are of the following structure:

$$\begin{cases} \hat{x}_{i,k+1} = A\hat{x}_{i,k} + A_d\hat{x}_{i,k-\tau} + K_{i,1}r_{i,k} + K_{i,2} \sum_{j \in \mathcal{N}_i} h_{ij}\tilde{r}_{j,k}^t, \\ \hat{z}_{i,k} = L\hat{x}_{i,k} \end{cases} \tag{9.9}$$

where $\hat{z}_{i,k} \in \mathbb{R}^{n_z}$ is the estimated output on sensor node i. Here, $K_{i,1}$ and $K_{i,2}$ are the estimator gain matrices on node i to be determined.

Remark 9.1 *For distributed state estimation problems, the information available on each node is not only from itself but also from its neighbors according to the given topology. From the engineering viewpoint, the event-triggered communication protocol is adopted to determine at what time the information needs to be broadcast. Hence, for a given node, the amount of data received from any neighboring sensors should be less than that of the data from the node itself due to the application of the event-triggered protocol. This explains why we divide the innovation into two parts in (9.9), i.e., $r_{i,k}$*

concerning the data from the node itself and $\sum_{j \in \mathcal{N}_i} h_{ij} \tilde{r}^t_{j,k}$ concerning the data from the neiboring nodes. Therefore, the proposed estimator model (9.9) can be utilized to effectively cope with the complicated coupling issues between any sensor and its neighboring sensors and also adequately fuse these two kinds of information (i.e. $r_{i,k}$ and $\tilde{r}^t_{j,k}$) to improve the estimation performance.

Remark 9.2 *For described state estimation issues, an event-triggered communication mechanism (9.7) is adopted to attempt to reduce the communication burden and the energy consumption, where the innovation on each sensor is broadcast to its neighbors only when the certain triggering condition in (9.8) is violated. In light of such a condition, it is not difficult to see that a smaller threshold δ_i gives rise to a heavier communication load, and therefore an adequate trade-off can be achieved between the threshold and the acceptable network load.*

For notational simplicity, we define

$$\xi_k = \mathbf{1}_n \otimes x_k - \hat{x}_k, \quad \hat{x}_k = [\hat{x}^T_{1,k} \quad \hat{x}^T_{2,k} \quad \cdots \quad \hat{x}^T_{n,k}]^T,$$
$$\tilde{z}_{i,k} = z_k - \hat{z}_{i,k}, \quad \tilde{z}_k = [\tilde{z}^T_{1,k} \quad \tilde{z}^T_{2,k} \quad \cdots \quad \tilde{z}^T_{n,k}]^T,$$
$$e_k = [e^T_{1,k} \quad e^T_{2,k} \quad \cdots \quad e^T_{n,k}]^T, \quad \tilde{f}(x_k, \vartheta_k) = \mathbf{1}_n \otimes f(x_k, \vartheta_k),$$
$$\mathcal{A} = \text{diag}_n\{A\}, \quad \mathcal{A}_d = \text{diag}_n\{A_d\},$$
$$\mathcal{C} = \text{diag}\{C_1, C_2 \cdots, C_n\}, \quad \mathcal{K}_1 = \text{diag}\{K_{1,1}, K_{2,1} \cdots, K_{n,1}\},$$
$$\mathcal{D} = [\, D^T_1, \, D^T_2, \, \cdots, \, D^T_n \,]^T, \quad \mathcal{K}_2 = \text{diag}\{K_{1,2}, K_{2,2} \cdots, K_{n,2}\},$$
$$\Xi = \text{diag}_n\{\bar{\alpha}\}, \quad \Xi_k = \text{diag}\{\alpha_{1,k} - \bar{\alpha}, \, \alpha_{2,k} - \bar{\alpha}, \, \cdots, \, \alpha_{n,k} - \bar{\alpha}\}.$$

Using the defined notations, the dynamics of the estimation errors can be obtained as follows:

$$\begin{aligned}
\xi_{k+1} = &\, \big(\mathcal{A} - \mathcal{K}_1\mathcal{C} - \mathcal{K}_2(\mathcal{H}\Xi \otimes I)\mathcal{C}\big)\xi_k + \mathcal{A}_d\xi_{k-\tau} \\
&- \mathcal{K}_2(\mathcal{H}\Xi \otimes I)e_k - \big(\mathcal{K}_1\mathcal{D} + \mathcal{K}_2(\mathcal{H}\Xi \otimes I)\mathcal{D}\big)v_k \\
&+ (\mathbf{1} \otimes B)w_k + \tilde{f}(x_k, \vartheta_k) - \mathcal{K}_2(\mathcal{H}\Xi_k \otimes I)\mathcal{C}(\mathbf{1}_n \otimes x_k) \\
&- \mathcal{K}_2(\mathcal{H}\Xi_k \otimes I)e_k - \mathcal{K}_2(\mathcal{H}\Xi_k \otimes I)\mathcal{D}v_k.
\end{aligned} \tag{9.10}$$

Setting $\eta_k = [x^T_k \ \xi^T_k]^T$ and $\tilde{w}_k = [w^T_k \ v^T_k]^T$, an augmented system can be derived from (9.1) and (9.10) as follows:

$$\begin{cases}
\eta_{k+1} = \bar{\mathcal{A}}\eta_k + \tilde{\mathcal{A}}_k\eta_k + \mathscr{F}(x_k, \vartheta_k) \\
\qquad\quad + \bar{\mathcal{A}}_d\eta_{k-\tau} + \bar{\mathcal{B}}e_k + \tilde{\mathcal{B}}_ke_k + \bar{\mathcal{D}}\tilde{w}_k + \tilde{\mathcal{D}}_k\tilde{w}_k, \\
\tilde{z}_k = \bar{\mathcal{L}}\eta_k
\end{cases} \tag{9.11}$$

where

$$\mathscr{F}(x_k, \vartheta_k) = \mathbf{1}_{n+1} \otimes f(x_k, \vartheta_k), \quad \bar{\mathcal{L}} = \text{diag}\{0, I \otimes L\}, \quad \bar{\mathcal{A}}_d = \text{diag}_{n+1}\{A_d\},$$

$$\bar{\mathcal{A}} = \begin{bmatrix} A & 0 \\ 0 & \mathcal{A} - \mathcal{K}_1\mathcal{C} - \mathcal{K}_2(\mathcal{H}\Xi \otimes I)\mathcal{C} \end{bmatrix}, \quad \bar{\mathcal{B}} = \begin{bmatrix} 0 \\ -\mathcal{K}_2(\mathcal{H}\Xi \otimes I) \end{bmatrix},$$

$$\tilde{\mathcal{A}}_k = \begin{bmatrix} 0 & 0 \\ -\mathcal{K}_2(\mathcal{H}\Xi_k \otimes I)\mathcal{C}(\mathbf{1}_n \otimes I) & 0 \end{bmatrix}, \quad \tilde{\mathcal{B}}_k = \begin{bmatrix} 0 \\ -\mathcal{K}_2(\mathcal{H}\Xi_k \otimes I) \end{bmatrix},$$

$$\bar{\mathcal{D}} = \begin{bmatrix} B & 0 \\ \mathbf{1}_n \otimes B & -\mathcal{K}_1\mathcal{D} - \mathcal{K}_2(\mathcal{H}\Xi \otimes I)\mathcal{D} \end{bmatrix}, \quad \tilde{\mathcal{D}}_k = \begin{bmatrix} 0 & 0 \\ 0 & -\mathcal{K}_2(\mathcal{H}\Xi_k \otimes I)\mathcal{D} \end{bmatrix}.$$

Before proceeding further, we introduce the following definition and assumption.

Definition 9.1 *The augmented system (9.11) with $v_k = 0$ is said to be exponentially mean-square stable if there exist constants $\varepsilon > 0$ and $0 < \hbar < 1$ such that*

$$\mathbb{E}\{||\eta_k||^2\} \leq \varepsilon \hbar^k \max_{i \in [-\tau, \, 0]} \mathbb{E}\{||\tilde{w}_i||^2\}, \ k \in \mathbb{N}.$$

Assumption 9.1 *The matrices Π_i and Γ_i $(i = 1, 2, \cdots, s)$ in (9.5) have the following decomposition*

$$\Pi_i = \bar{\pi}_i \bar{\pi}_i^T = \begin{bmatrix} \pi_{1i} \\ \pi_{2i} \end{bmatrix} \begin{bmatrix} \pi_{1i} \\ \pi_{2i} \end{bmatrix}^T, \quad \Gamma_i = \bar{\theta}_i \bar{\theta}_i^T$$

where $\bar{\pi}_i$, π_{1i}, π_{2i} and $\bar{\theta}_i$ are known vectors with appropriate dimensions.

The purpose of this chapter is to design a set of state estimators of form (9.9) for the discrete-time stochastic nonlinear system (9.1) through sensor networks. More specifically, we are interested in looking for the parameters $K_{i,1}$ and $K_{i,2}$ $(i = 1, 2, \cdots, n)$ such that the following requirements are met simultaneously:

R1) The augmented system (9.11) with $\tilde{w}_k = 0$ is exponentially mean-square stable.

R2) Under the zero-initial condition, for a given disturbance attenuation level $\gamma > 0$ and all nonzero $\tilde{w}_k$, the estimation error $\tilde{z}_k$ satisfies

$$\frac{1}{n} \sum_{k=0}^{\infty} \mathbb{E}\{||\tilde{z}_k||^2\} \leq \gamma^2 \sum_{k=0}^{\infty} ||\tilde{w}_k||^2. \tag{9.12}$$

9.2 $\mathcal{H}_\infty$ Performance Analysis

In this section, by resorting to stochastic analysis techniques, we shall provide the analysis result of the $\mathcal{H}_\infty$ performance for the augmented system (9.11), and then proceed with the subsequent design stage of event-triggered estimators.

Theorem 9.1 *Let the estimator parameters $K_{i,1}$ and $K_{i,2}$ ($i = 1, 2, \cdots, n$) as well as a prescribed disturbance attenuation level $\gamma > 0$ be given. The dynamics of the estimation errors (9.11) is exponentially mean-square stable and also satisfies the prespecified $\mathcal{H}_\infty$ performance constraint (9.12) if there exist two positive definite matrices P, Q and a positive scalar λ satisfying*

$$\mathcal{R} = \begin{bmatrix} \mathcal{R}_{11} & \mathcal{R}_{12} & \mathcal{R}_{13} & \mathcal{R}_{14} \\ * & \mathcal{R}_{22} & \mathcal{R}_{23} & \mathcal{R}_{24} \\ * & * & \mathcal{R}_{33} & \mathcal{R}_{34} \\ * & * & * & \mathcal{R}_{44} \end{bmatrix} < 0 \qquad (9.13)$$

where

$$\Theta = diag\{\delta_1, \delta_2, \cdots, \delta_n\}, \quad \mathcal{H}_i := diag\{h_{i,1}, h_{i,2}, \cdots, h_{i,n}\},$$

$$\Psi = \bar{\alpha}(1 - \bar{\alpha}) \sum_{i=1}^n (\mathcal{H}_i^T \mathcal{H}_i) \otimes (K_{i,2}^T P K_{i,2}),$$

$$\mathbf{I} = [\, I \ 0 \ 0 \ \cdots \ 0 \,], \quad \tilde{\mathbf{I}} = [\, 0 \ I \ I \ \cdots \ I \,], \quad \tilde{\mathcal{D}} = [\, 0 \ \mathcal{D} \,],$$

$$\mathcal{R}_{11} = \bar{\mathcal{A}}^T (I \otimes P)\bar{\mathcal{A}} + \Upsilon_1 + \sum_{i=1}^s (n+1) tr[P\Pi_i] \mathbf{I}^T \Gamma_i \mathbf{I}$$

$$+ \lambda (\mathbf{1}_n \otimes \tilde{\mathbf{I}})^T \mathcal{C}^T (\Theta \otimes I)\mathcal{C}(\mathbf{1}_n \otimes \tilde{\mathbf{I}}) - (I \otimes P) + (I \otimes Q) + \frac{1}{n}\bar{\mathcal{L}}^T \bar{\mathcal{L}},$$

$$\mathcal{R}_{12} = \bar{\mathcal{A}}^T (I \otimes P)\bar{\mathcal{A}}_d, \quad \mathcal{R}_{13} = \bar{\mathcal{A}}^T (I \otimes P)\bar{\mathcal{B}},$$

$$\mathcal{R}_{14} = \bar{\mathcal{A}}^T (I \otimes P)\bar{\mathcal{D}} + \lambda (\mathbf{1}_n \otimes \mathbf{I})^T \mathcal{C}^T (\Theta \otimes I)\tilde{\mathcal{D}},$$

$$\mathcal{R}_{22} = \bar{\mathcal{A}}_d^T (I \otimes P)\bar{\mathcal{A}}_d - (I \otimes Q), \ \mathcal{R}_{23} = \bar{\mathcal{A}}_d^T (I \otimes P)\bar{\mathcal{B}}, \ \mathcal{R}_{24} = \bar{\mathcal{A}}_d^T (I \otimes P)\bar{\mathcal{D}},$$

$$\mathcal{R}_{33} = \bar{\mathcal{B}}^T (I \otimes P)\bar{\mathcal{B}} + \Psi - \lambda I, \quad \mathcal{R}_{34} = \bar{\mathcal{B}}^T (I \otimes P)\bar{\mathcal{D}},$$

$$\mathcal{R}_{44} = \bar{\mathcal{D}}^T (I \otimes P)\bar{\mathcal{D}} + \Upsilon_2 + \lambda \tilde{\mathcal{D}}^T (\Theta \otimes I)\tilde{\mathcal{D}} - \gamma^2 I,$$

$$\Upsilon_1 = \begin{bmatrix} (\mathbf{1}_n \otimes I)^T \mathcal{C}^T \Psi \mathcal{C}(\mathbf{1}_n \otimes I) & 0 \\ 0 & 0 \end{bmatrix}, \quad \Upsilon_2 = \begin{bmatrix} 0 & 0 \\ 0 & \mathcal{D}^T \Psi \mathcal{D} \end{bmatrix}.$$

Proof *First, noting the stochastic matrix Ξ_k, one has*

$$\mathbb{E}\{(\mathcal{H}\Xi_k \otimes I)^T \mathcal{K}_2^T (I \otimes P)\mathcal{K}_2(\mathcal{H}\Xi_k \otimes I)\}$$

$$= \bar{\alpha}(1 - \bar{\alpha}) \sum_{i=1}^n (\mathcal{H}_i \otimes I)^T (I \otimes K_{i,2})^T (I \otimes P)(I \otimes K_{i,2})(\mathcal{H}_i \otimes I) \qquad (9.14)$$

$$= \bar{\alpha}(1 - \bar{\alpha}) \sum_{i=1}^n (\mathcal{H}_i^T \mathcal{H}_i) \otimes (K_{i,2}^T P K_{i,2}).$$

Then, by employing the property of matrix trace, it follows from (9.4) and (9.5) that

$$
\begin{aligned}
&\mathbb{E}\{\mathscr{F}^T(x_k,\vartheta_k)(I\otimes P)\mathscr{F}(x_k,\vartheta_k)\}\\
&= \mathbb{E}\{(1_{n+1}\otimes f(x_k,\vartheta_k))^T(I\otimes P)(1_{n+1}\otimes f(x_k,\vartheta_k))\}\\
&= \mathbb{E}\{(1_{n+1}^T 1_{n+1})\otimes(f^T(x_k,\vartheta_k)Pf(x_k,\vartheta_k))\}\\
&= (1_{n+1}^T 1_{n+1})\otimes\mathbb{E}\{f^T(x_k,\vartheta_k)Pf(x_k,\vartheta_k)\}\\
&= (1_{n+1}^T 1_{n+1})\otimes\mathbb{E}\left\{tr[Pf(x_k,\vartheta_k)f^T(x_k,\vartheta_k)]\right\}\\
&= (1_{n+1}^T 1_{n+1})\otimes\mathbb{E}\left\{x_k^T\sum_{i=1}^{s}tr[P\Pi_i]\Gamma_i x_k\right\}\\
&= \mathbb{E}\left\{x_k^T\sum_{i=1}^{s}(n+1)tr[P\Pi_i]\Gamma_i x_k\right\}.
\end{aligned}
\tag{9.15}
$$

In what follows, choose the Lyapunov function for system (9.11):

$$
V_k = \eta_k^T(I\otimes P)\eta_k + \sum_{i=k-\tau}^{k-1}\eta_i^T(I\otimes Q)\eta_i.
\tag{9.16}
$$

Calculating the difference of V_k along the trajectory of system (9.11) with $\tilde{w}_k = 0$ and taking the mathematical expectation, one has

$$
\begin{aligned}
&\mathbb{E}\{\Delta V_k\}\\
&:= \mathbb{E}\left\{V_{k+1}-V_k\right\}\\
&= \mathbb{E}\left\{\left(\bar{A}\eta_k+\tilde{A}_k\eta_k+\mathscr{F}(x_k,\vartheta_k)+\bar{A}_d\eta_{k-\tau}+\bar{B}e_k+\tilde{B}_k e_k\right)^T\right.\\
&\qquad\times(I\otimes P)\left(\bar{A}\eta_k+\tilde{A}_k\eta_k+\mathscr{F}(x_k,\vartheta_k)+\bar{A}_d\eta_{k-\tau}+\bar{B}e_k+\tilde{B}_k e_k\right)\\
&\qquad\left. -\eta_k^T(I\otimes P)\eta_k+\sum_{i=k-\tau+1}^{k}\eta_i^T(I\otimes Q)\eta_i-\sum_{i=k-\tau}^{k-1}\eta_i^T(I\otimes Q)\eta_i\right\}\\
&= \mathbb{E}\left\{\eta_k^T\bar{A}^T(I\otimes P)\bar{A}\eta_k+2\eta_k^T\bar{A}^T(I\otimes P)\bar{A}_d\eta_{k-\tau}+2\eta_k^T\bar{A}^T(I\otimes P)\bar{B}e_k\right.\\
&\qquad+\eta_k^T\Upsilon_1\eta_k+\eta_k^T\sum_{i=1}^{s}(n+1)tr[P\Pi_i]\mathbf{I}^T\Gamma_i\mathbf{I}\eta_k+\eta_{k-\tau}^T\bar{A}_d^T(I\otimes P)\bar{A}_d\eta_{k-\tau}\\
&\qquad+2\eta_{k-\tau}^T\bar{A}_d^T(I\otimes P)\bar{B}e_k+e_k^T\bar{B}^T(I\otimes P)\bar{B}e_k+e_k^T\Psi e_k-\eta_k^T(I\otimes P)\eta_k\\
&\qquad\left. +\eta_k^T(I\otimes Q)\eta_k-\eta_{k-\tau}^T(I\otimes Q)\eta_{k-\tau}\right\}.
\end{aligned}
\tag{9.17}
$$

Furthermore, it follows from the event-triggering condition (9.7) that

$$
\lambda e_k^T e_k - \lambda\eta_k^T(1_n\otimes\tilde{\mathbf{I}})^T\mathcal{C}^T(\Theta\otimes I)\mathcal{C}(1_n\otimes\tilde{\mathbf{I}})\eta_k \le 0.
\tag{9.18}
$$

Taking the above inequality into account, we have

$$
\begin{aligned}
\mathbb{E}\{\Delta V_k\} \leq \mathbb{E}\Big\{ &\eta_k^T \bar{\mathcal{A}}^T (I \otimes P) \bar{\mathcal{A}} \eta_k + 2\eta_k^T \bar{\mathcal{A}}^T (I \otimes P) \bar{\mathcal{A}}_d \eta_{k-\tau} \\
&+ 2\eta_k^T \bar{\mathcal{A}}^T (I \otimes P) \bar{\mathcal{B}} e_k + \eta_k^T \Upsilon_1 \eta_k \\
&+ \eta_k^T \sum_{i=1}^{s} (n+1) tr[P\Pi_i] \mathbf{I}^T \Gamma_i \mathbf{I} \eta_k + \eta_{k-\tau}^T \bar{\mathcal{A}}_d^T (I \otimes P) \bar{\mathcal{A}}_d \eta_{k-\tau} \\
&+ 2\eta_{k-\tau}^T \bar{\mathcal{A}}_d^T (I \otimes P) \bar{\mathcal{B}} e_k + e_k^T \bar{\mathcal{B}}^T (I \otimes P) \bar{\mathcal{B}} e_k + e_k^T \Psi e_k \\
&- \lambda e_k^T e_k + \lambda \eta_k^T (\mathbf{1}_n \otimes \tilde{\mathbf{I}})^T \mathcal{C}^T (\Theta \otimes I) \mathcal{C} (\mathbf{1}_n \otimes \tilde{\mathbf{I}}) \eta_k \\
&- \eta_k^T (I \otimes P) \eta_k + \eta_k^T (I \otimes Q) \eta_k - \eta_{k-\tau}^T (I \otimes Q) \eta_{k-\tau} \Big\}, \quad (9.19)
\end{aligned}
$$

which results in

$$
\mathbb{E}\{\Delta V_k\} \leq \mathbb{E}\big\{ \bar{\eta}_k^T \tilde{\mathcal{R}} \bar{\eta}_k \big\} \tag{9.20}
$$

where $\bar{\eta}_k = [\; \eta_k^T \;\; \eta_{k-\tau}^T \;\; e_k^T \;]^T$ and

$$
\tilde{\mathcal{R}} = \begin{bmatrix} \mathcal{R}_{11} - \frac{1}{n} \bar{\mathcal{L}}^T \bar{\mathcal{L}} & \mathcal{R}_{12} & \mathcal{R}_{13} \\ * & \mathcal{R}_{22} & \mathcal{R}_{23} \\ * & * & \mathcal{R}_{33} \end{bmatrix}.
$$

By considering (9.13), one has $\tilde{\mathcal{R}} < 0$ and, subsequently,

$$
\mathbb{E}\{\|\eta_k\|^2\} \leq -\lambda_{\min}(-\tilde{\mathcal{R}})\|\bar{\eta}_k\|^2. \tag{9.21}
$$

Finally, along the similar line of the proof of Theorem 1 in [26], one can prove that the augmented system (9.11) is exponentially mean-square stable.

To establish the $\mathcal{H}_\infty$ performance, we introduce the following:

$$
\begin{aligned}
\mathbb{E}\Big\{ &\Delta V_k + \frac{1}{n} \|\tilde{z}_k\|^2 - \gamma^2 \|\tilde{w}_k\|^2 \Big\} \\
\leq \mathbb{E}\Big\{ &\bar{\eta}_k^T \Omega \bar{\eta}_k + 2\eta_k^T \bar{\mathcal{A}}^T (I \otimes P) \bar{\mathcal{D}} \tilde{w}_k + 2\eta_{k-\tau}^T \bar{\mathcal{A}}_d^T (I \otimes P) \bar{\mathcal{D}} \tilde{w}_k \\
&+ 2 e_k^T \bar{\mathcal{B}}^T (I \otimes P) \bar{\mathcal{D}} \tilde{w}_k + \tilde{w}_k^T \bar{\mathcal{D}}^T (I \otimes P) \bar{\mathcal{D}} \tilde{w}_k \\
&+ \tilde{w}_k^T \Upsilon_2 \tilde{w}_k + \frac{1}{n} \eta_k^T \bar{\mathcal{L}}^T \bar{\mathcal{L}} \eta_k - \gamma^2 \tilde{w}_k^T \tilde{w}_k \\
&+ 2\lambda \eta_k^T (\mathbf{1}_n \otimes \tilde{\mathbf{I}})^T \mathcal{C}^T (\Theta \otimes I) \mathcal{D} v_k + \lambda v_k^T \mathcal{D}^T (\Theta \otimes I) \mathcal{D} v_k \Big\},
\end{aligned} \tag{9.22}
$$

which leads to

$$
\mathbb{E}\Big\{ \Delta V_k + \frac{1}{n} \|\tilde{z}_k\|^2 - \gamma^2 \|\tilde{w}_k\|^2 \Big\} \leq \mathbb{E}\{ \tilde{\eta}_k^T \mathcal{R} \tilde{\eta}_k \} < 0. \tag{9.23}
$$

It follows from the zero initial conditions and (9.23) that

$$
\frac{1}{n} \sum_{k=0}^{\infty} \mathbb{E}\{\|\tilde{z}_k\|^2\} \leq \gamma^2 \sum_{k=0}^{\infty} \|\tilde{w}_k\|^2
$$

and therefore the proof is now complete.

9.3 $\mathcal{H}_\infty$ Estimator Design

Having established the analysis results, we are now ready to handle the distributed estimator design problem with an event-trigged communication mechanism. For this purpose, we firstly need to deal with the trace operation in Theorem 9.1, and then establish a sufficient condition for the existence of the desired $\mathcal{H}_\infty$ estimator.

Theorem 9.2 *Let the estimator parameters $K_{i,1}$ and $K_{i,2}$ $(i = 1, 2, \cdots, n)$ as well as a prescribed disturbance attenuation level $\gamma > 0$ be given. The dynamics of estimation errors (9.11) is exponentially mean-square stable and also satisfies the prespecified $\mathcal{H}_\infty$ performance constraint (9.12) if there exist two positive definite matrices P and Q, and positive scalars λ and ϖ_i $(i = 1, 2, \cdots, s)$ satisfying*

$$\begin{bmatrix} -\varpi_i & \bar{\pi}^T P \\ * & -P \end{bmatrix} < 0, \quad i = 1, 2, \cdots, s, \tag{9.24}$$

$$\begin{bmatrix} \bar{\mathcal{R}} & \bar{\mathcal{R}}_5^T & \mathcal{W}^T \mathcal{Y}^T (I \otimes P) \\ * & -I \otimes P & 0 \\ * & * & -I \otimes P \end{bmatrix} < 0 \tag{9.25}$$

where

$$\bar{\mathcal{R}} = \begin{bmatrix} \bar{\mathcal{R}}_{11} & 0 & 0 & \bar{\mathcal{R}}_{14} \\ * & \bar{\mathcal{R}}_{22} & 0 & 0 \\ * & * & \bar{\mathcal{R}}_{33} & 0 \\ * & * & * & \bar{\mathcal{R}}_{44} \end{bmatrix},$$

$$\bar{\mathcal{R}}_5 = (I \otimes P) \begin{bmatrix} \bar{A} & \bar{A}_d & \bar{B} & \bar{D} \end{bmatrix}, \quad \mathcal{Y} = \begin{bmatrix} \mathcal{Y}_1 & \mathcal{Y}_2 & \cdots & \mathcal{Y}_n \end{bmatrix},$$

$$\mathcal{W} = \sqrt{\bar{\alpha}(1 - \bar{\alpha})} \, diag\Big\{ diag\{\mathcal{C}(\mathbf{1}_n \otimes I), 0\}, \, 0, \, I, \, diag\{0, \mathcal{D}\}\Big\},$$

$$\mathcal{Y}_i = diag\Big\{ diag\{\mathcal{H}_i \otimes K_{i,2}, 0\}, \, 0, \, \mathcal{H}_i \otimes K_{i,2}, \, diag\{0, \mathcal{H}_i \otimes K_{i,2}\}\Big\},$$

$$\bar{\mathcal{R}}_{11} = \lambda (\mathbf{1}_n \otimes \tilde{\mathbf{I}})^T \mathcal{C}^T (\Theta \otimes I) \mathcal{C} (\mathbf{1}_n \otimes \tilde{\mathbf{I}})$$

$$+ (n + 1) \sum_{i=1}^{s} \varpi_i \mathbf{I}^T \Gamma_i \mathbf{I} - I \otimes P + I \otimes Q + \frac{1}{n} \bar{\mathcal{L}}^T \bar{\mathcal{L}},$$

$$\bar{\mathcal{R}}_{14} = \lambda (\mathbf{1}_n \otimes \tilde{\mathbf{I}})^T \mathcal{C}^T (\Theta \otimes I) \tilde{\mathcal{D}}, \quad \bar{\mathcal{R}}_{22} = -I \otimes Q,$$

$$\bar{\mathcal{R}}_{33} = -\lambda I, \quad \bar{\mathcal{R}}_{44} = \lambda \tilde{\mathcal{D}}^T (\Theta \otimes I) \tilde{\mathcal{D}} - \gamma^2 I.$$

Proof *First, it is not difficult to see that (9.13) is equivalent to*

$$\mathcal{S} + (n + 1) \sum_{i=1}^{s} diag\Big\{ tr[P\Pi_i] \mathbf{I}^T \Gamma_i \mathbf{I} - \varpi_i, \, 0, \, 0, \, 0\Big\} < 0 \tag{9.26}$$

where $\mathcal{S}_{11} = \mathcal{R}_{11} - \sum_{i=1}^{s}(n+1)tr\big[P\Pi_i\big]\mathbf{I}^T\Gamma_i\mathbf{I} + (n+1)\sum_{i=1}^{s}\varpi_i$ and

$$\mathcal{S} = \begin{bmatrix} \mathcal{S}_{11} & \mathcal{R}_{12} & \mathcal{R}_{13} & \mathcal{R}_{14} \\ * & \mathcal{R}_{22} & \mathcal{R}_{23} & \mathcal{R}_{24} \\ * & * & \mathcal{R}_{33} & \mathcal{R}_{34} \\ * & * & * & \mathcal{R}_{44} \end{bmatrix}.$$

On the other hand, in light of the Schur complement lemma, (9.24) is equivalent to

$$\bar{\pi}_i^T P \bar{\pi}_i < \varpi_i, \quad (\varpi_i = 1, 2, \cdots, s),$$

which, by using the property of matrix trace, can be rewritten as

$$tr\big[P\Pi_i\big] < \varpi_i, \quad (\varpi_i = 1, 2, \cdots, s).$$

Therefore, if $\mathcal{S} < 0$, one has that (9.13) is true. Furthermore, by using the Schur complement lemma again, it follows that (9.25) is equivalent to $\mathcal{S} < 0$. Finally, according to Theorem 9.1, the design requirements R1) and R2) are simultaneously satisfied. The proof is complete.

Finally, by utilizing variable substitution, we have the following theorem whose proof is omitted to save space.

Theorem 9.3 *Let the disturbance attenuation level $\gamma > 0$ be given. Assume that there exist two positive definite matrices P and Q, matrices $\tilde{K}_{i,1}$ and $\tilde{K}_{i,2}$ $(i = 1, 2, \cdots, s)$, and positive scalars λ and ϖ_i $(i = 1, 2, \cdots, s)$ satisfying the following linear matrix inequalities*

$$\begin{bmatrix} -\varpi_i & \bar{\pi}^T P \\ * & -P \end{bmatrix} < 0, \quad i = 1, 2, \cdots, s, \tag{9.27}$$

$$\begin{bmatrix} \bar{\mathcal{R}} & \mathcal{Z}^T & \mathcal{W}^T \tilde{y}^T \\ * & -I \otimes P & 0 \\ * & * & -I \otimes P \end{bmatrix} < 0 \tag{9.28}$$

where

$$\mathcal{Z} = \begin{bmatrix} \mathcal{Z}_1 & \mathcal{Z}_2 & \mathcal{Z}_3 & \mathcal{Z}_4 \end{bmatrix}, \quad \tilde{y} = \begin{bmatrix} \tilde{y}_1 & \tilde{y}_2 & \cdots & \tilde{y}_n \end{bmatrix},$$

$$\tilde{y}_i = diag\Big\{ diag\{\mathcal{H}_i \otimes \tilde{K}_{i,2}, 0\}, 0, \mathcal{H}_i \otimes \tilde{K}_{i,2}, diag\{0, \mathcal{H}_i \otimes \tilde{K}_{i,2}\} \Big\},$$

$$\mathcal{Z}_1 = \begin{bmatrix} PA & 0 \\ 0 & (I \otimes P)\mathcal{A} - \tilde{\mathcal{K}}_1\mathcal{C} - \tilde{\mathcal{K}}_2(\mathcal{H}\Xi \otimes I)\mathcal{C} \end{bmatrix}, \quad \mathcal{Z}_2 = diag_{n+1}\{PA_d\},$$

$$\mathcal{Z}_3 = \begin{bmatrix} 0 \\ -\tilde{\mathcal{K}}_2(\mathcal{H}\Xi \otimes I) \end{bmatrix}, \quad \mathcal{Z}_4 = \begin{bmatrix} PB & 0 \\ 1_n \otimes (PB) & -\tilde{\mathcal{K}}_1\mathcal{D} - \tilde{\mathcal{K}}_2(\mathcal{H}\Xi \otimes I)\mathcal{D} \end{bmatrix}.$$

In this case, with the estimator gain matrices given by $K_{i,1} = P^{-1}\tilde{K}_{i,1}$ and $K_{i,2} = P^{-1}\tilde{K}_{i,2}$ $(i = 1, 2, \cdots, s)$, the dynamics of estimation errors (9.11) is exponentially mean-square stable while achieving the prespecified $\mathcal{H}_\infty$ performance constraint (9.12).

Remark 9.3 *In this chapter, a novel distributed estimator is first proposed in order to properly fuse two classes of information (i.e. the innovation for the node itself without the event-triggering mechanism and the innovation for neighboring nodes subject to the event-triggering mechanism). It can be seen that, in the main results in Theorems 9.1-9.3, the information about the given topology, the probability of packet dropouts and the threshold of event-triggering conditions on the estimation performance are all involved. The main technical contributions lie in that 1) a reasonable model is established to describe the event-triggered communication mechanism and the network-induced phenomena in an unified framework; and 2) the gains of proposed distributed estimators are obtained by solving a set of linear matrix inequalities reflecting both the threshold and the desired $\mathcal{H}_\infty$ performance.*

9.4 Simulation Examples

In this section, a simulation example is presented to illustrate the effectiveness of the proposed design scheme of distributed $\mathcal{H}_\infty$ estimators for discrete-time stochastic nonlinear systems with both event-triggered communication protocol and packet dropouts through sensor networks.

The considered target plant and sensor dynamics are, respectively, modeled by (9.1) and (9.2) with the following parameters:

$$A = \begin{bmatrix} 0.7 & 0.3 \\ 0.3 & -0.6 \end{bmatrix}, \quad A_d = \begin{bmatrix} 0.02 & 0 \\ 0 & 0.15 \end{bmatrix}, \quad B = \begin{bmatrix} 0.20 \\ -0.3 \end{bmatrix},$$

$$C_1 = \begin{bmatrix} -0.30 & 0.10 \end{bmatrix}, \quad C_2 = \begin{bmatrix} -0.25 & 0.12 \end{bmatrix}, \quad C_3 = \begin{bmatrix} -0.30 & 0.10 \end{bmatrix},$$

$$C_4 = \begin{bmatrix} -0.32 & 0.12 \end{bmatrix}, \quad C_5 = \begin{bmatrix} -0.25 & 0.09 \end{bmatrix}, \quad D_1 = 0.02,$$

$$D_2 = D_3 = 0.02, \quad D_4 = D_5 = 0.01, \quad L = \begin{bmatrix} 0.20 & 0.30 \end{bmatrix}$$

and the stochastic nonlinear function $f(x_k, \vartheta_k)$ is chosen as

$$f(x_k, \vartheta_k) = \left(0.1\mathrm{sign}(x_k^1)x_k^1\vartheta_k^1 + 0.13\mathrm{sign}(x_k^2)x_k^2\vartheta_k^2 \right) \begin{bmatrix} 0.06 \\ 0.09 \end{bmatrix}$$

where x_k^i ($i = 1, 2$) denotes the i-th element of the system state, and ϑ_k^1 and ϑ_k^2 are zero mean, uncorrelated Gaussian white noise sequences with unity covariance. It is not difficult to verify that the above stochastic nonlinear function satisfies

$$\mathbb{E}\left\{ f(x_k, \vartheta_k) | x_k \right\} = 0,$$

$$\mathbb{E}\left\{ f(x_k, \vartheta_k) f^T(x_k, \vartheta_k) | x_k \right\} = \begin{bmatrix} 0.06 \\ 0.09 \end{bmatrix} \begin{bmatrix} 0.06 \\ 0.09 \end{bmatrix}^T x_k^T \begin{bmatrix} 0.01 & 0 \\ 0 & 0.0169 \end{bmatrix} x_k.$$

$$\tag{9.29}$$

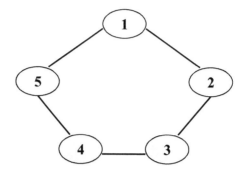

FIGURE 9.1
Topological structure of the sensor network.

The sensor network shown in Fig. 9.1 is represented by a graph $\mathcal{G} = (\mathcal{V}, \mathcal{E}, \mathcal{H})$ with the set of nodes $\mathcal{V} = \{1, 2, 3, 4, 5\}$, the set of edges $\mathcal{E} = \{(1, 2), (1, 5), (2, 1), (2, 3), (3, 2), (3, 4), (4, 3), (4, 5), (5, 1), (5, 4)\}$ and the following adjacency matrix

$$\mathcal{H} = \begin{bmatrix} 0 & 0.20 & 0 & 0 & 0.12 \\ 0.12 & 0 & 0.20 & 0 & 0 \\ 0 & 0.12 & 0 & 0.20 & 0 \\ 0 & 0 & 0.20 & 0 & 0.20 \\ 0.20 & 0 & 0 & 0.12 & 0 \end{bmatrix}.$$

The $\mathcal{H}_\infty$ performance level γ, the threshold δ_i ($i = 1, 2, \cdots, 5$), the time-delay τ, and the probability $\bar{\alpha}$ are taken as 0.98, 0.04, 3, and 0.95, respectively. Using the MATLAB® software (with SeDuMi 1.32), a set of solutions of linear matrix inequalities (9.27)-(9.28) in Theorem 9.3 is obtained as follows:

$$P = \begin{bmatrix} 0.4085 & 0.0277 \\ 0.0277 & 0.4134 \end{bmatrix}, \quad Q = \begin{bmatrix} 0.1023 & 0.0002 \\ 0.0002 & 0.1012 \end{bmatrix},$$

$$\tilde{K}_{1,1} = \begin{bmatrix} -0.7493 \\ -0.6945 \end{bmatrix}, \quad \tilde{K}_{1,2} = \begin{bmatrix} 0.0009 \\ 0.0007 \end{bmatrix},$$

$$\tilde{K}_{2,1} = \begin{bmatrix} -0.7628 \\ -0.8572 \end{bmatrix}, \quad \tilde{K}_{2,2} = \begin{bmatrix} 0.0003 \\ 0.0009 \end{bmatrix},$$

$$\tilde{K}_{3,1} = \begin{bmatrix} -0.7500 \\ -0.6953 \end{bmatrix}, \quad \tilde{K}_{3,2} = \begin{bmatrix} 0.0007 \\ 0.0004 \end{bmatrix},$$

$$\tilde{K}_{4,1} = \begin{bmatrix} -0.6724 \\ -0.6594 \end{bmatrix}, \quad \tilde{K}_{4,2} = \begin{bmatrix} 0.0006 \\ -0.0001 \end{bmatrix},$$

$$\tilde{K}_{5,1} = \begin{bmatrix} -0.8752 & -0.8412 \end{bmatrix}^T, \quad \lambda = 0.3924,$$

$$\tilde{K}_{5,2} = \begin{bmatrix} 0.0009 & 0.0001 \end{bmatrix}^T, \quad \varpi = 0.2134.$$

Furthermore, the desired estimator parameters are

$$K_{1,1} = \begin{bmatrix} -1.7279 \\ -1.5641 \end{bmatrix}, \quad K_{1,2} = \begin{bmatrix} 0.0021 \\ 0.0016 \end{bmatrix},$$

$$K_{2,1} = \begin{bmatrix} -1.7342 \\ -1.9574 \end{bmatrix}, \quad K_{2,2} = \begin{bmatrix} -0.0008 \\ 0.0022 \end{bmatrix},$$

$$K_{3,1} = \begin{bmatrix} -1.7295 \\ -1.5659 \end{bmatrix}, \quad K_{3,2} = \begin{bmatrix} 0.0018 \\ 0.0008 \end{bmatrix},$$

$$K_{4,1} = \begin{bmatrix} -1.5447 \\ -1.4916 \end{bmatrix}, \quad K_{4,2} = \begin{bmatrix} 0.0015 \\ -0.0001 \end{bmatrix},$$

$$K_{5,1} = \begin{bmatrix} -2.0133 & -1.8999 \end{bmatrix}, \quad K_{5,2} = \begin{bmatrix} 0.0022 & 0.0001 \end{bmatrix}^T.$$

In the simulation, the exogenous disturbance inputs are selected as

$$w_k = \frac{0.2\sin(0.2k)}{0.2k+1}, \quad v_k = 0.2\cos(0.25k)\exp(-0.1k).$$

The initial conditions are set as $x_{-3} = [-0.10\ 0.15]^T$, $x_{-2} = [0.20\ -0.27]^T$, $x_{-1} = [0.125\ -0.17]^T$, $x_0 = [0.25\ -0.55]^T$ and $\hat{x}_{i,k} = [0\ 0]^T$ ($k = -3,-2,-1,0$, $i = 1,2,\cdots,5$). Simulation results are shown in Figs. 9.2-9.3, where Fig. 9.2 depicts the estimation errors $\tilde{z}_{i,k}$ as well as the event-triggered times, and Fig. 9.3 plots the trajectories for the states and the estimates. The simulation results show that estimators have a satisfactory estimation performance, which confirms that the distributed estimation scheme presented in this chapter is indeed effective.

9.5 Summary

In this chapter, we have dealt with the event-triggered distributed $\mathcal{H}_\infty$ state estimation problem for a class of discrete-time stochastic nonlinear systems through sensor networks. To reduce the network burden and the energy consumption, we have considered the event-triggered communication mechanism, where the innovation on each sensor has been transmitted only when a certain triggering condition has been violated. By employing the Lyapunov stability theorem, some sufficient conditions have been established to ensure that the dynamics of the estimation error satisfies the desired $\mathcal{H}_\infty$ performance constraint. Finally, an illustrative example has been provided to confirm the usefulness of the developed state estimation approach.

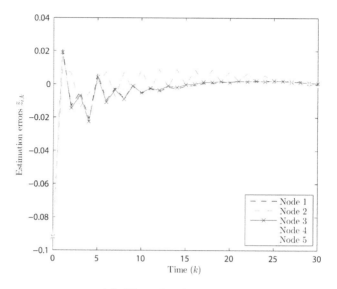

(a) The estimation errors.

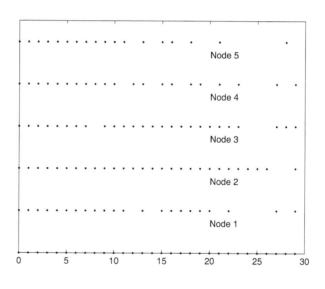

(b) The event-triggered times.

FIGURE 9.2
The estimation errors and event-triggered times.

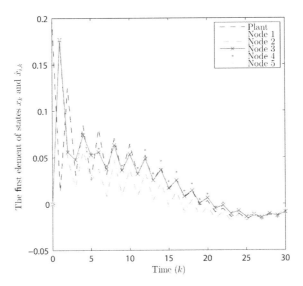

(a) The first element of dynamic trajectories.

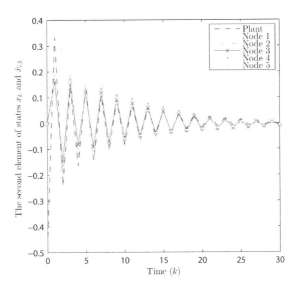

(b) The second element of dynamic trajectories.

FIGURE 9.3
The dynamic trajectories.

10

Event-Triggered Consensus Control for Multi-Agent Systems in the Framework of Input-to-State Stability in Probability

By resorting to the developed backward RDE technology, the $\mathcal{H}_\infty$ consensus control has been discussed in Chapter 3 for discrete time-varying multi-agent systems with missing measurements and parameter uncertainties. The effect on consensus from missing measurements has been characterized by the solvability of the derived RDEs. When the event triggering is the concern, the impact from sparse data should be thoroughly investigated in theory with the intent to achieve a balance between consensus performance and network service quality. Up to now, such an impact has been thoroughly discussed in Chapter 9 for the distributed filtering problem over sensor networks. On the other hand, almost all research focus has been on the steady-state behaviors of the consensus, i.e., the consensus performance when time tends to infinity. Unfortunately, transient consensus performance at specific times has been largely overlooked despite its significance in reflecting the consensus dynamics. As such, this chapter will deal with event-triggered consensus control for discrete-time stochastic multi-agent systems with state-dependent noises. The challenges we are going to cope with are identified as follows: 1) how can we establish a general analysis framework to analyze the dynamic behaviors in probability of the multi-agent systems, and 2) how can we design an appropriate threshold of the triggering condition based on the measurement outputs?

For this concern, a novel definition of *consensus in probability* is proposed to better describe the consensus dynamics of the addressed stochastic multi-agent systems. The measurement output available for the controller is not only from the individual agent but also from its neighboring ones according to the given topology. An event-triggered mechanism is adopted to attempt to reduce the communication burden, where the control input on each agent is updated only when a certain triggering condition is violated. For the addressed problem, first of all, a theoretical framework is established for analyzing the so-called *input-to-state stability in probability* (ISSiP) for general discrete-time nonlinear stochastic systems. Within such a theoretical framework, some sufficient conditions of the event-triggered control protocol are derived under which the consensus in probability is reached. Furthermore, both the controller parameter and the triggering threshold are obtained in terms of the solution to certain matrix inequalities involving the topology information and the desired consensus probability. Finally, a simulation example is utilized to illustrate the usefulness of the proposed control protocol.

10.1   Modeling and Problem Formulation

In this chapter, the multi-agent system has N agents which communicate with each other according to a fixed network topology represented by an undirected graph $\mathscr{G} = (\mathscr{V}, \mathscr{E}, \mathscr{H})$ of order N with the set of agents $\mathscr{V} = \{1, 2, \cdots, N\}$, the set of edges $\mathscr{E} \in \mathscr{V} \times \mathscr{V}$, and the weighted adjacency matrix $\mathscr{H} = [h_{ij}]$ with nonnegative adjacency element h_{ij}. If $(i, j) \in \mathscr{E}$, then $h_{ij} > 0$, else $h_{ij} = 0$. An edge of $\mathscr{G}$ is denoted by the ordered pair (i, j). The adjacency elements associated with the edges of the graph are positive, i.e., $h_{ij} > 0 \Longleftrightarrow (i, j) \in \mathscr{E}$, which means that agent i can obtain information from agent j. Furthermore, self-edges (i, i) are not allowed, i.e. $(i, i) \notin \mathscr{E}$ for any $i \in \mathscr{V}$. The neighborhood of agent i is denoted by $\mathscr{N}_i = \{j \in \mathscr{V} : (j, i) \in \mathscr{E}\}$. An element of $\mathscr{N}_i$ is called a neighbor of agent i. The in-degree of agent i is defined as $\deg_{in}^i = \sum_{j \in \mathscr{N}_i} h_{ij}$.

Consider a discrete-time stochastic multi-agent system consisting of N agents, in which the dynamics of agent i is described by

$$\begin{cases} x_i(k+1) = Ax_i(k) + Dx_i(k)w(k) + Bu_i(k), \\ \quad y_i(k) = Cx_i(k) + Ex_i(k)v(k) \end{cases} \tag{10.1}$$

where $x_i(k) \in \mathbb{R}^n$, $y_i(k) \in \mathbb{R}^q$ and $u_i(k) \in \mathbb{R}^p$ are, respectively, the states, the measurement outputs, and the control inputs of agent i. $w(k)$, $v(k) \in \mathbb{R}$ are zero-mean random sequences with $\mathbb{E}\{w^2(k)\} = \mathbb{E}\{v^2(k)\} = 1$ and $\mathbb{E}\{w(k)v(k)\} = \sigma^2$. A, B, C, D and E are known constant matrices with compatible dimensions. It is assumed that the rank of B is p.

A control protocol with the following form is adopted:

$$u_i(k) = K \sum_{j \in \mathscr{N}_i} h_{ij} \Big(y_j(k) - y_i(k) \Big) := K\phi_i(k) \tag{10.2}$$

where $K \in \mathbb{R}^{n \times q}$ is a feedback gain matrix to be determined and $\phi_i(k)$ is the input signal of the controller, which can be employed to describe the disagreement between agents' measurements.

In order to characterize the event-triggering mechanism, let the triggering time sequence of agent i be $0 \leq k_0^i < k_1^i < k_2^i < \cdots < k_s^i < \cdots$ and the event generator functions $\Upsilon_i(\cdot, \cdot) : \mathbb{R}^q \times \mathbb{R} \to \mathbb{R}$ $(i = 1, 2, \cdots, N)$ be selected as follows:

$$\Upsilon_i(e_i(k), \theta_i) = e_i^T(k)e_i(k) - \theta_i. \tag{10.3}$$

Here, $e_i(k) := \phi_i^u(k_s^i) - \phi_i(k)$ $(k > k_s^i)$ where $\phi_i^u(k_s^i)$ is the control input signal at the latest triggering time k_s^i and θ_i is a positive scalar *to be designed*. The executions are triggered as long as the condition

$$\Upsilon_i(e_i(k), \theta_i) > 0 \tag{10.4}$$

is satisfied. Obviously, the sequence of event-triggering instants is determined iteratively by

$$k_{s+1}^i = \inf\{k \in \mathbb{N} | k > k_s^i, \ \Upsilon_i(e_i(k), \theta_i) > 0\}.$$

In the event-triggered setup, the control law can be further described by

$$u_i(k) = \begin{cases} 0, & k \in [0, k_0^i), \\ K\phi_i^u(k), & k \in [k_0^i, \infty) \end{cases} \tag{10.5}$$

where $\phi_i^u(k) = 0$ for $k \in [0, k_0^i)$ and $\phi_i^u(k) = \phi_i(k_s^i)$ for $k \in [k_s^i, k_{s+1}^i)$. It is easy to see that the controller input remains a constant in the execution interval $[k_s, k_{s+1})$. Furthermore, such a control law can be rewritten as

$$u_i(k) = K(\phi_i(k) + e_i(k)), \quad \|e_i(k)\|^2 \le \theta_i.$$

The closed-loop system can be obtained as follows

$$\begin{aligned} x_{k+1} &= (I \otimes A + \mathcal{H} \otimes (BKC))x_k \\ &+ I \otimes Dx_k w_k + \mathcal{H} \otimes (BKE)x_k v_k + I \otimes (BK)e_k \end{aligned} \tag{10.6}$$

where

$$\begin{aligned} x_k &= [\ x_1^T(k) \quad x_2^T(k) \quad \cdots \quad x_N^T(k)\ ]^T, \\ e_k &= [\ e_1^T(k) \quad e_2^T(k) \quad \cdots \quad e_N^T(k)\ ]^T, \\ \mathcal{H} &= \begin{bmatrix} -\deg_{in}^1 & h_{1,2} & h_{1,3} & \cdots & h_{1,N} \\ h_{2,1} & -\deg_{in}^2 & h_{2,3} & \cdots & h_{2,N} \\ \vdots & \vdots & \vdots & \ddots & \vdots \\ h_{N,1} & h_{N,2} & h_{N,3} & \cdots & -\deg_{in}^N \end{bmatrix}. \end{aligned}$$

On the other hand, denote the average state of all agents by

$$\bar{x}_k = \frac{1}{N} \sum_{i=1}^N x_i(k) = \frac{1}{N}(\mathbf{1}^T \otimes I)x_k. \tag{10.7}$$

Then, taking $\mathbf{1}^T \mathcal{H} = 0$ into consideration, one has

$$\begin{aligned} \bar{x}_{k+1} &= \frac{1}{N}(\mathbf{1}^T \otimes I)x_{k+1} \\ &= \frac{1}{N}(\mathbf{1}^T \otimes I)\Big\{ (I \otimes A + \mathcal{H} \otimes (BKC))x_k \\ &\quad + I \otimes Dx_k w_k + \mathcal{H} \otimes (BKE)x_k v_k + I \otimes (BK)e_k \Big\} \\ &= A\bar{x}_k + D\bar{x}_k w_k + \frac{1}{N}(\mathbf{1}^T \otimes (BK))e_k. \end{aligned} \tag{10.8}$$

Furthermore, in order to look at the deviation of each state from the average state, we denote $\tilde{x}_i(k) := x_i(k) - \bar{x}_k$ and $\tilde{x}_k := [\tilde{x}_1^T(k) \ \tilde{x}_2^T(k) \ \cdots \ \tilde{x}_N^T(k)]$. Moreover, subtracting (10.8) from (10.6) leads to

$$
\begin{aligned}
\tilde{x}_{k+1} = {} & \left(I \otimes A + \mathcal{H} \otimes (BKC)\right)\tilde{x}_k \\
& + I \otimes D\tilde{x}_k w_k + \mathcal{H} \otimes (BKE)\tilde{x}_k v_k + \mathcal{N} \otimes (BK)e_k
\end{aligned}
\tag{10.9}
$$

where $\mathcal{N} = [a_{ij}]_{N \times N}$ with

$$
a_{ij} = \left\{ \begin{array}{ll} \frac{N-1}{N}, & i = j, \\ -\frac{1}{N}, & i \neq j. \end{array} \right.
$$

Similar to the method in [147], one can select $\varphi_i \in \mathbb{R}^n$ with $\varphi_i^T \mathcal{H} = \lambda_i \varphi_i^T$ $(i = 2, 3, \cdots, N)$ to form the unitary matrix $M = [1/\sqrt{N} \ \varphi_2 \ \cdots \ \varphi_N]$ and then transform $\mathcal{H}$ into a diagonal form: $\mathrm{diag}\{0, \lambda_2, \cdots, \lambda_N\} = M^T \mathcal{H} M$. It can be found that φ_i and λ_i are the ith eigenvalue and its corresponding eigenvector of the Laplacian matrix $\mathcal{H}$. Furthermore, we assume that eigenvalues of $\mathcal{H}$ in descending order are written as $0 = \lambda_1 > \lambda_2 \geq \lambda_3 \geq \cdots \geq \lambda_N$.

Define $\tilde{\eta}_k := (M \otimes I)^T \tilde{x}_k \in \mathbb{R}^{nN}$ and partition it into two parts, i.e., $\tilde{\eta}_k = [\eta_k^T \ \xi_k^T]^T$, where $\eta_k \in \mathbb{R}^n$ is a vector consisting of the first n elements of $\tilde{\eta}_k$. Due to $(\mathcal{H} \otimes (BKE))(\mathbf{1} \otimes \bar{x}_k) = 0$, it can be easily shown that

$$
\left\{ \begin{array}{ll} \eta_{k+1} = \dfrac{1}{\sqrt{N}} \displaystyle\sum_{i=1}^{N} \tilde{x}_i(k+1) = 0, & (10.10a) \\[2mm] \xi_{k+1} = (\mathcal{A} + \Xi \otimes (BKC))\xi_k + \mathcal{D}\xi_k w_k & \\ \qquad + \Xi \otimes (BKE)\xi_k v_k + (\mathcal{M}\mathcal{N}) \otimes (BK)e_k & (10.10b) \end{array} \right.
$$

where

$$
\begin{aligned}
\mathcal{A} &= I_{N-1} \otimes A, \quad \Xi = \mathrm{diag}\{\lambda_2, \lambda_3, \cdots, \lambda_N\}, \\
\mathcal{D} &= I_{N-1} \otimes D, \quad \mathcal{M} = [\ \varphi_2 \ \ \varphi_3 \ \ \cdots \ \ \varphi_N \ ]^T.
\end{aligned}
$$

Before proceeding further, we introduce the following definition.

Definition 10.1 *Let the undirected communication graph $\mathscr{G}$, the parameter $\varepsilon > 0$, and the bounded function $U(\|\tilde{x}_0\|)$ be given. The discrete-time stochastic multi-agent system (10.1) with the event-triggered control protocol (10.5) is said to reach the consensus with probability $1 - \varepsilon$ if the system dynamics, for $\forall \tilde{x}_0 \in \mathbb{R}^{nN} \backslash \{0\}$, satisfies*

$$
Prob\big\{\|x_i(k) - x_j(k)\| < U(\|\tilde{x}_0\|)\big\} \geq 1 - \varepsilon, \ \forall i, j \in \mathscr{V}, \ \forall k \geq 0. \tag{10.11}
$$

Definition 10.2 *Let the undirected communication graph $\mathscr{G}$, the parameter $\varepsilon > 0$, and the bounded function $U(\|\tilde{x}_0\|)$ be given. The discrete-time stochastic multi-agent system (10.1) is said to be ε-consensusable if there exist a controller gain K in (10.2) and thresholds θ_i $(i = 1, 2, \cdots, N)$ in the triggering condition (10.4) such that the closed-loop system (10.6) can reach the consensus with probability $1 - \varepsilon$.*

We are now in a position to state the main goal of this chapter as follows. We aim to design both the controller parameter K and the threshold θ_i ($i = 1, 2, \cdots, N$) such that the discrete-time stochastic multi-agent system (10.1) reaches the consensus with probability $1 - \varepsilon$.

Remark 10.1 *It can be seen from (10.1) that the process noise $w(k)$ and the measurement noise $v(k)$ are one-step cross-correlated, which typically occurs when both noises are dependent on the system state [12, 18, 64, 103, 115]. On the other hand, different from traditional consensus definitions focusing on the asymptotic (with time tending to infinity) behavior in the mean-square sense, the new consensus definition proposed in this chapter is associated with the transient (at any time point k) behavior in the probability sense, that is, the consensus with probability $1 - \varepsilon$. Our definition is more suitable when the transient and probabilistic performances of the consensus process are a concern, which is often the case in practice.*

Remark 10.2 *Another feature of this chapter is that the threshold θ_i in (10.4) is a parameter to be designed and this provides more flexibility to achieve the desired consensus performance. Note that, in most existing literature, such thresholds have been assumed to be fixed a priori. Obviously, the threshold of the triggering condition is closely related to the network traffic. A smaller threshold gives rise to a heavier network load, and therefore an adequate trade-off between the threshold and the acceptable network load. In the extreme case, when the threshold $\theta_i = 0$, the event-triggering mechanism reduces to the traditional time-triggering one.*

10.2   Analysis of Input-to-State Stability in Probability

In this section, a framework is established to deal with the ISSiP requirement for *general stochastic nonlinear systems*. Furthermore, within such a framework, the proposed ISSiP conditions are applied in the following section to the multi-agent system (10.1) in order to design both the controller gain matrix K and the thresholds θ_i ($i = 1, 2, \cdots, N$).

Let us consider the following discrete-time stochastic nonlinear systems

$$x_{k+1} = f(k, x_k, u_k) + g(k, x_k, u_k)w_k \tag{10.12}$$

where $x_k \in \mathbb{R}^n$ is the state vector, $u_k \in \mathbb{R}^p$ is the control input, $w_k \in \mathbb{R}$ is a zero-mean random sequence on a probability space $(\Omega, \mathscr{F}, \text{Prob})$ with $\mathbb{E}\{w_k^2\} = 1$, and $f, g : [0, +\infty) \times \mathbb{R}^n \times \mathbb{R}^p \to \mathbb{R}^n$ are two continuous functions satisfying $f(k, 0, 0) = g(k, 0, 0) = 0$ for all $k \geq 0$.

Definition 10.3 *The system (10.12) is said to be input-to-state stable with* *probability if, for any* $\varepsilon > 0$, *there exist functions* $\beta \in \mathscr{KL}$ *and* $\gamma \in \mathscr{K}$ *such that the system dynamic* x_k *satisfies*

$$Prob\{\|x_k\| < \beta(\|x_0\|) + \gamma(\|u_k\|_\infty)\} \geq 1 - \varepsilon, \ \forall k \geq 0, \ \forall x_0 \in \mathbb{R}^n \backslash \{0\} \quad (10.13)$$

where $\|u_k\|_\infty := \sup_k\{\|u_k\|\}$. *In this case, the system (10.12) is said to have input-to-state stability in probability (ISSiP).*

Remark 10.3 *The definition of input-to-state stability in probability (ISSiP) has been given in [76], and Definition 10.3 can be viewed as its discrete-time version. Since its inception by Sontag [116], the input-to-state stability (ISS) theory has been playing an important role in the analysis and synthesis of a large class of nonlinear systems. Accordingly, in the past decade, the ISS concept has been widely adopted to deal with the stochastic nonlinear systems leading to several new stability definitions including ISSiP [76], γ-ISS [129], p-th moment ISS [53], noise-to-state stability [25] and exponential ISS [130]. These new ISS-related concepts have been exclusively studied for continuous-time systems, and the corresponding investigation for the discrete-time cases remains an open problem despite its crucial importance in analyzing the dynamic behaviors of discrete-time stochastic nonlinear systems.*

Lemma 10.1 *[139] Let* $\mathscr{F}_0$ *be a* σ *sub-field of* $\mathscr{F}_1$ *and* X *be an integrable random variable. The following is true*

$$\mathbb{E}\{\mathbb{E}\{X|\mathscr{F}_0\}|\mathscr{F}_1\} = \mathbb{E}\{X|\mathscr{F}_0\} = \mathbb{E}\{\mathbb{E}\{X|\mathscr{F}_1\}|\mathscr{F}_0\}.$$

In the following theorem, a Lyapunov-like theorem is established to ensure the ISSiP for the general discrete-time stochastic nonlinear system (10.12), based on which the addressed consensus problem can be readily solved.

Theorem 10.1 *The stochastic system (10.12) is input-to-state stable with probability if there exist a positive definite function* $\mathcal{V} : [0, +\infty) \times \mathbb{R}^n \to \mathbb{R}$ *(called an ISSiP-Lyapunov function), two* $\mathscr{K}_\infty$ *class functions* $\underline{\alpha}$ *and* $\bar{\alpha}$, *and two* $\mathscr{K}$ *class functions* χ *and* α *such that, for all* $x_k \in \mathbb{R}^n \backslash \{0\}$, *the following two inequalities hold*

$$\underline{\alpha}(\|x_k\|) \leq \mathcal{V}(k, x_k) \leq \bar{\alpha}(\|x_k\|), \quad (10.14)$$
$$\mathbb{E}\{\mathcal{V}(k+1, x_{k+1})|\mathscr{F}_k\} - \mathcal{V}(k, x_k) \leq \chi(\|u\|_\infty) - \alpha(\|x_k\|). \quad (10.15)$$

Furthermore, if (10.14) and (10.15) hold, then the functions φ *and* γ *in Definition 10.3 can be selected as* $\beta(\cdot) = \underline{\nu}^{-1}(\varepsilon^{-1}\bar{\nu}(\cdot))$ *and* $\gamma(\cdot) = \underline{\nu}^{-1}\big(\varepsilon^{-1}\big(\bar{\nu}(\tilde{\nu}^{-1}(\tilde{\chi}(\cdot))) + \tilde{\chi}(\cdot)\big)\big)$, *respectively.*

Proof *Along the similar line of the proof of Theorem 2 in [76], we denote*

$$\mathscr{B} = \{x : \|x\| < \alpha^{-1}(\chi(\|u\|_\infty))\}, \quad \mathscr{B}^c = \mathbb{R}^n \backslash \mathscr{B}. \quad (10.16)$$

Then, a sequence of stopping times $\{\tau_i\}_{i \geq 1}$ can be defined by

$$\tau_0 = 0,$$

$$\tau_1 = \begin{cases} \inf\{k \geq \tau_0 : x_k \in \mathscr{B}\}, & \text{if } \{k \geq \tau_0 : x_k \in \mathscr{B}\} \neq \emptyset, \\ \infty, & \text{otherwise}, \end{cases}$$

$$\tau_{2i} = \begin{cases} \inf\{k \geq \tau_{2i-1} : x_k \in \mathscr{B}^c\}, & \text{if } \{k \geq \tau_{2i-1} : x_k \in \mathscr{B}^c\} \neq \emptyset, \\ \infty, & \text{otherwise}, \end{cases} \quad (10.17)$$

$$\tau_{2i+1} = \begin{cases} \inf\{k \geq \tau_{2i} : x_k \in \mathscr{B}\}, & \text{if } \{k \geq \tau_{2i} : x_k \in \mathscr{B}\} \neq \emptyset, \\ \infty, & \text{otherwise} \end{cases}$$

where τ_{2i+1} and τ_{2i} $(i = 0, 1, 2, \cdots)$ are the times at which the trajectory enters the sets $\mathscr{B}$ and $\mathscr{B}^c$, respectively. It is not difficult to see that $x_k \in \mathscr{B}^c$ when $k \in \{\tau_{2i}, \tau_{2i} + 1, \cdots, \tau_{2i+1} - 1\} := [\tau_{2i}, \tau_{2i+1})$ and $x_k \in \mathscr{B}$ when $k \in \{\tau_{2i+1}, \tau_{2i+1} + 1, \cdots, \tau_{2i+2} - 1\} := [\tau_{2i+1}, \tau_{2i+2})$.

According to the definitions of stopping times τ_{2i} and τ_{2i+1}, for any $k \in [\tau_{2i}, \tau_{2i+1})$, one has

$$\|x_k\| \geq \alpha^{-1}(\chi(\|u\|_\infty)), \quad a.s. \quad (10.18)$$

It follows from (10.15) that

$$\mathbb{E}\{\mathcal{V}(k+1, x_{k+1}) | \mathscr{F}_k\} - \mathcal{V}(k, x_k) \leq \chi(\|u\|_\infty) - \alpha(\|x_k\|) \leq 0, \ a.s., \quad (10.19)$$

which implies

$$\begin{aligned} \mathbb{E}\{\mathcal{V}(\tau_{2i+1}, x_{\tau_{2i+1}}) | \mathscr{F}_{\tau_{2i}}\} &\leq \mathbb{E}\{\mathcal{V}(\tau_{2i+1} - 1, x_{\tau_{2i+1}-1}) | \mathscr{F}_{\tau_{2i}}\} \\ &\leq \cdots \leq \mathbb{E}\{\mathcal{V}(\tau_{2i}, x_{\tau_{2i}}) | \mathscr{F}_{\tau_{2i}}\}. \end{aligned} \quad (10.20)$$

It can be easily found that the sequence $\mathcal{V}_k^i := \mathcal{V}((k \vee \tau_{2i}) \wedge \tau_{2i+1}, x_{(k \vee \tau_{2i}) \wedge \tau_{2i+1}})$ is a supermartingale.

In what follows, the process of the proof can be divided into two cases, namely, $x_0 \in \mathscr{B}^c$ and $x_0 \in \mathscr{B} \backslash \{0\}$ for the purpose of convenience.

Case a) *$x_0 \in \mathscr{B}^c$. In this case, we have $x_k \in \mathscr{B}^c$ for any $k \in [\tau_0, \tau_1)$. Therefore, the inequality (10.20) holds. By using Chebyshev's inequality, we have*

$$\text{Prob}\left\{\mathcal{V}((k \vee \tau_0) \wedge \tau_1, x_{(k \vee \tau_0) \wedge \tau_1}) \geq \varepsilon^{-1} \mathcal{V}(\tau_0, x_{\tau_0}) | \mathscr{F}_{\tau_0}\right\}$$

$$= \text{Prob}\left\{\mathcal{V}_k^0 \geq \varepsilon^{-1} \mathcal{V}_{\tau_0}^0 | \mathscr{F}_{\tau_0}\right\} \leq \frac{\mathbb{E}\{\mathcal{V}_k^0 | \mathscr{F}_{\tau_0}\}}{\varepsilon^{-1} \mathcal{V}_{\tau_0}^0} \leq \varepsilon.$$

Furthermore, defining $\beta(r) = \underline{\alpha}^{-1}(\varepsilon^{-1} \bar{\alpha}(r))$, it follows that

$$\text{Prob}\left\{x_k < \beta(\|x_0\|) | \mathscr{F}_0\right\} \geq 1 - \varepsilon, \quad \forall k \in [\tau_0, \tau_1). \quad (10.21)$$

Defining $\mathscr{A} = \bigcup\limits_{i=0}^{+\infty} [\tau_{2i+1}, \tau_{2i+2})$ *and* $\mathscr{C} = \bigcup\limits_{i=1}^{+\infty} [\tau_{2i}, \tau_{2i+1})$ *for* $k \in [\tau_1, +\infty)$, *it follows that* $\mathscr{A} \cap \mathscr{C} = \emptyset$, $\mathscr{A} \cup \mathscr{C} = [\tau_1, +\infty)$ *and*

$$
\mathbb{E}\{\mathcal{V}(k \vee \tau_1, x_{k\vee\tau_1}) \big| \mathscr{F}_0\}
$$
$$
= \mathbb{E}\{\mathcal{V}(k \vee \tau_1, x_{k\vee\tau_1})\mathbb{I}_{\{k\in\mathscr{A}\}} \big| \mathscr{F}_0\} + \mathbb{E}\{\mathcal{V}(k \vee \tau_1, x_{k\vee\tau_1})\mathbb{I}_{\{k\in\mathscr{C}\}} \big| \mathscr{F}_0\}
$$
$$
= \sum_{i=0}^{+\infty} \mathbb{E}\{\mathcal{V}(k, x_k)\mathbb{I}_{\{k\in[\tau_{2i+1}, \tau_{2i+2})\}} \big| \mathscr{F}_0\} \qquad (10.22)
$$
$$
+ \sum_{i=1}^{+\infty} \mathbb{E}\{\mathcal{V}(k, x_k)\mathbb{I}_{\{k\in[\tau_{2i}, \tau_{2i+1})\}} \big| \mathscr{F}_0\}.
$$

Noticing that $x_k \in \mathscr{B}$ *for* $k \in [\tau_{2i+1}, \tau_{2i+2})$, *it follows from (10.14) that*

$$
\mathcal{V}(k, x_k) \leq \bar{\alpha} \circ \alpha^{-1}(\chi(\|u\|_\infty)),
$$

which means

$$
\sum_{i=0}^{+\infty} \mathbb{E}\{\mathcal{V}(k, x_k)\mathbb{I}_{\{k\in[\tau_{2i+1}, \tau_{2i+2})\}} \big| \mathscr{F}_0\}
$$
$$
\qquad (10.23)
$$
$$
\leq \bar{\alpha} \circ \alpha^{-1}(\chi(\|u\|_\infty)) \sum_{i=0}^{+\infty} Prob\{k \in [\tau_{2i+1}, \tau_{2i+2})\}.
$$

Similarly, due to $x_{\tau_{2i}-1} \in \mathscr{B}$, *it follows from (10.15) that*

$$
\mathbb{E}\{\mathcal{V}(\tau_{2i}, x_{\tau_{2i}}) \big| \mathscr{F}_{\tau_{2i}-1}\} \leq \mathcal{V}(\tau_{2i} - 1, x_{\tau_{2i}-1}) + \chi(\|u\|_\infty) - \alpha(\|x_{\tau_{2i}-1}\|)
$$
$$
\leq \bar{\alpha} \circ \alpha^{-1}(\chi(\|u\|_\infty)) + \chi(\|u\|_\infty).
$$
$$
\qquad (10.24)
$$

Furthermore, taking (10.20), (10.24) and Lemma 10.1 into consideration, one has

$$
\sum_{i=0}^{+\infty} \mathbb{E}\{\mathcal{V}(k, x_k)\mathbb{I}_{\{k\in[\tau_{2i}, \tau_{2i+1})\}} \big| \mathscr{F}_0\}
$$
$$
= \sum_{i=0}^{+\infty} \mathbb{E}\{\mathcal{V}((k \vee \tau_{2i}) \wedge (\tau_{2i+1} - 1), x_{(k\vee\tau_{2i})\wedge(\tau_{2i+1}-1)})\mathbb{I}_{\{k\in[\tau_{2i}, \tau_{2i+1})\}} \big| \mathscr{F}_0\}
$$
$$
= \sum_{i=0}^{+\infty} \mathbb{E}\Big\{\mathbb{E}\{\mathcal{V}((k \vee \tau_{2i}) \wedge (\tau_{2i+1} - 1), x_{(k\vee\tau_{2i})\wedge(\tau_{2i+1}-1)})
$$
$$
\times \mathbb{I}_{\{k\in[\tau_{2i}, \tau_{2i+1})\}} \big| \mathscr{F}_{\tau_{2i}}\} \big| \mathscr{F}_0\Big\} \qquad (10.25)
$$
$$
\leq \sum_{i=0}^{+\infty} \mathbb{E}\{\mathcal{V}(\tau_{2i}, x_{\tau_{2i}})\mathbb{I}_{\{k\in[\tau_{2i}, \tau_{2i+1})\}} \big| \mathscr{F}_0\}
$$
$$
\leq \sum_{i=0}^{+\infty} Prob\{k \in [\tau_{2i}, \tau_{2i+1})\}\big(\bar{\alpha} \circ \alpha^{-1}(\chi(\|u\|_\infty)) + \chi(\|u\|_\infty)\big).
$$

Finally, substituting (10.23) and (10.25) into (10.22) yields

$$\mathbb{E}\{\mathcal{V}(k \vee \tau_1, x_{k\vee\tau_1})|\mathscr{F}_0\}$$
$$\leq \bar{\alpha} \circ \underline{\alpha}^{-1}(\chi(\|u\|_\infty)) + \chi(\|u\|_\infty) := \psi(\chi(\|u\|_\infty)). \tag{10.26}$$

On the other hand, recalling that $\mathcal{V}(k, x_k)$ *is nonnegative, one has*

$$\mathbb{E}\{\mathcal{V}(k \vee \tau_1, x_{k\vee\tau_1})|\mathscr{F}_0\}$$
$$\geq \mathbb{E}\{\mathcal{V}(k \vee \tau_1, x_{k\vee\tau_1})\mathbb{I}_{\{\mathcal{V}(k\vee\tau_1, x_{k\vee\tau_1})\geq\varphi\circ\psi(\chi(\|u\|_\infty))\}}|\mathscr{F}_0\} \tag{10.27}$$
$$\geq (\varphi \circ \psi(\chi(\|u\|_\infty)))Prob\{\mathcal{V}(k \vee \tau_1, x_{k\vee\tau_1}) \geq \varphi \circ \psi(\chi(\|u\|_\infty))|\mathscr{F}_0\}$$

where $\varphi \in \mathscr{K}_\infty$ *satisfies*

$$\frac{\psi(\chi(\|u\|_\infty))}{\varphi \circ \psi(\chi(\|u\|_\infty))} \leq \varepsilon.$$

Therefore, it is easy to obtain from (10.26) and (10.27) that

$$Prob\{\mathcal{V}(k \vee \tau_1, x_{k\vee\tau_1}) \geq \varphi \circ \psi(\chi(\|u\|_\infty))|\mathscr{F}_0\} \leq \frac{\psi(\chi(\|u\|_\infty))}{\varphi \circ \psi(\chi(\|u\|_\infty))} \leq \varepsilon$$

and, furthermore,

$$Prob\left\{x_{k\vee\tau_1} < \underline{\alpha}^{-1} \circ \varphi \circ \psi(\chi(\|u\|_\infty))|\mathscr{F}_0\right\} \geq 1 - \varepsilon. \tag{10.28}$$

Let $\gamma(s) = \underline{\alpha}^{-1} \circ \varphi(\bar{\alpha} \circ \underline{\alpha}^{-1}(\chi(s)) + \chi(s))$. *Then, it can be verified from (10.21) and (10.28) that for any* $k \geq 0$ *and* $x_0 \in \mathcal{B}^c$

$$Prob\{\|x_k\| < \beta(\|x_0\|) + \gamma(\|u_k\|_\infty)\}$$
$$= Prob\{\|x_k\| < \beta(\|x_0\|)$$
$$+ \underline{\alpha}^{-1} \circ \varphi(\bar{\alpha} \circ \underline{\alpha}^{-1}(\chi(\|u_k\|_\infty)) + \chi(\|u_k\|_\infty))\}$$
$$\geq Prob\Big\{\{\|x_{k\wedge\tau_1}\| < \beta(\|x_0\|)\} \tag{10.29}$$
$$\cup \{\|x_{k\vee\tau_1}\| < \underline{\alpha}^{-1} \circ \varphi(\bar{\alpha} \circ \underline{\alpha}^{-1}(\chi(\|u_k\|_\infty)) + \chi(\|u_k\|_\infty))\}\Big\}$$
$$\geq 1 - \varepsilon.$$

Case b) $x_0 \in \mathcal{B}\backslash\{0\}$. *In this case, one has* $\tau_0 = \tau_1$ *and* $[\tau_0, \tau_1) = \emptyset$. *From the proof of Case a), we can obtain that (10.28) still holds for* $\forall k \in [\tau_1, +\infty) = [0, +\infty)$.

In conclusion, according to Case a) and Case b), one has

$$Prob\{\|x_k\| < \beta(\|x_0\|) + \gamma(\|u_k\|_\infty)\} \geq 1 - \varepsilon, \quad \forall k \geq 0 \text{ and } x_0 \neq 0,$$

which completes the proof.

Remark 10.4 *The sufficient condition provided in Theorem 10.1 can be referred to as a "discrete-time" version of the stability condition for ISSiP that can be used to analyze the stability performance for general discrete-time stochastic nonlinear systems. As with the traditional literature dealing with the ISS issues, a combination of the stopping times, the properties of conditional expectation, and Chebyshev's inequality are employed in the proof. Nevertheless, the proof of Theorem 10.1 exhibits its distinctive features as compared to those for the continuous-time systems. For example, $x(\tau_{2i})$ $(x(\tau_{2i+1}))$ may not lie in the boundary of the set $\mathcal{B}^c$ $(\mathcal{B})$. This makes it impossible to directly obtain the relationship among $x(\tau_{2i})$, $x(\tau_{2i+1})$ and $\chi(\|u_k\|_\infty)$, as required when analyzing ISSiP for continuous-time systems [76].*

10.3   Event-Triggered Consensus Control for Multi-Agent Systems

So far, we have developed a general framework for analyzing the ISSiP for a class of general discrete-time stochastic nonlinear systems, and let us now show that the established stability conditions can be applied to deal with the consensus (in probability) problem addressed for the multi-agent system (10.1). In fact, in Theorem 10.1, the relationship is described between the desired stability and the bound of inputs. For the event-triggering mechanism proposed in this chapter, when $e_i(k)$ outweighs a certain threshold θ_i for the system (10.1), the control input is updated and $e_i(k)$ returns to zero at the same time. As such, $e_i(k)$ is a bounded input and the stability result in Theorem 10.1 can be directly utilized to analyze the event-triggered consensus control problem, based on which both the controller gain matrix K and the thresholds θ_i $(i = 1, 2, \cdots, N)$ can be designed.

Before proceeding further, we introduce the following lemma which will be needed for the derivation of our main results.

Lemma 10.2 *[66] Let Ψ_1 and $\Psi_2 \in \mathbb{R}^{n \times n}$ be two symmetric matrices with their eigenvalues listed as follows*

$$\lambda_1(\Psi_1) \geq \lambda_2(\Psi_1) \geq \cdots \geq \lambda_n(\Psi_1), \quad \lambda_1(\Psi_2) \geq \lambda_2(\Psi_2) \geq \cdots \geq \lambda_n(\Psi_2),$$
$$\lambda_1(\Psi_1 + \Psi_2) \geq \lambda_2(\Psi_1 + \Psi_2) \geq \cdots \geq \lambda_n(\Psi_1 + \Psi_2).$$

Then, we have $\lambda_n(\Psi_1 + \Psi_2) \leq \lambda_n(\Psi_1) + \lambda_1(\Psi_2)$. Furthermore, if $\Psi_1 \geq \Psi_2$, then $\lambda_n(\Psi_1) \geq \lambda_n(\Psi_2)$.

Theorem 10.2 *Let the positive scalar ε and the bounded function $U(|\tilde{x}_0|)$ be given. The discrete-time stochastic multi-agent system (10.1) with the undirected graph $\mathscr{G}$ and the event-triggered control law (10.5) is ε-consensusable if there exist a matrix K, two positive definite matrices $\mathcal{Q}$ and*

$\mathcal{P}$, *and a positive scalar* δ *satisfying*

$$
\begin{aligned}
\mathcal{Q} = \mathcal{P} &- (1+\delta)(\mathcal{A} + \Xi \otimes (BKC))^T \mathcal{P}(\mathcal{A} + \Xi \otimes (BKC)) \\
&- \mathcal{D}^T \mathcal{P} \mathcal{D} - (\Xi \otimes (BKE))^T \mathcal{P}(\Xi \otimes (BKE)) \\
&- \sigma^2 \mathcal{D}^T \mathcal{P}(\Xi \otimes (BKE)) - \sigma^2(\Xi \otimes (BKE))^T \mathcal{P} \mathcal{D}
\end{aligned}
\tag{10.30}
$$

and the following triggering condition

$$
\|e_i(k)\|^2 \le \theta_i := \frac{deg_{in}^i(U(\|\tilde{x}_0\|) - 2\tilde{\beta}\|\tilde{x}_0\|^2)}{2\tilde{\gamma} \sum_{j=1}^N deg_{in}^i}
\tag{10.31}
$$

where

$$
\begin{aligned}
\Gamma &= (\delta^{-1} + 1)\big(\mathcal{N} \otimes (BK)\big)^T \mathcal{P}\big(\mathcal{N} \otimes (BK)\big), \\
\tilde{\gamma} &= \sqrt{\varepsilon^{-1}\lambda_{\min}^{-1}(\mathcal{P})\big(\lambda_{\max}(\mathcal{P})\lambda_{\min}^{-1}(\mathcal{Q}) + 1\big)\lambda_{\max}(\Gamma)}, \\
\tilde{\beta} &= \sqrt{\varepsilon^{-1}\lambda_{\min}^{-1}(\mathcal{P})\lambda_{\max}(\mathcal{P})}.
\end{aligned}
$$

Proof *Constructing the Lyapunov function* $\mathcal{V}(k) = \xi_k^T \mathcal{P} \xi_k$ *and calculating its difference along the trajectory of system (10.10b), it follows that*

$$
\begin{aligned}
\mathbb{E}\{\Delta V(k)\} &:= \mathbb{E}\{V(k+1) - V(k)|\xi_k\} \\
&= \mathbb{E}\{\xi_{k+1}^T \mathcal{P} \xi_{k+1}|\xi_k\} - \xi_k^T \mathcal{P} \xi_k \\
&= \xi_k^T \Big((\mathcal{A} + \Xi \otimes (BKC))^T \mathcal{P}(\mathcal{A} + \Xi \otimes (BKC)) + \mathcal{D}^T \mathcal{P} \mathcal{D} \\
&\quad + (\Xi \otimes (BKE))^T \mathcal{P}(\Xi \otimes (BKE)) + 2\sigma^2 \mathcal{D}^T \mathcal{P}(\Xi \otimes (BKE)) - \mathcal{P}\Big)\xi_k \\
&\quad + 2\xi_k^T (\mathcal{A} + \Xi \otimes (BKC))^T \mathcal{P}\big((\mathcal{M}\mathcal{N}) \otimes (BK)\big)e_k \\
&\quad + e_k^T \big((\mathcal{M}\mathcal{N}) \otimes (BK)\big)^T \mathcal{P}\big((\mathcal{M}\mathcal{N}) \otimes (BK)\big)e_k.
\end{aligned}
\tag{10.32}
$$

Considering $\mathcal{M}^T \mathcal{M} = I$ *and the elementary inequality* $2a^T b \le \delta a^T a + \delta^{-1}b^T b$, *we have*

$$
\begin{aligned}
\mathbb{E}\{\Delta V(k)\} &\le -\xi_k^T \mathcal{Q} \xi_k + e_k^T \Gamma e_k \\
&\le -\lambda_{\min}(\mathcal{Q})\|\xi_k\|^2 + \lambda_{\max}(\Gamma)\|e_k\|^2 \\
&\le -\lambda_{\min}(\mathcal{Q})\|\xi_k\|^2 + \lambda_{\max}(\Gamma)\|e_k\|_\infty^2.
\end{aligned}
\tag{10.33}
$$

Selecting $\chi(\|e\|_\infty) = \lambda_{\max}(\Gamma)\|e_k\|_\infty^2$ *and* $\alpha(\|\xi_k\|) = \lambda_{\min}(Q)\|\xi_k\|^2$, *it follows from Theorem 10.1 that the system modeled by (10.10a) and (10.10b) is input-to-state stable with probability. Since* $\tilde{\eta}_k = (M \otimes I)^T \tilde{x}_k$ *and* $(M \otimes$

$I)(M \otimes I)^T = I$, *we have*

$$
\begin{aligned}
&Prob\{\|\xi_k\| < \beta(\|\xi_0\|) + \gamma(\|e_k\|_\infty)\} \\
&= Prob\{\|(M \otimes I)^T \tilde{x}_k\| < \beta(|(M \otimes I)^T \tilde{x}_0|) + \gamma(\|e_k\|_\infty)\} \\
&= Prob\{\|\tilde{x}_k\| < \beta(\|\tilde{x}_0\|) + \gamma(\|e_k\|_\infty)\} \\
&\geq 1 - \varepsilon, \quad \forall \tilde{x}_0 \in \mathbb{R}^{nN} \setminus \{0\}
\end{aligned}
\tag{10.34}
$$

where $\beta(z) = \tilde{\beta}z$ *and* $\gamma(z) = \tilde{\gamma}z$ *for* $\forall z \in \mathbb{R}$.

On the other hand, in terms of the triggering condition (10.31), one has

$$
\|e_k\| \leq \frac{1}{2\tilde{\gamma}}(U(\|\tilde{x}_0\|) - 2\tilde{\beta}\|\tilde{x}_0\|).
\tag{10.35}
$$

In addition, it is easy to see that

$$
\|x_i(k) - x_j(k)\| = \|x_i(k) - \bar{x}_k - (x_j(k) - \bar{x}_k)\| \leq 2\|\tilde{x}_k\|.
\tag{10.36}
$$

Finally, it follows from (10.34)-(10.36) that

$$
\begin{aligned}
&Prob\{\|x_i(k) - x_j(k)\| < U(\|\tilde{x}_0\|)\} \\
&\geq Prob\{\|\tilde{x}_k\| < \beta(\|\tilde{x}_0\|) + \gamma(\|e_k\|_\infty)\} \geq 1 - \varepsilon, \quad \forall i, j \in \mathscr{V},
\end{aligned}
$$

which completes the proof.

In what follows, a sufficient condition is provided in order to reduce the computation complexity by using the second and Nth eigenvalues of $\mathcal{H}$.

Theorem 10.3 *Let the positive scalar* ε *and the bounded function* $U(|\tilde{x}_0|)$ *be given. The discrete-time stochastic multi-agent system (10.1) with the undirected graph* $\mathscr{G}$ *and the event-triggered control law (10.5) is* ε-*consensusable, if there exist a matrix* K, *positive definite matrix* P, *and two positive scalars* δ *and* ϖ *satisfying*
a) when $\lambda_2 \neq \lambda_N$

$$
\begin{cases}
(1 - \sigma^4)D^T PD - P + \varpi(\lambda_2 - \lambda_N)I \\
\quad + (1 + \delta)(A + \lambda_2 BKC))^T P(A + \lambda_2 BKC) \\
\quad + (\sigma^2 D + \lambda_2 BKE)^T P(\sigma^2 D + \lambda_2 BKE) < 0, & (10.37a) \\
(1 + \delta)\big((BKC)^T PA + A^T PBKC \\
\quad + (\lambda_2 + \lambda_N)(BKC)^T PBKC\big) + \varpi I + \big(\sigma^2 (BKE)^T PD \\
\quad + \sigma^2 D^T PBKE + (\lambda_2 + \lambda_N)(BKE)^T PBKE\big) > 0, & (10.37b)
\end{cases}
$$

b) when $\lambda_2 = \lambda_N$

$$
\begin{aligned}
(1 - \sigma^4)D^T PD - P &+ (1 + \delta)(A + \lambda_2 BKC))^T P(A + \lambda_2 BKC) \\
&+ (\sigma^2 D + \lambda_2 BKE)^T P(\sigma^2 D + \lambda_2 BKE) < 0
\end{aligned}
\tag{10.38}
$$

and the triggering condition

$$\|e_i(k)\|^2 \le \theta_i := \frac{deg_{in}^i\big(U(\|\tilde{x}_0\|) - 2\bar{\beta}\|\tilde{x}_0\|^2\big)}{2\bar{\gamma}\sum_{j=1}^N deg_{in}^i} \tag{10.39}$$

where

$$Q_2 = P - (1 - \sigma^4)D^T PD - (1 + \delta)(A + \lambda_2 BKC))^T P(A + \lambda_2 BKC)$$
$$\quad - (\sigma^2 D + \lambda_2 BKE)^T P(\sigma^2 D + \lambda_2 BKE),$$
$$\Gamma = (\delta^{-1} + 1)\big((\mathcal{MN}) \otimes (BK)\big)^T (I \otimes P)\big((\mathcal{MN}) \otimes (BK)\big),$$
$$\bar{\beta} = \sqrt{\varepsilon^{-1}\lambda_{\min}^{-1}(P)\lambda_{\max}(P)}, \quad \varpi = \lambda_{\min}(Q_2) - \varpi(\lambda_2 - \lambda_N) > 0,$$
$$\bar{\gamma} = \sqrt{\varepsilon^{-1}\lambda_{\min}^{-1}(P)\big(q\pi^{-1}\lambda_{\max}(P) + 1\big)\lambda_{\max}(\Gamma)}.$$

Proof *Based on Theorem 10.2, it suffices to show that (10.30)-(10.31) are satisfied for both cases of $\lambda_2 \ne \lambda_N$ and $\lambda_2 = \lambda_N$.*

Case a) $\lambda_2 \ne \lambda_N$. *First, it follows from (10.37a) that $Q_2 > \varpi(\lambda_2 - \lambda_N)I$. For $1 < i < N$, denote*

$$Q_i = P - (1 - \sigma^4)D^T PD - (1 + \delta)(A + \lambda_i BKC))^T P(A + \lambda_i BKC) \tag{10.40}$$
$$\quad - (\sigma^2 D + \lambda_i BKE)^T P(\sigma^2 D + \lambda_i BKE).$$

As $\lambda_N < \lambda_i = \lambda_2$ $(1 < i < N)$, one has $Q_i = Q_2$. Furthermore, it follows from $\lambda_N < \lambda_i < \lambda_2 < 0$ that

$$Q_i - Q_2 + \varpi(\lambda_2 - \lambda_N)I$$
$$= (1 + \delta)(A + \lambda_2 BKC))^T P(A + \lambda_2 BKC)$$
$$\quad - (1 + \delta)(A + \lambda_i BKC)^T P(A + \lambda_i BKC)$$
$$\quad + (\sigma^2 D + \lambda_2 BKE)^T P(\sigma^2 D + \lambda_2 BKE)$$
$$\quad - (\sigma^2 D + \lambda_i BKE)^T P(\sigma^2 D + \lambda_i BKE) + \varpi(\lambda_2 - \lambda_N)I$$
$$= (\lambda_2 - \lambda_i)(1 + \delta)\Big((BKC)^T PA + A^T PBKC$$
$$\quad + (\lambda_2 + \lambda_i)(BKC)^T PBKC\Big) + (\lambda_2 - \lambda_i)\Big(\sigma^2 (BKE)^T PD$$
$$\quad + \sigma^2 D^T PBKE + (\lambda_2 + \lambda_i)(BKE)^T PBKE + \varpi I\Big)$$
$$\quad + (\lambda_2 - \lambda_i)\varpi\Big(\frac{\lambda_2 - \lambda_N}{\lambda_2 - \lambda_i} - 1\Big)I$$
$$\ge (\lambda_2 - \lambda_i)(1 + \delta)\Big((BKC)^T PA + A^T PBKC$$
$$\quad + (\lambda_2 + \lambda_N)(BKC)^T PBKC\Big) + (\lambda_2 - \lambda_i)\Big(\sigma^2 (BKE)^T PD$$

$$+ \sigma^2 D^T PBKE + (\lambda_2 + \lambda_N)(BKE)^T PBKE + \varpi I\Big)$$

$$+ (\lambda_2 - \lambda_i)\Big((\lambda_2 + \lambda_i) - (\lambda_N + \lambda_2)\Big)$$

$$\times \Big((1 + \delta)(BKC)^T PBKC + (BKE)^T PBKE\Big)$$

$$> 0, \tag{10.41}$$

which implies $Q_i \geq Q_2 - \varpi(\lambda_2 - \lambda_N)I > 0$. *Therefore, by selecting* $\Psi_1 = Q_i$, $\Psi_2 = \varpi(\lambda_2 - \lambda_N)I$ *and exploiting Lemma 10.2, it can be found that*

$$\lambda_{\min}(Q_i) + \lambda_{\max}(\varpi(\lambda_2 - \lambda_N)I) \geq \lambda_{\min}(Q_i + \varpi(\lambda_2 - \lambda_N)) \geq \lambda_{\min}(Q_2),$$

which results in

$$\lambda_{\min}(Q_i) \geq \lambda_{\min}(Q_2) - \varpi(\lambda_2 - \lambda_N)$$

and therefore

$$\lambda_{\min}(\mathcal{Q}) := \lambda_{\min}\big(diag\{Q_2, Q_3, \cdots, Q_N\}\big)$$
$$\geq \lambda_{\min}(Q_2) - \varpi(\lambda_2 - \lambda_N) > 0. \tag{10.42}$$

By selecting $\mathcal{P} = I \otimes P$, *it can be easily checked that (10.31) in Theorem 10.2 can be guaranteed.*

On the other hand, in terms of (10.42) and the selected $\mathcal{P}$, *one has* $\bar{\beta} = \tilde{\beta}$ *and* $\bar{\gamma} > \tilde{\gamma}$. *Therefore, it follows that*

$$\|e_i(k)\|^2 \leq \frac{deg_{in}^i(U(\|\tilde{x}_0\|) - 2\bar{\beta}\|\tilde{x}_0\|^2)}{2\bar{\gamma}\sum_{j=1}^N deg_{in}^i} \leq \frac{deg_{in}^i(U(\|\tilde{x}_0\|) - 2\tilde{\beta}\|\tilde{x}_0\|^2)}{2\tilde{\gamma}\sum_{j=1}^N deg_{in}^i} \tag{10.43}$$

and therefore (10.30) in Theorem 10.2 holds.

Case b) $\lambda_2 = \lambda_N$. *Using a procedure similar to the proof of Case a), one can find from (10.38) and (10.40) that* $Q_N = Q_{N-1} = \cdots = Q_2 > 0$. *Therefore,* $\lambda_{\min}(\mathcal{Q}) = \lambda_{\min}(Q_2) > 0$ *which, in turn, implies that (10.30)-(10.31) in Theorem 10.2 are true.*

So far, it can be concluded that, for both Case a) and Case b), (10.30)-(10.31) are satisfied and it then follows from Theorem 10.2 that the discrete-time stochastic multi-agent system (10.1) is ε-*consensusable with the event-triggered control law (10.5).*

Having obtained the analysis results, we are now in a position to handle the design problem of both the threshold θ_i and the controller gain matrix K. First, in terms of the Schur complement lemma, (10.37a), (10.37b) and

(10.38) are, respectively, equivalent to

$$
\begin{bmatrix}
\Pi_{11} + \varpi(\lambda_2 - \lambda_N)I & * & * \\
\delta\Pi_{21} & -P^{-1} & * \\
\Pi_{31} & 0 & -P^{-1}
\end{bmatrix} < 0,
$$

$$
\begin{bmatrix}
\bar{\Pi}_{11} & * & * \\
\tilde{\delta}\bar{\Pi}_{21} & -P^{-1} & * \\
\bar{\Pi}_{31} & 0 & -P^{-1}
\end{bmatrix} < 0, \tag{10.44}
$$

$$
\begin{bmatrix}
\Pi_{11} & * & * \\
\tilde{\delta}\Pi_{21} & -P^{-1} & * \\
\Pi_{31} & 0 & -P^{-1}
\end{bmatrix} < 0
$$

where

$$
\begin{aligned}
&\Pi_{11} = (1 - \sigma^4)D^T PD - P, \quad \Pi_{21} = A + \lambda_2 BKC, \\
&\Pi_{31} = \sigma^2 D + \lambda_2 BKE, \quad \bar{\lambda} = \sqrt{-(\lambda_2 + \lambda_N)}, \\
&\bar{\Pi}_{11} = -\varpi I - \tilde{\delta}\bar{\lambda}^{-2} A^T PA - \sigma^4 \bar{\lambda}^{-2} D^T PD, \quad \tilde{\delta} = \sqrt{1 + \delta}, \\
&\bar{\Pi}_{21} = \bar{\lambda}^{-1} A - \bar{\lambda} BKC, \quad \bar{\Pi}_{31} = \sigma^2 \bar{\lambda}^{-1} D - \bar{\lambda} BKE.
\end{aligned} \tag{10.45}
$$

Then, denote $W = [B((B^T B)^{-1})^T \ B^\perp]^T$ where $B^\perp$ stands for an orthogonal basis of the null space for B^T. Similar to the method in [32], by introducing a free matrix

$$
R := \begin{bmatrix} R_{11} & R_{12} \\ 0 & R_{22} \end{bmatrix}
$$

where $R_{11} \in \mathbb{R}^{p \times p}$, $R_{12} \in \mathbb{R}^{n \times (n-p)}$ and $R_{22} \in \mathbb{R}^{(n-p) \times (n-p)}$ are arbitrary matrices, we conclude that (10.44) is satisfied if

$$
\begin{bmatrix}
\Pi_{11} + \varpi(\lambda_2 - \lambda_N)I & * & * \\
\tilde{\delta}(RWA + \lambda_2 RWBKC) & P_w & * \\
\sigma^2 RWD + \lambda_2 RWBKE & 0 & P_w
\end{bmatrix} < 0,
$$

$$
\begin{bmatrix}
\bar{\Pi}_{11} & * & * \\
\tilde{\delta}(\bar{\lambda}^{-1} RWA - \bar{\lambda} RWBKC) & P_w & * \\
\sigma^2 \bar{\lambda}^{-1} RWD - \bar{\lambda} RWBKE & 0 & P_w
\end{bmatrix} < 0,
$$

$$
\begin{bmatrix}
\Pi_{11} & * & * \\
\tilde{\delta}(RWA + \lambda_2 RWBKC) & P_w & * \\
\sigma^2 RWD + \lambda_2 RWBKE & 0 & P_w
\end{bmatrix} < 0.
$$

with $P_w = P - RW - W^T R^T$. Finally, by utilizing the variable substitution $\tilde{K} := [\bar{K}^T \ 0]^T := RWBK$, one has the following theorem.

Theorem 10.4 *Let the positive scalar ε and the bounded function $U(|\tilde{x}_0|)$ be given. The discrete-time stochastic multi-agent system (10.1) with the undirected graph $\mathcal{G}$ and the event-triggered control law (10.5) is ε-consensusable if there exist two matrices $\tilde{K}$ and R, positive definite matrix*

P, and positive scalars ϖ and $\tilde{\delta} > 1$ satisfying

a) when $\lambda_2 \neq \lambda_N$

$$
\begin{bmatrix}
\Pi_{11} + \varpi(\lambda_2 - \lambda_N)I & * & * \\
\tilde{\delta}(RWA + \lambda_2\tilde{K}C) & P_w & * \\
\sigma^2 RWD + \lambda_2\tilde{K}E & 0 & P_w
\end{bmatrix} < 0, \tag{10.46}
$$

$$
\begin{bmatrix}
\bar{\Pi}_{11} & * & * \\
\tilde{\delta}(\bar{\lambda}^{-1}RWA - \bar{\lambda}\tilde{K}C) & P_w & * \\
\sigma^2\bar{\lambda}^{-1}RWD - \bar{\lambda}\tilde{K}E & 0 & P_w
\end{bmatrix} < 0, \tag{10.47}
$$

b) when $\lambda_2 = \lambda_N$

$$
\begin{bmatrix}
\Pi_{11} & * & * \\
\tilde{\delta}(RWA + \lambda_2\tilde{K}C) & P_w & * \\
\sigma^2 RWD + \lambda_2\tilde{K}E & 0 & P_w
\end{bmatrix} < 0 \tag{10.48}
$$

where the related parameters are defined as in (10.45) and Theorem 10.3. Furthermore, if (10.46)-(10.47) hold for Case a) or (10.48) holds for Case b), then the output feedback controller gain matrix is given by $K = R_{11}^{-1}\tilde{K}$ and the threshold θ_i of the triggering condition (10.4) is determined by (10.39).

Remark 10.5 *In Theorem 10.4, the system parameters, the desired consensus probability and the eigenvalues of the Laplacian matrix $\mathcal{H}$ corresponding to the given topology are all reflected in the set of parameter-dependent matrix inequalities. Obviously, when the parameter $\tilde{\delta}$ is fixed, (10.46)-(10.48) reduce to linear matrix inequalities that can be solved by standard algorithm packages, and such a parameter offers additional flexibility with the possibility of improving the consensus performance.*

Remark 10.6 *In this chapter, a novel definition of consensus in probability is first proposed in order to 1) characterize the phenomenon where all agents are close to each other within a given bounded area with a given probability; and 2) describe the transient dynamics of the consensus process at each step for the addressed stochastic multi-agent systems. Then, we examine the impact from the given topology, the consensus probability, and the state-dependent noises on the consensus performance. The main technical contributions are that 1) the discrete-time version of ISSiP for general discrete-time stochastic nonlinear systems is established in order to facilitate later analysis of the consensus performance; and 2) both the output feedback controller and the threshold of the triggering condition are obtained by solving a set of parameter-dependent matrix inequalities reflecting both the topology information and the desired consensus probability.*

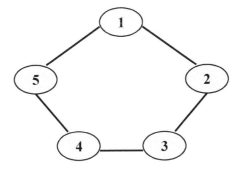

FIGURE 10.1
Communication graph $\mathcal{G}$.

10.4 Simulation Examples

In this section, a simulation example is presented to illustrate the effectiveness of the proposed event-triggered consensus protocol for discrete-time stochastic multi-agent systems with state-dependent noise.

Consider the system (10.1) with

$$A = \begin{bmatrix} 1 & 0.1 \\ 0 & 1 \end{bmatrix}, \ B = \begin{bmatrix} 0.30 \\ 0.25 \end{bmatrix}, \ D = \begin{bmatrix} 0 & 0.01 \\ 0 & 0 \end{bmatrix},$$
$$C = \begin{bmatrix} -0.4 & 0 \end{bmatrix}, \ E = \begin{bmatrix} 0.01 & 0 \end{bmatrix}.$$

Suppose that there are five agents with an undirected communication graph $\mathcal{G}$ shown in Fig. 10.1. The associated adjacency matrix $\mathcal{H}$ is selected as follows

$$\mathcal{H} = \begin{bmatrix} -8 & 4 & 0 & 0 & 4 \\ 4 & -8 & 4 & 0 & 0 \\ 0 & 4 & -8 & 4 & 0 \\ 0 & 0 & 4 & -8 & 4 \\ 4 & 0 & 0 & 4 & -8 \end{bmatrix}.$$

For such a Laplacian matrix, the eigenvalues are $\lambda_2 = \lambda_3 = -4.1459$, $\lambda_4 = \lambda_5 = -10.8541$. Select $U(|\tilde{x}_0|) = 80|\tilde{x}_0|^2$, $\sigma^2 = 0.0025$ and $\varepsilon = 0.25$. By linearly searching $\tilde{\delta}$ over the interval $[1.01 \ 2]$ and solving the resulting linear matrix inequalities using MATLAB® software (with YALMIP 3.0), a set of feasible solutions is obtained as follows:

$$P = \begin{bmatrix} 2.7723 & -1.2632 \\ -1.2632 & 1.3873 \end{bmatrix}, \ R = \begin{bmatrix} 0.4996 & -4.7485 \\ 0 & 4.5002 \end{bmatrix},$$
$$\tilde{K} = -0.2875, \quad \varpi = 8.0676 \times 10^{-4}, \quad \tilde{\delta} = 1.15$$

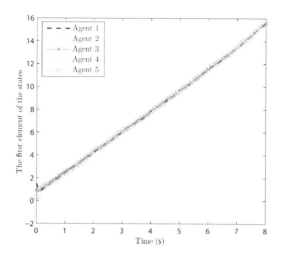

FIGURE 10.2
The trajectories for the 1st state of closed-loop systems.

and therefore

$$K = -0.5754, \; \lambda_{\max}(P) = 4.6942, \; \lambda_{\min}(P) = 0.4759,$$
$$\lambda_{\min}(Q_2) = 0.0461, \; \lambda_{\max}(\Gamma) = 0.5980, \; \bar{\gamma} = 2.7842 \times 10^2, \; \tilde{\beta} = 39.4584.$$

In the simulation, $x_s(0)$ $(s = 1, 2, \cdots, 5)$ are generated that obey the Gaussian distribution $\mathcal{N}(\hat{x}_0^s, \bar{\Lambda}^s)$ with $\hat{x}_0^s = [0.6 + 0.1s \quad 0.8 + 0.13s]^T$ and $\bar{\Lambda}^s = \mathrm{diag}\{0.25 + 0.1s, 0.15 + 0.08s\}$. According to the inequality (10.39), the triggering condition is $\|e_s\| \geq 0.0257$ $(s = 1, 2, \cdots, 5)$.

Simulation results are shown in Figs. 10.2-10.9. Figs. 10.2-10.5 plot the state trajectories of $x_s(k)$ $(s = 1, 2, \cdots, 5)$ for the discrete-time stochastic multi-agent system (10.1), respectively, with and without the event-triggered consensus controller (10.5). Obviously, the open-loop multi-agent system cannot reach the desired consensus and the closed-loop system does so. Furthermore, the update times and update control signals are shown in Figs. 10.6-10.7, from which we can easily find that the proposed consensus control scheme can effectively reduce the update frequency as compared with the traditional time-triggered mechanism. Finally, the consensus errors of the first agent are shown in Figs. 10.8-10.9, and the results for the other four agents are similar. Furthermore, by utilizing 2000 independent simulation trials, the average event-triggered law for each agent is exhibited in Table 10.1, which shows that the updated number is effectively reduced. The simulation results have confirmed that the designed event-triggered consensus controller performs very well.

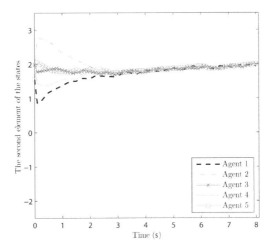

FIGURE 10.3
The trajectories for the 2nd state of closed-loop systems.

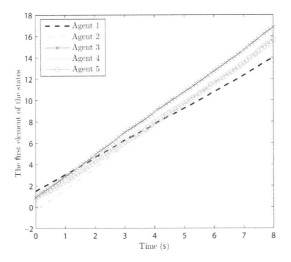

FIGURE 10.4
The trajectories for the 1st state of open-loop systems.

10.5 Summary

In this chapter, we have dealt with the event-triggered consensus control problem for a class of discrete-time stochastic multi-agent systems with

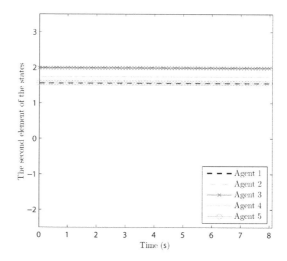

FIGURE 10.5
The trajectories for the 2nd state of open-loop systems.

TABLE 10.1
The average event-triggered law.

Agent	1	2	3	4	5
The event-triggered law	21.67%	22.02%	23.07%	22.68%	20.81%

state-dependent noises. By using intensive stochastic analysis approaches, a discrete-time version of ISSiP has been derived for a class of general discrete-time stochastic nonlinear systems. Within the established framework, some sufficient conditions have been proposed for the existence of desired event-triggered consensus controllers. Furthermore, the output feedback controller gain matrix and the threshold of the triggering condition have been designed by solving a set of parameter-dependent matrix inequalities. Finally, a numerical simulation example has been exploited to show the effectiveness of the event-triggered consensus control scheme.

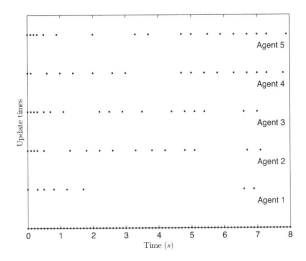

FIGURE 10.6
Update times of controllers.

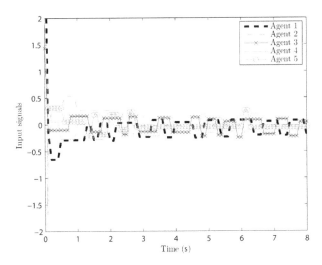

FIGURE 10.7
Controller inputs ϕ_s^u.

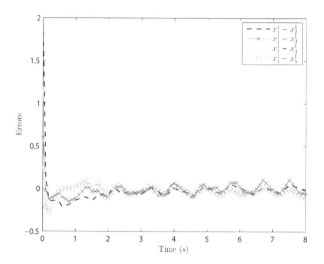

FIGURE 10.8
Consensus errors of the first element of the states.

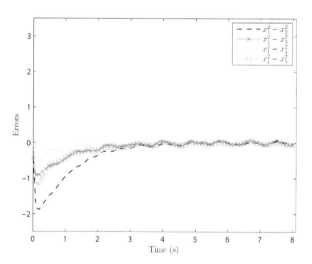

FIGURE 10.9
Consensus errors of the second element of the states.

11

Event-Triggered Security Control for Discrete-Time Stochastic Systems Subject to Cyber-Attacks

Recent years have witnessed rapid development of cyber-physical systems due to advances in computing, communication, and related hardware technologies. In a cyber-physical system, the resource scheduling via various shared or own networks plays an important role. Due to physical constraints or technological limitations, data among sensors, actuators and other networked components may be transmitted over networks without proper security protections. On the one hand, the interconnection of large-scale networked components makes it complicated to protect against inherent physical vulnerabilities therein. On the other hand, however, the cyber-integration usually sets up an underscore on the security and the resilience against unforeseen patterns or threats from cyberspace. In Chapter 8, we have identified the main challenges from cyber-attacks and investigated a class of distributed filtering issue over sensor networks. However, security control for cyber-physical systems remains an open yet challenging issue, especially when the event-triggering scheme addressed in Chapter 10 is also a concern.

This chapter is concerned with the event-based security control problem for a class of discrete-time stochastic systems with multiplicative noises subject to both randomly occurring denial-of-service (DoS) attacks and randomly occurring deception attacks. An event-triggered mechanism is adopted to attempt to reduce the communication burden, where the measurement signal is transmitted only when a certain triggering condition is violated. A novel attack model is proposed to reflect the randomly occurring behaviors of the DoS attacks as well as the deception attacks within a unified framework via two sets of Bernoulli distributed white sequences with known conditional probabilities. A new concept of mean-square security domain is put forward to quantify the security degree. By using the stochastic analysis techniques, some sufficient conditions are established to guarantee the desired security requirement and the control gain is obtained by solving some linear matrix inequalities with nonlinear constraints. A simulation example is utilized to illustrate the usefulness of the proposed controller design scheme.

11.1   Problem Formulation

In this chapter, consider the following discrete-time stochastic system with multiplicative noises in both the system and measurement equations:

$$
\begin{cases}
x_{k+1} = \left(A_0 + \displaystyle\sum_{i=1}^{r} \omega_{i,k} A_i\right) x_k + B u_k, \\[2mm]
\tilde{y}_k = \left(C_0 + \displaystyle\sum_{i=1}^{s} \varpi_{i,k} C_i\right) x_k
\end{cases}
\tag{11.1}
$$

where $x_k \in \mathbb{R}^{n_x}$, $\tilde{y}_k \in \mathbb{R}^{n_y}$ and $u_k \in \mathbb{R}^{n_u}$ are the state vector, the sensor measurement, and the controller input, respectively. A_i $(i = 0, 1, \cdots, r)$, B and C_i $(i = 0, 1, \cdots, s)$ are known constant matrices with appropriate dimensions. $\omega_{i,k} \in \mathbb{R}$ $(i = 1, 2, \cdots, r)$ and $\varpi_{i,k} \in \mathbb{R}$ $(i = 1, 2, \cdots, s)$ are multiplicative noises with zero means and unity variances, and are mutually uncorrelated in k and i, r and s are known positive integers. It is assumed that the rank of B is n_u.

In this chapter, an event-triggered communication mechanism is taken into consideration in order to reduce the communication burden. Define the event generator function $\psi(\cdot, \cdot) : \mathbb{R}^{n_y} \times \mathbb{R} \to \mathbb{R}$ as follows:

$$
\psi(e_k, \delta) = e_k^T e_k - \delta_1^2
\tag{11.2}
$$

where

$$
e_k := \tilde{y}_{k_s}^t - \tilde{y}_k
$$

with $\tilde{y}_{k_s}^t$ being the *transmitted* information at the *latest* event instant and δ_1 being a given positive scalar. The executions are triggered as long as the condition

$$
\psi(e_k, \delta_1) > 0
$$

is satisfied. Therefore, the sequence of event-triggered instants

$$
0 \le s_0 < s_1 < \cdots < s_l < \cdots
$$

is determined iteratively by

$$
s_{l+1} = \inf\{k \in \mathbb{N} | k > s_l, \ \psi(e_k, \delta_1) > 0\}.
$$

Furthermore, it is assumed that the attackers only destroy the transmitted data and have the ability to carry out both DoS attacks and deception attacks with certain success probabilities, in other words, both kinds of attack can be launched in a random way. To reflect such a situation that is of practical significance, a new attack model is proposed as follows:

$$
y_{k_s}^t = \alpha_{k_s}^t (\tilde{y}_{k_s}^t + \beta_{k_s}^t v_{k_s}^t) + (1 - \alpha_{k_s}^t) y_{k_s-1}^t
\tag{11.3}
$$

where $y_{k_s}^t$ is the *register information* on the controller and $v_{k_s}^t \in \mathbb{R}^{n_y}$ stands for the signals sent by attackers. In addition, $v_{k_s}^t$ is modeled as

$$v_{k_s}^t = -\tilde{y}_{k_s}^t + \xi_{k_s}^t$$

for deception attacks where the non-zero $\xi_{k_s}^t$ satisfying

$$\|\xi_{k_s}^t\| \le \delta_2$$

is an arbitrary bounded energy signal. The stochastic variables α_{k_s} and β_{k_s} are Bernoulli distributed white sequences taking values on 0 or 1 with the following probabilities

$$\mathrm{Prob}\{\alpha_{k_s} = 0\} = 1 - \bar{\alpha}, \quad \mathrm{Prob}\{\alpha_{k_s} = 1\} = \bar{\alpha},$$
$$\mathrm{Prob}\{\beta_{k_s} = 0\} = 1 - \bar{\beta}, \quad \mathrm{Prob}\{\beta_{k_s} = 1\} = \bar{\beta}$$

where $\bar{\alpha} \in [0, 1)$ and $\bar{\beta} \in [0, 1)$ are two known constants.

Remark 11.1 *Generally speaking, network attacks can be divided into DoS attacks and deception attacks. For DoS attacks, the adversary prevents the controller from receiving sensor measurements. For deception attacks, the adversary sends false information to controllers. Due to network-induced phenomena and the application of safety protection devices, there is a nonzero probability for each attack launched via networks to be unsuccessful at a certain time. The model proposed in (11.3) provides a novel unified framework to account for the phenomena of both randomly occurring DoS attacks and randomly occurring deception attacks.*

Remark 11.2 *Three cases can be observed from (11.3) as follows: a) the systems suffer from deception attacks when $\alpha_{k_s}^t = 1$ and $\beta_{k_s}^t = 1$; b) the systems are subject to DoS attacks when $\alpha_{k_s}^t = 0$ and the register information on the controller cannot be updated in this case; and c) the controller receives the normal sensor measurements when $\alpha_{k_s}^t = 1$ and $\beta_{k_s}^t = 0$. Furthermore, it is worth mentioning that Case a) (with $\xi_{k_s}^t = 0$) describes the traditional phenomenon of packet dropouts and Case b) can also reflect the time-delays. Therefore, the proposed attack model covers time delays and packet dropouts as its special cases.*

Remark 11.3 *Due to limited energy, the adversaries could not arbitrarily launch the attacks and, from the defenders' perspective, the cyber-attack could be intermittent and the injected signal is bounded. Furthermore, the injected signal sent by adversaries is a kind of invalid information to achieve the control task and therefore it can be viewed as an energy-bounded noise. The main purpose of the present research is to improve security by enhancing the insensitivity to certain types of bounded and random cyber-attacks.*

For $k \in [k_s, k_{s+1})$, the *register information* (11.3) can be rewritten as

$$y_k = \alpha_{k_s}^t (\tilde{y}_{k_s}^t + \beta_{k_s}^t v_{k_s}^t) + (1 - \alpha_{k_s}^t) y_{k-1} \tag{11.4}$$

with $y_{k_s} = y_{k_s}^t$. Furthermore, taking both the event-triggering condition and the attack bound into consideration, and applying the output-feedback control

$$u_k = K y_k$$

where K is the control parameter to be determined, one has the following closed-loop system

$$
\begin{aligned}
\tilde{x}_{k+1} = {} & \mathcal{A}_1 \tilde{x}_k + (\bar{\alpha} - \alpha_{k_s}^t) \mathcal{A}_2 \tilde{x}_k + ((\alpha_{k_s}^t - \bar{\alpha}) - \chi_{k_s}) \mathcal{A}_3 \tilde{x}_k \\
& + \mathcal{A}_{4,k_s} \tilde{x}_k + \bar{\alpha} \bar{\beta} \mathcal{A}_5 \xi_{k_s}^t + \bar{\alpha}(1 - \bar{\beta}) \mathcal{A}_5 e_k \\
& + \chi_{k_s} \mathcal{A}_5 \xi_{k_s}^t + ((\alpha_{k_s}^t - \bar{\alpha}) - \chi_{k_s}) \mathcal{A}_5 e_k
\end{aligned}
\tag{11.5}
$$

where

$$
\begin{aligned}
\tilde{x}_k &= \begin{bmatrix} x_k^T & y_{k-1}^T \end{bmatrix}^T, \\
\chi_{k_s} &= (\alpha_{k_s}^t - \bar{\alpha})\bar{\beta} + (\beta_{k_s}^t - \bar{\beta})\bar{\alpha} + (\alpha_{k_s}^t - \bar{\alpha})(\beta_{k_s}^t - \bar{\beta}), \\
\mathcal{A}_1 &= \begin{bmatrix} A_0 + \bar{\alpha}(1-\bar{\beta})BKC_0 & (1-\bar{\alpha})BK \\ \bar{\alpha}(1-\bar{\beta})C_0 & (1-\bar{\alpha})I \end{bmatrix}, \\
\mathcal{A}_2 &= \begin{bmatrix} 0 & BK \\ 0 & I \end{bmatrix}, \quad \mathcal{A}_3 = \begin{bmatrix} BKC_0 & 0 \\ C_0 & 0 \end{bmatrix}, \quad \mathcal{A}_5 = \begin{bmatrix} BK \\ I \end{bmatrix}, \\
\mathcal{A}_{4,k_s} &= \begin{bmatrix} \sum_{i=1}^r \varpi_{i,k} A_i + \alpha_{k_s}^t (1 - \beta_{k_s}^t) \sum_{i=1}^s \varpi_{i,k} BKC_i & 0 \\ \alpha_{k_s}^t (1 - \beta_{k_s}^t) \sum_{i=1}^s \varpi_{i,k} C_i & 0 \end{bmatrix}.
\end{aligned}
$$

Remark 11.4 *A schematic structure of the addressed control problem is shown in Fig. 11.1. The adversary can detect the transmitted data from plants and then try to destroy them to attain certain goals. Furthermore, in order to increase the success ratio and enhance the covertness of attacks, the adversary could randomly adopt any type of attack at each time. On the other hand, it is easy to find from (11.2) that the time interval (also called the inter-event time) between adjacent event-triggering instants is generally not a constant, which is determined by the event generator function ψ. Therefore, the closed-loop system (11.5) cannot be regarded as a discrete-time expression with a constant sampling interval.*

Before proceeding further, we introduce the following definitions.

Definition 11.1 *For a given positive scalar δ_3 representing the desired security level, the set*

$$\mathscr{D} = \left\{ \tilde{x}_0 \in \mathbb{R}^{n_x} : \mathbb{E}\|\tilde{x}_k\|^2 \leq \delta_3^2, \ \forall k \right\} \tag{11.6}$$

is said to be the mean-square security domain of the origin of the closed-loop system (11.5).

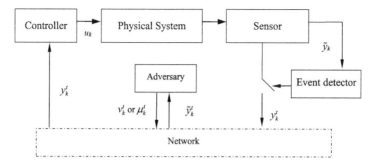

FIGURE 11.1
Attacks on an event-based control system.

Definition 11.2 *Let R be a positive definite matrix and δ_1, δ_2, δ_3 and δ_4 be given constants. The closed-loop system (11.5) is said to be secure with respect to $(\delta_1, \delta_2, \delta_3, \delta_4, R)$ if, when $\psi(e_k, \delta_1) \leq 0$, $\|\xi_k\| \leq \delta_2$ and $\tilde{x}_0^T R \tilde{x}_0 \leq \delta_4^2$, one has $\mathbb{E}\|\tilde{x}_k\|^2 \leq \delta_3^2$ for all k.*

Remark 11.5 *The five parameters δ_1, δ_2, δ_3, δ_4 and R do have their own engineering insights. To be specific, δ_1 is the triggering threshold that governs the transmission frequency for the benefit of energy saving, δ_2 is the energy bound for the false signals that the adversary likes to impose on the measurement output from the attacked system, δ_3 is associated with the desired security level (i.e., the upper bound for the dynamics evolution of the attached system in the mean square sense), δ_4 is about the energy of the initial system state, and R is the weighting matrix for the initial system state. Obviously, these five parameters play crucial roles in system security performance evaluation and design.*

Our aim in this chapter is to design an output feedback controller for system (11.5) with an event-triggering communication mechanism and randomly occurring cyber-attacks. In other words, we are going to determine the controller gain K such that the closed-loop system (11.5) is secure with respect to $(\delta_1, \delta_2, \delta_3, \delta_4, R)$.

11.2   Security Performance Analysis

In this section, security is analyzed for the closed-loop system (11.5) with the event-triggering communication mechanism and randomly occurring cyber-attacks. A sufficient condition is provided to guarantee that the closed-loop system (11.5) is secure with respect to $(\delta_1, \delta_2, \delta_3, \delta_4, R)$. Then, the explicit expression of the desired controller gain is proposed in terms of the solution to certain matrix inequalities subject to nonlinear constraints.

Let us start by giving the following lemma that will be used in the proof of our main result in this chapter.

Lemma 11.1 *Given constant matrices* Σ_1, Σ_2, Σ_3, *where* $\Sigma_1 = \Sigma_1^T$ *and* $\Sigma_2 = \Sigma_2^T > 0$, $\Sigma_1 + \Sigma_3^T \Sigma_2^{-1} \Sigma_3 < 0$ *if and only if*

$$\begin{bmatrix} \Sigma_1 & \Sigma_3^T \\ \Sigma_3 & -\Sigma_2 \end{bmatrix} < 0 \quad or \quad \begin{bmatrix} -\Sigma_2 & \Sigma_3 \\ \Sigma_3^T & \Sigma_1 \end{bmatrix} < 0.$$

Theorem 11.1 *Let the positive scalars* $\delta_1, \delta_2, \delta_3, \delta_4$, *the positive definite matrix* R, *and the controller gain* K *be given. The closed-loop system (11.5) is secure with respect to* $(\delta_1, \delta_2, \delta_3, \delta_4, R)$ *if there exist two positive definite matrices* P_1 *and* P_2, *and three positive scalars* ε_1, ε_2 *and* π *satisfying the following inequalities*

$$\begin{cases} \Xi_1 = \begin{bmatrix} \Xi_{11} & \Xi_{12} & \Xi_{13} \\ * & \Xi_{22} & 0 \\ * & * & \Xi_{33} \end{bmatrix} < 0, & (11.7a) \\[4mm] \Xi_2 = \max \left\{ \dfrac{\lambda_{\max}(P_R)\delta_4^2}{\lambda_{\min}(P)}, \dfrac{\theta^2 \gamma}{\lambda_{\min}(P)(\gamma - 1)} \right\} \le \delta_3^2 & (11.7b) \end{cases}$$

where

$$\mathcal{A}_4 = diag\{ \sum_{i=1}^{r} A_i^T P_1 A_i + \bar{\alpha}(1 - \bar{\beta}) \sum_{i=1}^{s} ((BKC_i)^T P_1 BKC_i + C_i^T P_2 C_i), 0 \},$$

$$P = diag\{P_1, P_2\}, \quad \bar{\chi} = \tilde{\alpha}\bar{\beta}^2 + \tilde{\beta}\bar{\alpha}^2 + \tilde{\alpha}\tilde{\beta}, \quad \theta = \sqrt{\varepsilon_1 \delta_1^2 + \varepsilon_2 \delta_2^2},$$

$$\Xi_{11} = \mathcal{A}_1^T P \mathcal{A}_1 + \tilde{\alpha}\mathcal{A}_2^T P \mathcal{A}_2 - 2(\tilde{\alpha} - \tilde{\alpha}\bar{\beta})\mathcal{A}_2^T P \mathcal{A}_3$$
$$\qquad + (\tilde{\alpha} - 2\tilde{\alpha}\bar{\beta} + \bar{\chi})\mathcal{A}_3^T P \mathcal{A}_3 + \mathcal{A}_4 - P + \pi I,$$

$$\Xi_{12} = \bar{\alpha}\bar{\beta}\mathcal{A}_1^T P \mathcal{A}_5 - \tilde{\alpha}\bar{\beta}\mathcal{A}_2^T P \mathcal{A}_5 + (\tilde{\alpha}\bar{\beta} - \bar{\chi})\mathcal{A}_3^T P \mathcal{A}_5,$$

$$\Xi_{13} = \bar{\alpha}(1 - \bar{\beta})\mathcal{A}_1^T P \mathcal{A}_5 - (\tilde{\alpha} - \tilde{\alpha}\bar{\beta})\mathcal{A}_2^T P \mathcal{A}_5 + (\tilde{\alpha} - 2\tilde{\alpha}\bar{\beta} + \bar{\chi})\mathcal{A}_3^T P \mathcal{A}_5,$$

$$\Xi_{22} = ((\bar{\alpha}\bar{\beta})^2 + \bar{\chi})\mathcal{A}_5^T P \mathcal{A}_5 - \varepsilon_2 I, \quad \Xi_{23} = (\bar{\alpha}^2 \bar{\beta} + \tilde{\alpha}\bar{\beta} - \bar{\chi})\mathcal{A}_5^T P \mathcal{A}_5,$$

$$\Xi_{33} = (\bar{\alpha}^2(1 - \bar{\beta})^2 + \tilde{\alpha} - 2\tilde{\alpha}\bar{\beta} + \bar{\chi})\mathcal{A}_5^T P \mathcal{A}_5 - \varepsilon_1 I, \quad \tilde{\alpha} = \bar{\alpha}(1 - \bar{\alpha}),$$

$$P_R = R^{-1/2} P R^{-1/2}, \quad \rho = \lambda_{\max}(P), \quad \gamma = \frac{\rho}{\rho - \pi}, \quad \tilde{\beta} = \bar{\beta}(1 - \bar{\beta}).$$

Proof: Construct the Lyapunov function

$$V_k = \tilde{x}_k^T P \tilde{x}_k.$$

By calculating the difference of V_k along the trajectory of system (11.5) and taking the mathematical expectation on $\omega_{i,k}$, $\varpi_{j,k}$ ($i = 1, 2, \cdots, r$, $j = 1, 2, \cdots, s$), $\alpha_{k_s}^t$ and $\beta_{k_s}^t$, one has

$$
\begin{aligned}
\mathbb{E}\{\Delta V_k | \tilde{x}_k\} &= \mathbb{E}\{\tilde{x}_{k+1}^T P \tilde{x}_{k+1} - \tilde{x}_k^T P \tilde{x}_k | \tilde{x}_k\} \\
&= \mathbb{E}\left\{\mathbb{E}\{\tilde{x}_{k+1}^T P \tilde{x}_{k+1} - \tilde{x}_k^T P \tilde{x}_k | \alpha_{k_s}^t, \beta_{k_s}^t\} \Big| \tilde{x}_k\right\} \\
&= \mathbb{E}\Big\{\tilde{x}_k^T \Big(\mathcal{A}_1^T P \mathcal{A}_1 + \tilde{\alpha}\mathcal{A}_2^T P \mathcal{A}_2 - 2(\tilde{\alpha} - \tilde{\alpha}\bar{\beta})\mathcal{A}_2^T P \mathcal{A}_3 \\
&\quad + (\tilde{\alpha} - 2\tilde{\alpha}\bar{\beta} + \bar{\chi})\mathcal{A}_3^T P \mathcal{A}_3 + \mathcal{A}_4 - P\Big)\tilde{x}_k \\
&\quad + \tilde{x}_k^T \big(2\bar{\alpha}\bar{\beta}\mathcal{A}_1^T P \mathcal{A}_5 - 2\tilde{\alpha}\bar{\beta}\mathcal{A}_2^T P \mathcal{A}_5 + 2(\tilde{\alpha}\bar{\beta} - \bar{\chi})\mathcal{A}_3^T P \mathcal{A}_5\big)\xi_{k_s}^t \\
&\quad + \tilde{x}_k^T \big(2\bar{\alpha}(1 - \bar{\beta})\mathcal{A}_1^T P \mathcal{A}_5 - 2(\tilde{\alpha} - \tilde{\alpha}\bar{\beta})\mathcal{A}_2^T P \mathcal{A}_5 \\
&\quad + 2(\tilde{\alpha} - 2\tilde{\alpha}\bar{\beta} + \bar{\chi})\mathcal{A}_3^T P \mathcal{A}_5\big)e_k + (\xi_{k_s}^t)^T\big(((\bar{\alpha}\bar{\beta})^2 + \bar{\chi})\mathcal{A}_5^T P \mathcal{A}_5\big)\xi_{k_s}^t \\
&\quad + 2(\xi_{k_s}^t)^T\big((\bar{\alpha}^2\bar{\beta} + \tilde{\alpha}\bar{\beta} - \bar{\chi})\mathcal{A}_5^T P \mathcal{A}_5\big)e_k \\
&\quad + e_k^T\big((\bar{\alpha}^2(1 - \bar{\beta})^2 + \tilde{\alpha} - 2\tilde{\alpha}\bar{\beta} + \bar{\chi})\mathcal{A}_5^T P \mathcal{A}_5\big)e_k \Big| \tilde{x}_k\Big\} \\
&= \eta_k^T \tilde{\Xi} \eta_k
\end{aligned}
\tag{11.8}
$$

where

$$
\eta_k = [\ \tilde{x}_k^T \quad (\xi_{k_s}^t)^T \quad e_k^T\ ]^T,
$$

$$
\tilde{\Xi} = \begin{bmatrix} \Xi_{11} - \pi I & \Xi_{12} & \Xi_{13} \\ * & \Xi_{22} + \varepsilon_2 I & 0 \\ * & * & \Xi_{33} + \varepsilon_1 I \end{bmatrix}.
$$

Subsequently, taking $\psi(e_k, \delta_1) \le 0$ and $\|\xi_k^t\| \le \delta_2$ into consideration, one has

$$
\begin{aligned}
\mathbb{E}\{\Delta V(k)\} \\
\le \mathbb{E}\{\eta_k^T \tilde{\Xi} \eta_k + \varepsilon_1(\delta_1^2 - e_k^T e_k) + \varepsilon_2(\delta_2^2 - (\xi_{k_s}^t)^T \xi_{k_s}^t)\} \\
= \mathbb{E}\{\eta_k^T \Xi_1 \eta_k - \eta_k^T \operatorname{diag}\{\pi I, \ 0, \ 0\}\eta_k\} + \varepsilon_1\delta_1^2 + \varepsilon_2\delta_2^2 \\
\le -\mathbb{E}\{\eta_k^T \operatorname{diag}\{\pi I, \ 0, \ 0\}\eta_k\} + \theta^2,
\end{aligned}
\tag{11.9}
$$

which implies

$$
\mathbb{E}\{\Delta V_k\} \le -\pi \mathbb{E}\{\|\tilde{x}_k\|^2\} + \theta^2.
\tag{11.10}
$$

For any scalar $\gamma > 1$, it follows from the above inequality that

$$
\begin{aligned}
\mathbb{E}\{\gamma^{k+1} V_{k+1}\} - \mathbb{E}\{\gamma^k V_k\} \\
= \gamma^{k+1}\mathbb{E}\{V_{k+1} - V_k\} + \gamma^k(\gamma - 1)\mathbb{E}\{V_k\} \\
\le \gamma^k((\gamma - 1)\rho - \gamma\pi)\mathbb{E}\{\|\tilde{x}_k\|^2\} + \gamma^{k+1}\theta^2.
\end{aligned}
\tag{11.11}
$$

Selecting $\gamma = \frac{\rho}{\rho - \pi}$, one has from (11.11) that

$$
\mathbb{E}\{\gamma^k V_k\} - \mathbb{E}\{V_0\} \le (\gamma^k + \gamma^{k-1} + \cdots + \gamma)\theta^2,
\tag{11.12}
$$

which implies

$$\begin{aligned}
\mathbb{E}\{V_k\} &\leq \gamma^{-k}\mathbb{E}\{V_0\} + (1 + \gamma^{-1} + \cdots + \gamma^{-k+1})\theta^2 \\
&= \gamma^{-k}\mathbb{E}\{V_0\} + \frac{[1 - (1/\gamma)^k]\theta^2}{1 - 1/\gamma} \\
&= \gamma^{-k}\left[\mathbb{E}\{V_0\} - \frac{\theta^2\gamma}{\gamma - 1}\right] + \frac{\theta^2\gamma}{\gamma - 1} \\
&\leq \gamma^{-k}\left(\lambda_{\max}(R^{-1/2}PR^{-1/2})\delta_4^2 - \frac{\theta^2\gamma}{\gamma - 1}\right) + \frac{\theta^2\gamma}{\gamma - 1} \\
&\leq \max\left\{\lambda_{\max}(R^{-1/2}PR^{-1/2})\delta_4^2, \frac{\theta^2\gamma}{\gamma - 1}\right\}.
\end{aligned} \tag{11.13}$$

Finally, it can be concluded from (11.7b) that the closed-loop system (11.5) is secure with respect to $(\delta_1, \delta_2, \delta_3, \delta_4, R)$, which completes the proof.

11.3   Security Controller Design

Having obtained the analysis results, we are now in a position to handle the design problem of the controller gain matrix K. First, in terms of Lemma 11.1 and the inequality

$$2(\mathcal{A}_2 - \mathcal{A}_3)^T P \mathcal{A}_3 \leq (\mathcal{A}_2 - \mathcal{A}_3)^T P(\mathcal{A}_2 - \mathcal{A}_3) + \mathcal{A}_3^T P \mathcal{A}_3,$$

(11.7a) is true if the following holds

$$\Pi_1 = \begin{bmatrix} \Pi_{11} & * & * \\ \bar{\Pi}_{12} & -\bar{P}^{-1} & * \\ \Pi_{17} & 0 & -I \otimes P_1^{-1} \end{bmatrix} < 0 \tag{11.14}$$

where

$$\mathcal{B}_1 = [\ \mathcal{A}_3 \quad -\mathcal{A}_5 \quad \mathcal{A}_5\ ], \quad \mathcal{B}_2 = [\ \mathcal{A}_2 - \mathcal{A}_3 \quad \mathcal{A}_5 \quad \mathcal{A}_5\ ],$$

$$\mathcal{S}_1 = \operatorname{diag}\{I, 0, -I\}, \quad \mathcal{S}_2 = \operatorname{diag}\{I, -I, I\},$$

$$\bar{P} = \operatorname{diag}\{P, P, P, P, P\}, \quad \Upsilon = [\ C_1^T \quad C_2^T \quad \cdots \quad C_s^T\ ]^T,$$

$$\Pi_{00} = \sum_{i=1}^{r} A_i^T P A_i + \bar{\alpha}(1 - \bar{\beta})\sum_{i=1}^{s} C_i^T P_2 C_i - P_1 + \pi I,$$

$$\Pi_{11} = \operatorname{diag}\left\{\Pi_{00}, -P_2 + \pi I, -\varepsilon_2 I, -\varepsilon_1 I\right\},$$

$$\bar{\Pi}_{12} = [\ \Pi_{12}^T \quad \Pi_{13}^T \quad \Pi_{14}^T \quad \Pi_{15}^T \quad \Pi_{16}^T\ ]^T,$$

$$\Pi_{12} = [\ \mathcal{A}_1 \quad \bar{\alpha}\bar{\beta}\mathcal{A}_5 \quad \bar{\alpha}(1 - \bar{\beta})\mathcal{A}_5\ ], \quad \Pi_{13} = \sqrt{\bar{\chi}}\mathcal{B}_1,$$

$$\Pi_{14} = \sqrt{\bar{\alpha}}\mathcal{B}_2\mathcal{S}_1, \quad \Pi_{15} = \sqrt{\bar{\alpha}\bar{\beta}}\mathcal{B}_2\mathcal{S}_2, \quad \Pi_{16} = \sqrt{\bar{\alpha}\bar{\beta}}\mathcal{B}_1\mathcal{S}_1,$$

$$\Pi_{17} = [\ \sqrt{\bar{\alpha}(1 - \bar{\beta})}(I \otimes BK)\Upsilon \quad 0 \quad 0 \quad 0\ ].$$

In what follows, we introduce a free matrix

$$\Theta = \left[\begin{array}{cc} \Theta_{11} & \Theta_{12} \\ 0 & \Theta_{22} \end{array} \right]$$

and denote

$$W =[\ B((B^T B)^{-1})^T \quad B^\perp\ ]^T,$$
$$\bar{K} =\Theta_{11}K,\ \mathcal{K} = [\bar{K}^T\ 0]^T,$$
$$\Gamma =\Theta W + W^T\Theta^T - P_1,\ \Psi = \Theta W P_1^{-1} W^T \Theta^T$$

where $\Theta_{11} \in \mathbb{R}^{n_x \times p}$, $\Theta_{12} \in \mathbb{R}^{n_x \times (n_x-p)}$ and $\Theta_{22} \in \mathbb{R}^{(n_x-p)\times(n_x-p)}$, $B^\perp$ stands for an orthogonal basis of the null space for B^T.

Pre- and post-multiplying the inequality (11.14) by

$$\mathrm{diag}\{I, I_5 \otimes \mathrm{diag}\{\Theta W, P_2\}, I \otimes (\Theta W)\}$$

and

$$\mathrm{diag}\{I, I_5 \otimes \mathrm{diag}\{(\Theta W)^T, P_2\}, I \otimes (\Theta W)^T\}$$

yields

$$\Pi_2 = \left[\begin{array}{ccc} \Pi_{11} & * & * \\ \tilde{\Pi}_{12}^* & -I_5 \otimes \mathrm{diag}\{\Psi, P_2\} & * \\ \tilde{\Pi}_{17} & 0 & -I \otimes \Psi \end{array} \right] \le 0 \qquad (11.15)$$

where

$$\tilde{\mathcal{B}}_1 = \left[\begin{array}{cccc} \mathcal{K}C_0 & 0 & -\mathcal{K} & \mathcal{K} \\ P_2 C_0 & 0 & -P_2 & P_2 \end{array} \right],\quad \tilde{\mathcal{B}}_2 = \left[\begin{array}{cccc} -\mathcal{K}C_0 & \mathcal{K} & \mathcal{K} & \mathcal{K} \\ -P_2 C_0 & P_2 & P_2 & P_2 \end{array} \right],$$
$$\tilde{\Pi}_{12} = \left[\begin{array}{cccc} \Theta W A_0 + \bar{\alpha}(1-\bar{\beta})\mathcal{K}C_0 & (1-\bar{\alpha})\mathcal{K} & \bar{\alpha}\bar{\beta}\mathcal{K} & \bar{\alpha}(1-\bar{\beta})\mathcal{K} \\ \bar{\alpha}(1-\bar{\beta})P_2 C_0 & (1-\bar{\alpha})P_2 & \bar{\alpha}\bar{\beta}P_2 & \bar{\alpha}(1-\bar{\beta})P_2 \end{array} \right],$$
$$\tilde{\Pi}_{13} = \sqrt{\bar{\chi}}\tilde{\mathcal{B}}_1,\quad \tilde{\Pi}_{14} = \sqrt{\bar{\alpha}}\tilde{\mathcal{B}}_2\mathcal{S}_1,\quad \tilde{\Pi}_{15} = \sqrt{\bar{\alpha}\bar{\beta}}\tilde{\mathcal{B}}_2\mathcal{S}_2,$$
$$\tilde{\Pi}_{16} = \sqrt{\bar{\alpha}\bar{\beta}}\tilde{\mathcal{B}}_2\mathcal{S}_2,\quad \tilde{\Pi}_{17} = [\ \sqrt{\bar{\alpha}(1-\bar{\beta})}(I \otimes \mathcal{K})\Upsilon\ \ 0\ \ 0\ \ 0\ ],$$
$$\tilde{\Pi}_{12}^* = [\ \tilde{\Pi}_{12}^T\ \ \tilde{\Pi}_{13}^T\ \ \tilde{\Pi}_{14}^T\ \ \tilde{\Pi}_{15}^T\ \ \tilde{\Pi}_{16}^T\ ]^T.$$

It is apparent from (11.15) that

$$\Pi_2 = \left[\begin{array}{ccc} \Pi_{11} & * & * \\ \tilde{\Pi}_{12}^* & -I_5 \otimes \mathrm{diag}\{\Gamma, P_2\} & * \\ \tilde{\Pi}_{17} & 0 & -I \otimes \Gamma \end{array} \right] \qquad (11.16)$$
$$+ \mathrm{diag}\Big\{0, I_5 \otimes \mathrm{diag}\{\Gamma - \Psi, 0\}, I \otimes (\Gamma - \Psi)\Big\}.$$

Finally, in light of

$$\Theta W + (\Theta W)^T - \Theta W P_1(\Theta W)^T - P_1$$
$$= -(P_1 - \Theta W)P_1(P_1 - \Theta W)^T \le 0,$$

one has

$$\Pi_2 \le \Pi_3 := \begin{bmatrix} \Pi_{11} & * & * \\ \tilde{\Pi}_{12}^* & -I_5 \otimes \mathrm{diag}\{\Gamma, P_2\} & * \\ \tilde{\Pi}_{17} & 0 & -I \otimes \Gamma \end{bmatrix}. \tag{11.17}$$

It should be pointed out that the matrix ΘW is invertible if $\Pi_2 \le \Pi_3 < 0$. It can be seen that $\Pi_2 < 0$ is equivalent to (11.14), and therefore $\Pi_2 < 0$ is equivalent to (11.7a) in Theorem 11.1. Finally, according to the analysis conducted above, the following theorem is easily accessible from Theorem 11.1 and its proof is therefore omitted.

Theorem 11.2 *Let the positive scalars $\delta_1, \delta_2, \delta_3, \delta_4$ and the positive definite matrix R be given. Assume that there exist two positive definite matrices P_1 and P_2, two matrices Θ and $\bar{K}$ and three positive scalars $\varepsilon_1, \varepsilon_2$ and π satisfying the matrix inequalities $\Pi_3 < 0$, and the condition (11.7b). In this case, with the controller gain matrix given by*

$$K = \Theta_{11}^{-1}\bar{K},$$

the closed-loop system (11.5) is secure with respect to $(\delta_1, \delta_2, \delta_3, \delta_4, R)$.

Controller design algorithm

Step 1.	Denote a positive scalar τ for linearly searching step size.
Step 2.	Let $\pi = 0$ and then check the solvability of inequality (11.7a). If it is solvable, go to next step, else go to *Step 6*.
Step 3.	Let $\pi = \pi + \tau$, solve the following optimal problem:

$$\mathbf{OP}: \quad \min_{\pi, \tilde{\rho}_{\max}} \tilde{\rho}_{\max} + \varepsilon_1 + \varepsilon_2 \quad \text{s. t. (11.7a) and } P < \tilde{\rho}_{\max} I.$$

Step 4.	If the optimal problem **OP** is solvable, calculate ρ, γ, $\lambda_{\max}(P_R)$, $\lambda_{\min}(P)$ and θ^2. In what follows, obtain Ξ_2 and go to the next step. If it is not solvable, go to *Step 6*.
Step 5.	Check the condition (11.7b). If $\Xi_2 \le \delta_3^2$, $K = \Theta_{11}^{-1}\bar{K}$ is the desired controller gain, the calculation stops. Else go to *Step 3*.
Step 6.	The algorithm is infeasible. Stop.

If the threshold $\delta_1 = 0$, it is not difficult to see that the triggering rules are always fulfilled, that is, the event-based approach reduces to a time-driven one. Consequently, we have the following corollary.

Corollary 11.1 *Let the positive scalars $\delta_2, \delta_3, \delta_4$ and the positive definite matrix R be given. Assume that there exist two positive definite matrices P_1*

and P_2, two matrices Θ and $\bar{K}$ and two positive scalars ε_1 and π satisfying the following inequalities

$$
\begin{cases}
\Xi_3 = \begin{bmatrix} \Xi_{11}^* & * & * \\ \tilde{\Xi}_{12}^* & -I_5 \otimes \bar{\Gamma} & * \\ \Xi_{17}^* & 0 & -I \otimes \Gamma \end{bmatrix} < 0, & (11.18\text{a}) \\[4mm]
\Xi_4 = \max\left\{ \dfrac{\lambda_{\max}(P_R)\delta_4^2}{\lambda_{\min}(P)}, \dfrac{\theta^2\gamma}{\lambda_{\min}(P)(\gamma-1)} \right\} \le \delta_3^2 & (11.18\text{b})
\end{cases}
$$

where

$$
\bar{B}_1 = \begin{bmatrix} \mathcal{K}C_0 & 0 & -\mathcal{K} \\ P_2 C_0 & 0 & -P_2 \end{bmatrix}, \quad
\bar{B}_2 = \begin{bmatrix} -\mathcal{K}C_0 & \mathcal{K} & \mathcal{K} \\ -P_2 C_0 & P_2 & P_2 \end{bmatrix},
$$

$$
\Xi_{11}^* = diag\left\{ \Pi_{00}, -P_2 + \pi I, \ -\varepsilon_1 I \right\}, \quad \theta = \sqrt{\varepsilon_1}\delta_2,
$$

$$
\Xi_{12}^* = \begin{bmatrix} \Theta W A_0 + \bar{\alpha}(1-\bar{\beta})\mathcal{K}C_0 & (1-\bar{\alpha})\mathcal{K} & \bar{\alpha}\bar{\beta}\mathcal{K} \\ \bar{\alpha}(1-\bar{\beta})P_2 C_0 & (1-\bar{\alpha})P_2 & \bar{\alpha}\bar{\beta}P_2 \end{bmatrix},
$$

$$
\Xi_{13}^* = \sqrt{\bar{\chi}}\bar{B}_1, \quad \Xi_{14}^* = \sqrt{\tilde{\alpha}}\bar{B}_2\bar{S}_1, \quad \Xi_{15}^* = \sqrt{\bar{\alpha}\bar{\beta}}\bar{B}_2\bar{S}_2,
$$

$$
\Xi_{16}^* = \sqrt{\tilde{\alpha}\bar{\beta}}\bar{B}_1\bar{S}_1, \quad \Pi_{17}^* = \begin{bmatrix} \sqrt{\bar{\alpha}(1-\bar{\beta})}(I \otimes \mathcal{K})\Upsilon & 0 & 0 \end{bmatrix},
$$

$$
\tilde{\Xi}_{12}^* = \begin{bmatrix} \Xi_{12}^{*T} & \Xi_{13T}^* & \Xi_{14}^{*T} & \Xi_{15}^{*T} & \Xi_{16}^{*T} \end{bmatrix}^T,
$$

$$
\bar{S}_1 = diag\{I, 0\}, \quad \bar{S}_2 = diag\{-I, I\}.
$$

In this case, with the controller gain matrix given by $K = \Theta_{11}^{-1}\bar{K}$, the closed-loop system (11.5) is secure with respect to $(0, \delta_2, \delta_3, \delta_4, R)$.

Remark 11.6 *In this chapter, the event-based security control problem is investigated for a class of discrete-time stochastic systems with multiplicative noises and cyber-attacks. By utilizing two sets of Bernoulli distributed white sequences, a novel attack model is proposed to account for the phenomenon of both randomly occurring DoS attacks and randomly occurring deception attacks. Based on such a model, the result established in Theorem 11.2 contains all the information about the threshold of the event-triggered communication mechanism, the security requirements, and the statistical information of cyber-attacks. It is worth mentioning that the research approaches are general and independent of the system model. Therefore, the obtained results can been easily extended to more complex cases including the case that attacks between controllers and physical systems, and between sensors and controllers, are taken into account simultaneously.*

11.4   Simulation Examples

Following [151], we consider the security control problem for a geared DC motor, which is a component of the MS150 modular servo system. Setting the sampling time $T = 0.01s$, we obtain the following discretized nominal system matrix and measurement matrix:

$$A_0 = \begin{bmatrix} 1 & 0.0098 \\ 0 & 0.9653 \end{bmatrix}, \quad C_0 = \begin{bmatrix} 1 \\ 0 \end{bmatrix}^T.$$

In the MS150 modular servo system, the control input u_k is a voltage and the measurement output $\tilde{y}_k$ is a rotary angle of the extended shaft, which is also called the position of the shaft. The movement of the motor is affected by the stochastic disturbance $w_{i,k}$ $(i = 1)$. To this end, other parameters are given as

$$A_1 = \begin{bmatrix} 0.01 & 0 \\ 0 & 0.075 \end{bmatrix}, \quad B = \begin{bmatrix} 1.20 \\ 0.01 \end{bmatrix}, \quad C_1 = \begin{bmatrix} 0.12 \\ -0.14 \end{bmatrix}^T.$$

Assume that the attack probabilities are $\bar{\alpha} = 0.90$ and $\bar{\beta} = 0.25$. Moreover, the parameters δ_1, δ_2, δ_3, δ_4 and R are, respectively, 0.004, 0.10, 0.32, 0.30 and diag$\{0.95, 0.95\}$.

By using MATLAB® software (with YALMIP 3.0) where the solver is selected as "solvesdp", a set of feasible solutions of Theorem 11.2 is obtained as follows:

$$\varepsilon_1 = 25.22, \quad \varepsilon_2 = 22.48, \quad \pi = 0.4189, \quad \bar{K} = -7.1898, \quad P_2 = 5.1811,$$

$$P_1 = \begin{bmatrix} 9.4457 & 0.5763 \\ 0.5763 & 9.0843 \end{bmatrix}, \Theta = \begin{bmatrix} 12.80 & 2.8631 \\ 0 & 18.91 \end{bmatrix}.$$

Furthermore, the desired control parameter is obtained as $K = -0.5615$.

In the simulation, the disturbance signal of attackers ξ_k^t is selected as $\delta_2 \sin(k)$ and the initial value is set as $x_0 = [0.20 \ -0.18]^T$. The simulation results are shown in Figs. 11.2-11.4, where Fig. 11.2 plots the norm of states for the open-loop system and the closed-loop system without attacks. Since the spectral radius of the matrix $A_0 \otimes A_0 + \sum_{i=1}^p A_i \otimes A_i$ is 1.0015, we can see that the open-loop system is unstable.

Fig. 11.3 depicts the norm of states for the closed-loop system with attacks. The curve experiences three unexpected jumps at $k = 75$, 78, 81 and 114 since the system suffers from deception attacks at $k = 73$, 74, 77, 80 and 113. In comparison with the dotted line in Fig. 11.2, the control performance is degraded by attacks.

The event-triggered time and the successful time of attacks are shown in Fig. 11.4, from which we can easily find that the number of event-triggered communication is quite small and the communication burden is effectively reduced.

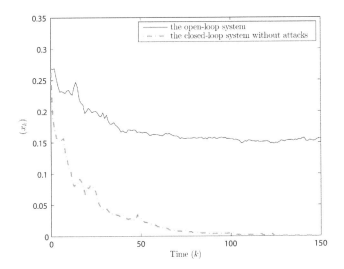

FIGURE 11.2
The norm of states for open- and closed-loop systems without attacks.

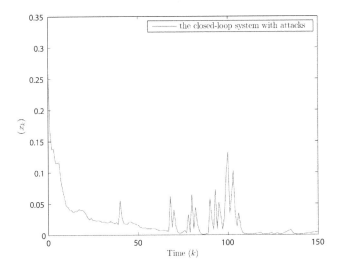

FIGURE 11.3
The norm of states for closed-loop systems with attacks.

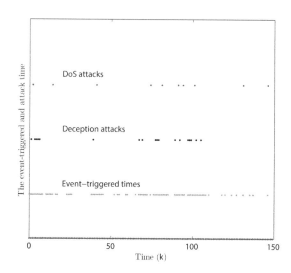

FIGURE 11.4
The event-triggered time and the attack time.

11.5   Summary

In this chapter, the event-based security control problem has been discussed for a class of discrete-time stochastic systems with multiplicative noises. The plant under consideration is subject to randomly occurring DoS attacks and randomly occurring deception attacks. First, a novel attack model has been provided to describe two such attacks within a unified framework. Then, an event-based communication mechanism has been utilized to reduce the communication burden. Furthermore, the output feedback controller gain matrix has been obtained by solving a linear matrix inequality with nonlinear constraints. Finally, a simulation example has been exploited to show the effectiveness of the event-triggered security control scheme proposed in this chapter.

12

Event-Triggered Consensus Control for Multi-Agent Systems Subject to Cyber-Attacks in the Framework of Observers

It has been recognized that an event-triggered control strategy is helpful in reducing the resource consumption while maintaining acceptable system performance. In addition, the adopted sensors collecting agent's information are made of low-cost devices with low computing capacity. Such limited computing capacity could become a stumbling block to the use of sophisticated encryption for secure transmission, and this makes it possible for the attackers to extract information from sensor transmissions. Therefore, it is of practical importance to deal with event-triggering consensus control subject to cyber-attacks. The challenges we are going to cope with are identified as follows: 1) How do we make mild assumptions on statistical behaviors of network- or detection-induced deception attacks? 2) How do we design a distributed observer effectively fusing the unreliable data and realizing output-feedback-based consensus control? 3) What kind of method can be developed to reduce the computation burden as the number of agents increases?

Benefiting from the developed analysis method in Chapter 10 and the proposed attack model in Chapter 11, the observer-based event-triggering consensus control problem is investigated for a class of discrete-time multi-agent systems with lossy sensors and cyber-attacks. A novel distributed observer is proposed to estimate the relative full states, and the estimated states are then used in the feedback protocol in order to achieve overall consensus. An event-triggered mechanism with a state-independent threshold is adopted to update the control input signals so as to reduce unnecessary data communications. The success ratio of the launched attacks is taken into account to reflect the probabilistic failures of the attacks passing through the protection devices subject to limited resources and network fluctuations. By making use of eigenvalues and eigenvectors of the Laplacian matrix, the closed-loop system is transformed into an easy-to-analyze setting and then a sufficient condition is derived to guarantee the desired consensus. Furthermore, the controller gain is obtained in terms of the solution to a certain matrix inequality, which is independent of the number of agents. An algorithm is provided to optimize the consensus bound. Finally, a simulation example is utilized to illustrate the usefulness of the proposed controller design scheme.

12.1   Modeling and Problem Formulation

In this chapter, it is assumed that the multi-agent system has N agents which communicate with each other according to a fixed network topology represented by an undirected graph $\mathscr{G} = (\mathscr{V}, \mathscr{E}, \mathscr{H})$ of order N with the set of agents $\mathscr{V} = \{1, 2, \cdots, N\}$, the set of edges $\mathscr{E} \in \mathscr{V} \times \mathscr{V}$, and the weighted adjacency matrix $\mathscr{H} = [h_{ij}]$ with nonnegative adjacency element h_{ij}. An edge of $\mathscr{G}$ is denoted by the ordered pair (i, j). The adjacency elements associated with the edges of the graph are positive, i.e., $h_{ij} > 0 \iff (i, j) \in \mathscr{E}$, which means that agent i can obtain information from agent j. The neighborhood of agent i is denoted by $\mathscr{N}_i = \{j \in \mathscr{V} : (j, i) \in \mathscr{E}\}$. An element of $\mathscr{N}_i$ is called a neighbor of agent i. In this chapter, self-edges (i, i) are not allowed, i.e., $(i, i) \notin \mathscr{E}$ for any $i \in \mathscr{V}$. The in-degree of agent i is defined as $\deg_{in}^i = \sum_{j \in \mathscr{N}_i} h_{ij}$.

Consider a multi-agent system consisting of N agents, in which the dynamics of agent i is described by the following discrete-time stochastic system:

$$\begin{cases} x_i(k+1) = Ax_i(k) + Dx_i(k)\varpi(k) + Bu_i(k), \\ y_{ij}(k) = \alpha_{ij}(k)C(x_j(k) - x_i(k)), \quad j \in \mathscr{N}_i \end{cases} \tag{12.1}$$

where $x_i(k) \in \mathbb{R}^n$, $y_{ij}(k) \in \mathbb{R}^q$ and $u_i(k) \in \mathbb{R}^p$ are states, relative measurement outputs to its neighbor j, and control inputs of agent i, respectively. $\varpi(k) \in \mathbb{R}$ is a multiplicative noise with zero mean and unity variance. A, B, C and D are known constant matrices with compatible dimensions. The stochastic variable $\alpha_{ij}(k)$, which accounts for the probabilistic packet dropouts (lossy behaviors) of the sensor outputs/communications, is a Bernoulli distributed white sequence taking values of 0 or 1 with the following probabilities:

$$\text{Prob}\{\alpha_{ij}(k) = 0\} = 1 - \bar{\alpha}, \quad \text{Prob}\{\alpha_{ij}(k) = 1\} = \bar{\alpha}$$

where $\bar{\alpha} \in [0, 1)$ is a known constant.

In this chapter, we assume that the information $\sum_{j \in \mathscr{N}_i} h_{ij} y_{ij}(k)$ on sensor i, when being transmitted to observer i, could be subject to deception attacks which occur in a random way (from the defenders' perspective) because of the success ratio of the launched attacks passing through the protection devices subject to limited resources and network fluctuations. A new model reflecting such a situation is proposed as follows:

$$\tilde{y}_i(k) = \sum_{j \in \mathscr{N}_i} h_{ij} y_{ij}(k) + \beta_i(k)\Big(\delta_i(k) - \sum_{j \in \mathscr{N}_i} h_{ij} y_{ij}(k)\Big) \tag{12.2}$$

where $\tilde{y}_i(k)$ is the received signal by observer i subject to cyber-attacks, and $\delta_i(k)$ satisfying $\|\delta_i(k)\|^2 \leq \delta_i$ stands for the signal sent by attackers. Here, δ_i is a known positive constant and the stochastic variable $\beta_i(k)$ is a Bernoulli distributed white sequence taking values of 0 or 1 with the probabilities

$$\text{Prob}\{\beta_i(k) = 0\} = \bar{\beta}, \quad \text{Prob}\{\beta_i(k) = 1\} = 1 - \bar{\beta}$$

where $\bar{\beta} \in [0, 1)$ is also a known positive constant.

Denoting $\xi_i(k) = \sum_{j \in \mathcal{N}_i} h_{ij}\big(x_j(k) - x_i(k)\big)$, a control protocol based on the estimation of $\xi_i(k)$ is designed as follows:

$$\begin{cases} \hat{\xi}_i(k+1) = A_c\hat{\xi}_i(k) + L_c\tilde{y}_i(k), \\ u_i(k) = K_c\hat{\xi}_i(k) \end{cases} \tag{12.3}$$

where A_c, L_c and K_c are the controller gain matrices to be determined.

We now introduce the event-triggering mechanism, as shown in Fig. 11.1, which is adopted to attempt to reduce the update frequency of controllers with a zero-order hold manner. In order to characterize such a mechanism, let the triggering time sequence of agent i be $0 = k_0^i < k_1^i < k_2^i < \cdots < k_{s_i}^i < \cdots$ and the event generator functions $\Upsilon_i(\cdot, \cdot) : \mathbb{R}^q \times \mathbb{R} \to \mathbb{R} \ (i = 1, 2, \cdots, N)$ be selected as follows:

$$\Upsilon_i(e_i(k), \theta_i) = e_i^T(k)e_i(k) - \theta_i. \tag{12.4}$$

Here $e_i(k) \triangleq \hat{\xi}_i(k_{s_i}^i) - \hat{\xi}_i(k) \ (k \in [k_{s_i}^i, k_{s_i+1}^i))$ where $\hat{\xi}_i(k_{s_i}^i)$ is the control input signal at the latest triggering time $k_{s_i}^i$ and $\theta_i \ (i = 1, 2, \cdots, N)$ are given positive scalars. The executions are triggered as long as the condition

$$\Upsilon_i(e_i(k), \theta_i) > 0 \tag{12.5}$$

is satisfied. Obviously, the sequence of event-triggering instants is determined iteratively by

$$k_{s+1}^i = \inf\{k \in \mathbb{N}|k > k_s^i, \ \Upsilon_i(e_i(k), \theta_i) > 0\}.$$

Therefore, in the event-triggered setup for $k \in [k_{s_i}^i, \ k_{s_i+1}^i)$, the observation-based control law can be written as

$$\begin{cases} \hat{\xi}_i(k+1) = A_c\hat{\xi}_i(k) + L_c\tilde{y}_i(k), \\ u_i(k) = K_c\hat{\xi}_i(k_{s_i}^i). \end{cases} \tag{12.6}$$

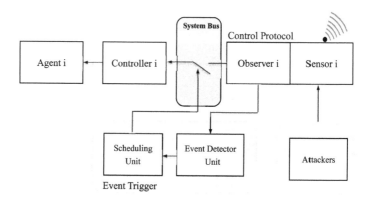

FIGURE 12.1
Event-triggering control scheme on Agent i.

For the purpose of facilitating these discussions, denote

$$x_k \triangleq [\ x_1^T(k) \quad x_2^T(k) \quad \cdots \quad x_N^T(k)\ ]^T, \quad \hat{\xi}_{i,k} \triangleq \hat{\xi}_i(k),$$

$$\hat{\xi}_k \triangleq [\ \hat{\xi}_1^T(k) \quad \hat{\xi}_2^T(k) \quad \cdots \quad \hat{\xi}_N^T(k)\ ]^T, \quad \hat{\xi}_{i,k}^* \triangleq \hat{\xi}_i(k_{s_i}^i),$$

$$\hat{\xi}_k^* \triangleq [\ \hat{\xi}_{1,k}^{*T} \quad \hat{\xi}_{2,k}^{*T} \quad \cdots \quad \hat{\xi}_{N,k}^{*T}\ ]^T, \quad \Upsilon(k) \triangleq [r_{ij}(k)]_{N \times N},$$

$$\tilde{\Upsilon}(k) \triangleq \operatorname{diag}\{\tilde{r}_1(k), \tilde{r}_2(k), \cdots, \tilde{r}_N(k)\},$$

$$\Lambda_k = \operatorname{diag}\{\beta_1(k), \beta_2(k), \cdots, \beta_N(k)\},$$

$$\delta_k = [\ \delta_1^T(k) \quad \delta_2^T(k) \quad \cdots \quad \delta_N^T(k)\ ]^T,$$

$$\mathcal{H} \triangleq \begin{bmatrix} -\deg_{in}^1 & h_{1,2} & h_{1,3} & \cdots & h_{1,N} \\ h_{2,1} & -\deg_{in}^2 & h_{2,3} & \cdots & h_{2,N} \\ \vdots & \vdots & \vdots & \ddots & \vdots \\ h_{N,1} & h_{N,2} & h_{N,3} & \cdots & -\deg_{in}^N \end{bmatrix}$$

with $\tilde{r}_i(k) = \sum_{j \in \mathcal{N}_i} h_{ij}\alpha_{ij}(k)$ and

$$r_{ij}(k) = \begin{cases} \alpha_{ij}(k), & j \in \mathcal{N}_i, \\ 0, & j \notin \mathcal{N}_i. \end{cases}$$

Then, for $k \in [k_{s_i}^i,\ k_{s_i+1}^i)$, the closed-loop system with event-triggering can be derived from (12.1) and (12.6) as follows:

$$\begin{cases} x_{k+1} = (I \otimes A)x_k + (I \otimes D)x_k \varpi_k + (I \otimes BK_c)\hat{\xi}_k^*, \\ \hat{\xi}_{i,k+1} = A_c \hat{\xi}_{i,k} + \left[(I - \Lambda_k)(\mathcal{H} \circ \Upsilon_k - \tilde{\Upsilon}_k) \right]_i \otimes (L_c C)x_k \\ \qquad\quad + ([\Lambda_k]_i \otimes L_c)\delta_k \end{cases} \qquad (12.7)$$

where $[\bullet]_i$ stands for a row vector whose elements come from the ith row of matrix $\bullet$.

Before proceeding further, we introduce the following definition.

Definition 12.1 *Let U be a known positive scalar and the initial value satisfy $\sum_{i=1}^{N} \|x_i(0) - \frac{1}{N} \sum_{j=1}^{N} x_j(0)\|^2 \leq \chi_0^2$. The discrete-time multi-agent system (12.1) with a given undirected communication graph $\mathscr{G}$ is said to reach consensus with bound U in a mean-square sense if the system dynamics satisfies*

$$\mathbb{E}\{\|x_i(k) - x_j(k)\|^2\} \leq U, \quad \forall i, j \in \mathscr{V}, \quad k \geq 0. \tag{12.8}$$

Remark 12.1 *Note that the multiplicative noise (also called state-dependent noise) $\varpi(k) \in \mathbb{R}$ is considered in system (12.1) to reflect the random perturbations of the system parameters, see [12, 18, 64, 115] for more details. On the other hand, for the addressed security-constrained consensus issue, the bound U is essentially a function with respect to χ_0, $\delta = \begin{bmatrix} \delta_1 & \cdots & \delta_N \end{bmatrix}^T$ and $\theta = \begin{bmatrix} \theta_1 & \cdots & \theta_N \end{bmatrix}^T$, and can therefore be denoted as $U(\chi_0, \delta, \theta)$. From the engineering viewpoint, a suitable bound function can be obtained by optimizing the control parameters. As such, in this chapter, we aim to design the parameters A_c, L_c and K_c such that the discrete-time multi-agent system (12.1) with the observer-based event-triggered scheme (12.3) is consensus with bound $U(\chi_0, \delta, \theta)$ in a mean-square sense.*

Remark 12.2 *In this chapter, an event-based control scheme shown in Fig. 12.1 is adopted to attempt to reduce both the update frequency of controllers and the communication burden via system buses. The triggered condition used in this chapter is based on the absolute condition changes where the threshold is state-independent. On the other hand, for deception attacks, the adversary destroys the real data and sends false information to controllers. Due to the application of safety protection devices/techniques, the communication protocol as well as the fluctuating network conditions, there is a nonzero probability for each launched cyber-attack to be unsuccessful at a certain time. Therefore, as discussed in the introduction, a set of Bernoulli-distributed white sequences with known conditional probabilities is adopted to reflect the success ratio of the launched cyber-attacks from the defenders' perspective.*

For the purpose of consensus control design, some suitable transformations are necessary by making use of the specific characteristics of the multi-agent systems. We start by defining the average state of all agents as

$$\bar{x}_k \triangleq \frac{1}{N} \sum_{i=1}^{N} x_i(k) = \frac{1}{N}(\mathbf{1}^T \otimes I)x_k \tag{12.9}$$

and the deviation of each state from the average state as

$$\tilde{x}_i(k) \triangleq x_i(k) - \bar{x}_k$$

where "**1**" is a compatible dimension vector with each element of one. It is easy to obtain that

$$
\begin{aligned}
\bar{x}_{k+1} &= \frac{1}{N}(\mathbf{1}^T \otimes I)x_{k+1} \\
&= \frac{1}{N}(\mathbf{1}^T \otimes I)\Big\{(I \otimes A)x_k + (I \otimes D)x_k \varpi_k + (I \otimes BK_c)\hat{\xi}_k^*\Big\} \\
&= A\bar{x}_k + D\bar{x}_k w_k + \frac{1}{N}(\mathbf{1}^T \otimes (BK_c))\hat{\xi}_k^*, \\
\tilde{x}_{k+1} &= x_{k+1} - (\mathbf{1} \otimes I)\bar{x}_{k+1} \\
&= (I \otimes A)\tilde{x}_k + (I \otimes D)\tilde{x}_k \varpi_k + (\mathcal{N} \otimes (BK_c))\hat{\xi}_k^*
\end{aligned}
\tag{12.10}
$$

where $\mathcal{N} = [n_{ij}]_{N \times N}$ with

$$
n_{ij} = \begin{cases} \frac{N-1}{N}, & i = j, \\ -\frac{1}{N}, & i \neq j. \end{cases}
$$

Similarly, define $\bar{\xi}_k \triangleq \frac{1}{N}\sum_{i=1}^{N}\hat{\xi}_{i,k}$ and $\tilde{\xi}_{i,k} \triangleq \hat{\xi}_{i,k} - \bar{\xi}_k$. Then, taking $\mathbf{1}^T(I - \Lambda_k)(\mathcal{H} \circ \Upsilon_k - \tilde{\Upsilon}_k)\mathbf{1} = 0$ into consideration, one has

$$
\begin{aligned}
\bar{\xi}_{k+1} &= \frac{1}{N}(\mathbf{1}^T \otimes I)\hat{\xi}_{k+1} \\
&= A_c\bar{\xi}_k + \frac{1}{N}\big(\mathbf{1}^T(I - \Lambda_k)(\mathcal{H} \circ \Upsilon_k - \tilde{\Upsilon}_k)\big) \otimes (L_cC)\tilde{x}_k \\
&\quad + \frac{1}{N}\big((\mathbf{1}^T\Lambda_k) \otimes L_c\big)\delta_k, \\
\tilde{\xi}_{k+1} &= \hat{\xi}_{k+1} - (\mathbf{1} \otimes I)\bar{\xi}_{k+1} \\
&= (I \otimes A_c)\tilde{\xi}_k + (\mathcal{N}(I - \Lambda_k)(\mathcal{H} \circ \Upsilon_k - \tilde{\Upsilon}_k)) \otimes (L_cC)\tilde{x}_k \\
&\quad + ((\mathcal{N}\Lambda_k) \otimes L_c)\delta_k.
\end{aligned}
\tag{12.11}
$$

With the aid of (12.10) and (12.11), the closed-loop system (12.7) can be transformed into

$$
\begin{cases}
\tilde{x}_{k+1} = (I \otimes A)\tilde{x}_k + (I \otimes D)\tilde{x}_k \varpi_k + \mathcal{N} \otimes (BK_c)\hat{\xi}_k^*, \\
\tilde{\xi}_{k+1} = (I \otimes A_c)\tilde{\xi}_k + (\mathcal{N}(I - \Lambda_k)(\mathcal{H} \circ \Upsilon_k - \tilde{\Upsilon}_k)) \otimes (L_cC)\tilde{x}_k \\
\qquad\quad + ((\mathcal{N}\Lambda_k) \otimes L_c)\delta_k.
\end{cases}
\tag{12.12}
$$

In what follows, by introducing the variables

$$
e_{i,k} \triangleq \hat{\xi}_{i,k} - \hat{\xi}_i(k_{s_i}^i), \quad e_k \triangleq [\ e_{1,k}^T \quad e_{2,k}^T \quad \cdots \quad e_{N,k}^T\ ]^T,
$$

one has

$$
\begin{cases}
\tilde{x}_{k+1} = (I \otimes A)\tilde{x}_k + (I \otimes D)\tilde{x}_k \varpi_k + I \otimes (BK_c)\tilde{\xi}_k - \mathcal{N} \otimes (BK_c)e_k, \\
\tilde{\xi}_{k+1} = (\mathcal{N}(I - \Lambda_k)(\mathcal{H} \circ \Upsilon_k - \tilde{\Upsilon}_k)) \otimes (LC)\tilde{x}_k \\
\qquad\quad + (I \otimes A_c)\tilde{\xi}_k + ((\mathcal{N}\Lambda_k) \otimes L)\delta_k
\end{cases}
\tag{12.13}
$$

where the relation $\mathcal{N} \otimes (BK)\hat{\xi}_k = \mathcal{N} \otimes (BK)\tilde{\xi}_k = I \otimes (BK)\tilde{\xi}_k$ has been utilized. Furthermore, subtracting their expectations from the stochastic matrices Υ_k, $\tilde{\Upsilon}_k$ and Λ_k yields

$$
\begin{cases}
\tilde{x}_{k+1} = (I \otimes A)\tilde{x}_k + (I \otimes D)\tilde{x}_k \varpi_k + I \otimes (BK_c)\tilde{\xi}_k - \mathcal{N} \otimes (BK_c)e_k, \\
\tilde{\xi}_{k+1} = (I \otimes A_c)\tilde{\xi}_k + \bar{\alpha}\bar{\beta}\mathcal{H} \otimes (L_cC)\tilde{x}_k + (1 - \bar{\beta})(\mathcal{N} \otimes L_c)\delta_k \\
\qquad + (\mathcal{N}\tilde{\Lambda}_k\tilde{\mathcal{H}}_k) \otimes (L_cC)\tilde{x}_k + \bar{\alpha}(\mathcal{N}\tilde{\Lambda}_k\mathcal{H}) \otimes (L_cC)\tilde{x}_k \\
\qquad + \bar{\beta}(\mathcal{N}\tilde{\mathcal{H}}_k) \otimes (L_cC)\tilde{x}_k - ((\mathcal{N}\tilde{\Lambda}_k) \otimes L_c)\delta_k
\end{cases}
$$

$$(12.14)$$

where

$$
\tilde{\Lambda}_k = (1 - \bar{\beta})I - \Lambda_k, \quad \tilde{\mathcal{H}}_k = \mathcal{H} \circ \Upsilon_k - \tilde{\Upsilon}_k - \bar{\alpha}\mathcal{H}.
$$

On the other hand, taking the coupling nature of the multi-agent systems into consideration, we select $\varphi_i \in \mathbb{R}^n$ with $\varphi_i^T\mathcal{H} = \lambda_i\varphi_i^T$ $(i = 2, 3, \cdots, N)$ to form the unitary matrix $M = [1/\sqrt{N} \ \varphi_2 \ \cdots \ \varphi_N]$. Then, $\mathcal{H}$ can be transformed into the following diagonal form:

$$
\text{diag}\{0, \lambda_2, \cdots, \lambda_N\} = M^T\mathcal{H}M.
$$

It can be found that φ_i and λ_i are, respectively, the ith eigenvector and eigenvalue of the Laplacian matrix $\mathcal{H}$. Furthermore, in this chapter, we assume that eigenvalues of $\mathcal{H}$ in a descending order are written as $0 = \lambda_1 > \lambda_2 \geq \lambda_3 \geq \cdots \geq \lambda_N$.

In what follows, let us define $\eta_{1,k} \triangleq (M \otimes I)^T\tilde{x}_k$ and $\eta_{2,k} \triangleq (M \otimes I)^T\tilde{\xi}_k$, and then transform system (12.20) into

$$
\begin{cases}
\eta_{1,k+1} = (I \otimes A)\eta_{1,k} + (I \otimes D)\eta_{1,k}\varpi_k \\
\qquad + I \otimes (BK_c)\eta_{2,k} - (M^T\mathcal{N}) \otimes (BK_c)e_k, \\
\eta_{2,k+1} = (I \otimes A_c)\eta_{2,k} + \bar{\alpha}\bar{\beta}(M^T\mathcal{H}M) \otimes (L_cC)\eta_{1,k} \\
\qquad + (1 - \bar{\beta})(M^T\mathcal{N} \otimes L_c)\delta_k \\
\qquad + (M^T\mathcal{N}\tilde{\Lambda}_k\tilde{\mathcal{H}}_kM) \otimes (L_cC)\eta_{1,k} \\
\qquad + \bar{\alpha}(M^T\mathcal{N}\tilde{\Lambda}_k\mathcal{H}M) \otimes (L_cC)\eta_{1,k} \\
\qquad + \bar{\beta}(M^T\mathcal{N}\tilde{\mathcal{H}}_kM) \otimes (L_cC)\eta_{1,k} \\
\qquad - ((M^T\mathcal{N}\tilde{\Lambda}_k) \otimes L_c)\delta_k.
\end{cases}
$$

$$(12.15)$$

Finally, partition $\eta_{1,k}$ and $\eta_{2,k}$ into two parts, i.e., $\eta_{1,k} = \begin{bmatrix} \zeta_{1,k}^T & \tilde{\eta}_{1,k}^T \end{bmatrix}^T$ and $\eta_{2,k} = \begin{bmatrix} \zeta_{2,k}^T & \tilde{\eta}_{2,k}^T \end{bmatrix}^T$, where $\zeta_{1,k}, \zeta_{2,k} \in \mathbb{R}^n$ are, respectively, the vectors consisting of the first n elements of the aforementioned two vectors. It is not difficult to verify that $\zeta_{1,k} = \zeta_{2,k} = 0$. In order to simplify the subsequent analysis, we denote $\Xi \triangleq \text{diag}\{\lambda_2, \cdots, \lambda_N\}$, $S \triangleq \begin{bmatrix} 0 & I \end{bmatrix}$, and then have the following

$$
\begin{cases}
\tilde{\eta}_{1,k+1} = (I \otimes A)\tilde{\eta}_{1,k} + (I \otimes D)\tilde{\eta}_{1,k}\varpi_k \\
\qquad\quad + I \otimes (BK_c)\tilde{\eta}_{2,k} - (SM^T\mathcal{N}) \otimes (BK_c)e_k, \\
\tilde{\eta}_{2,k+1} = (I \otimes A_c)\tilde{\eta}_{2,k} + \bar{\alpha}\bar{\beta}\Xi \otimes (L_cC)\tilde{\eta}_{1,k} \\
\qquad\quad + (1 - \bar{\beta})((SM^T\mathcal{N}) \otimes L_c)\delta_k \\
\qquad\quad + (SM^T\mathcal{N}\tilde{\Lambda}_k\tilde{\mathcal{H}}_k MS^T) \otimes (L_cC)\tilde{\eta}_{1,k} \\
\qquad\quad + \bar{\alpha}(SM^T\mathcal{N}\tilde{\Lambda}_k\mathcal{H}MS^T) \otimes (L_cC)\tilde{\eta}_{1,k} \\
\qquad\quad + \bar{\beta}(SM^T\mathcal{N}\tilde{\mathcal{H}}_k MS^T) \otimes (L_cC)\tilde{\eta}_{1,k} \\
\qquad\quad - ((SM^T\mathcal{N}\tilde{\Lambda}_k) \otimes L_c)\delta_k.
\end{cases}
\tag{12.16}
$$

Remark 12.3 *Obviously, the addressed system (12.7) is bounded consensus in a mean-square sense if the states of the equivalent system (12.16) are bounded. As is well known, the eigenvalues of the Laplacian matrix describing the topology of multi-agent systems play crucial roles in the consensus analysis. In this subsection, an equivalent dynamic system is established by removing zero dynamics ($\zeta_{1,k} = \zeta_{2,k} = 0$) with the aid of the eigenvectors and eigenvalues of the Laplacian matrix as well as the properties of the Kronecker product. Such a transformation shows how the consensus performance and controller design are influenced by the network topology.*

12.2   Consensus Analysis

After the preliminaries in the above subsection, we are now in a position to consider the consensus analysis for the multi-agent system (12.7) (or, equivalently, (12.16)) with the event-triggering mechanism. By utilizing the Lyapunov stability theorem combined with stochastic analysis approaches, a sufficient condition is proposed to guarantee that the multi-agent system (12.7) is bound consensus in a mean-square sense.

Theorem 12.1 *Let the undirected communication graph $\mathcal{G}$ and the matrices A_c, L_c and K_c be given. If there exist a positive definite matrix $\bar{\mathcal{P}}$ and two positive scalars γ_0 and ε satisfying*

$$
(1 + \varepsilon)\mathcal{A}_1^T\mathcal{P}\mathcal{A}_1 + \mathcal{A}_2^T\mathcal{P}\mathcal{A}_2 + \mathcal{F} - \mathcal{P} + \gamma_0 I < 0,
\tag{12.17}
$$

then the controlled multi-agent system (12.7) is consensus in a mean-square sense with bound

$$
U(\chi_0, \delta, \theta) = \max\left\{ \frac{2\rho_{\max}\chi_0^2}{\rho_{\min}}, \frac{\mu\Theta}{\mu - 1} \right\}
\tag{12.18}
$$

where

$$\mathcal{P} = \begin{bmatrix} I \otimes P_1 & * \\ I \otimes P_0 & I \otimes P_2 \end{bmatrix}, \quad \rho_{\max} = \lambda_{\max}(\mathcal{P}), \quad \rho_{\min} = \lambda_{\min}(\mathcal{P}),$$

$$\mathcal{A}_1 = \begin{bmatrix} I \otimes A & I \otimes (BK_c) \\ \bar{\alpha}\bar{\beta}\Xi \otimes (L_cC) & I \otimes A_c \end{bmatrix}, \quad \mathcal{A}_2 = \begin{bmatrix} I \otimes D & 0 \\ 0 & 0 \end{bmatrix},$$

$$\mathcal{D}_1 = \begin{bmatrix} -(SM^T\mathcal{N}) \otimes (BK_c) \\ 0 \end{bmatrix}, \quad \mathcal{D}_2 = \begin{bmatrix} 0 \\ (1-\bar{\beta})(SM^T\mathcal{N}) \otimes L_c \end{bmatrix},$$

$$\Psi_1 = (1+\varepsilon^{-1})\mathcal{D}_2^T\mathcal{P}\mathcal{D}_2 + \frac{\bar{\beta}(1-\bar{\beta})(N-1)}{N} I \otimes (L_c^T P_2 L_c),$$

$$\Psi = \begin{bmatrix} \Psi_1 & \mathcal{D}_1^T\mathcal{P}\mathcal{D}_2 \\ * & (1+\varepsilon^{-1})\mathcal{D}_1^T\mathcal{P}\mathcal{D}_1 \end{bmatrix}, \quad \Theta = \lambda_{\max}(\Psi)\sum_{i=1}^{N}(\theta_i + \delta_i),$$

$$\mathcal{F} = diag\left\{ \tilde{\mathcal{F}} \otimes (L_cC)^T P_2(L_cC), 0 \right\}, \quad \mu = \frac{\rho_{\max}}{\rho_{\max} - \gamma_0},$$

$$\tilde{\mathcal{F}} = \frac{N-1}{N}\left(2\bar{\alpha}(1-\bar{\alpha})(1+\bar{\beta}(1-\bar{\beta}))SM^T\tilde{\mathcal{H}}MS^T + \bar{\beta}(1-\bar{\beta})\Xi^T\Xi \right),$$

$$\tilde{\mathcal{H}} = [\tilde{h}_{ij}]_{N\times N}, \quad with \quad \tilde{h}_{ij} = \begin{cases} \sum_{j\in\mathcal{N}_i} h_{ij}^2, & i=j, \\ -h_{ij}^2, & i\neq j. \end{cases}$$

Proof: Before proceeding to the consensus analysis, let us calculate some important expectations of matrix functions on $\tilde{\Lambda}_k$ or/and $\tilde{\mathcal{H}}_k$. Firstly, for $\mathcal{N}\tilde{\mathcal{H}}_k \triangleq [\tilde{h}_{ij}(k)]_{N\times N}$, one has

$$\tilde{h}_{ij}(k) = \begin{cases} -\sum_{s\in\mathcal{N}_i} h_{is}(\alpha_{is}(k)-\bar{\alpha}) + m_j(k), & i=j, \\ h_{ij}(\alpha_{ij}(k)-\bar{\alpha}) + m_j(k), & i\neq j, \quad j\in\mathcal{N}_i, \\ m_j(k), & i\neq j, \quad j\notin\mathcal{N}_i \end{cases}$$

with

$$m_j(k) = \frac{1}{N}\sum_{s\in\mathcal{N}_i} h_{is}(\alpha_{is}(k)-\bar{\alpha}) - \frac{1}{N}\sum_{s\in\mathcal{N}_i} h_{si}(\alpha_{si}(k)-\bar{\alpha}).$$

Then, the expectation of $\left(SM^T\mathcal{N}\tilde{\mathcal{H}}_kMS^T\right)^T\left(SM^T\mathcal{N}\tilde{\mathcal{H}}_kMS^T\right)$ is calculated as

$$\begin{aligned} &\mathbb{E}\left\{\left(SM^T\mathcal{N}\tilde{\mathcal{H}}_kMS^T\right)^T\left(SM^T\mathcal{N}\tilde{\mathcal{H}}_kMS^T\right)\right\} \\ &= \mathbb{E}\left\{SM^T\tilde{\mathcal{H}}_k^T\mathcal{N}^T\mathcal{N}\tilde{\mathcal{H}}_kMS^T\right\} \\ &\quad - \mathbb{E}\left\{SM^T\tilde{\mathcal{H}}_k^T\mathcal{N}^TM(I-S^TS)M^T\mathcal{N}\tilde{\mathcal{H}}_kMS^T\right\} \\ &= \mathbb{E}\left\{SM^T\tilde{\mathcal{H}}_k^T\mathcal{N}^T\mathcal{N}\tilde{\mathcal{H}}_kMS^T\right\} \\ &= SM^T\mathbb{E}\left\{\tilde{\mathcal{H}}_k^T\mathcal{N}^T\mathcal{N}\tilde{\mathcal{H}}_k\right\}MS^T \\ &= \frac{2\bar{\alpha}(1-\bar{\alpha})(N-1)}{N}SM^T\tilde{\mathcal{H}}MS^T. \end{aligned} \tag{12.19}$$

Similarly, one can calculate the following expectations:

$$
\begin{aligned}
&\mathbb{E}\left\{\left(SM^T\mathcal{N}\tilde{\Lambda}_k\tilde{\mathcal{H}}_k MS^T\right)^T\left(SM^T\mathcal{N}\tilde{\Lambda}_k\tilde{\mathcal{H}}_k MS^T\right)\right\}\\
&= \frac{\bar{\beta}(1-\bar{\beta})(N-1)}{N}SM^T\mathbb{E}\{\tilde{\mathcal{H}}_k^T\tilde{\mathcal{H}}_k\}MS^T\\
&= \frac{2\bar{\alpha}\bar{\beta}(1-\bar{\alpha})(1-\bar{\beta})(N-1)}{N}SM^T\tilde{\mathcal{H}}MS^T
\end{aligned}
\tag{12.20}
$$

and

$$
\begin{aligned}
\mathbb{E}\left\{\left(SM^T\mathcal{N}\tilde{\Lambda}_k\right)^T\left(SM^T\mathcal{N}\tilde{\Lambda}_k\right)\right\} &= \frac{\bar{\beta}(1-\bar{\beta})(N-1)}{N}I,\\
\mathbb{E}\left\{\left(SM^T\mathcal{N}\tilde{\Lambda}_k\mathcal{H}MS^T\right)^T\left(SM^T\mathcal{N}\tilde{\Lambda}_k\mathcal{H}MS^T\right)\right\}\\
&= \frac{\bar{\beta}(1-\bar{\beta})(N-1)}{N}\Xi^T\Xi.
\end{aligned}
\tag{12.21}
$$

In what follows, construct the Lyapunov function:

$$
V_k \triangleq \tilde{\eta}_k^T\mathcal{P}\tilde{\eta}_k
$$

where $\tilde{\eta}_k = \begin{bmatrix} \tilde{\eta}_{1,k}^T & \tilde{\eta}_{2,k}^T \end{bmatrix}^T$. Then, calculating the difference of V_k along the trajectory of (12.16) and taking the mathematical expectation, one has

$$
\begin{aligned}
&\mathbb{E}\{\Delta V_k|\tilde{\eta}_k\} \triangleq \mathbb{E}\{V_{k+1}|\tilde{\eta}_k\} - V_k\\
&= \mathbb{E}\Bigg\{\Big(\big(\mathcal{A}_1 + \varpi_k\mathcal{A}_2 + \mathcal{F}_{1,k} + \mathcal{F}_{2,k} + \mathcal{F}_{3,k}\big)\tilde{\eta}_k\\
&\quad + \mathcal{D}_1 e_k + (\mathcal{D}_{3,k} + \mathcal{D}_2)\delta_k\Big)^T\mathcal{P}\Big(\big(\mathcal{A}_1 + \varpi_k\mathcal{A}_2 + \mathcal{F}_{1,k}\\
&\quad + \mathcal{F}_{2,k} + \mathcal{F}_{3,k}\big)\tilde{\eta}_k + \mathcal{D}_1 e_k + (\mathcal{D}_{3,k} + \mathcal{D}_2)\delta_k\Big)\Bigg\} - V_k\\
&= \tilde{\eta}_k^T\big(\mathcal{A}_1^T\mathcal{P}\mathcal{A}_1 + \mathcal{A}_2^T\mathcal{P}\mathcal{A}_2 + \mathcal{F} - \mathcal{P}\big)\tilde{\eta}_k + 2\tilde{\eta}_k^T\mathcal{A}_1^T\mathcal{P}\mathcal{D}_1 e_k\\
&\quad + 2\tilde{\eta}_k^T\mathcal{A}_1^T\mathcal{P}\mathcal{D}_2\delta_k + e_k^T\mathcal{D}_1^T\mathcal{P}\mathcal{D}_1 e_k + 2e_k^T\mathcal{D}_1^T\mathcal{P}\mathcal{D}_2\delta_k\\
&\quad + \delta_k\Big(\mathcal{D}_2^T\mathcal{P}\mathcal{D}_2 + \frac{\bar{\beta}(1-\bar{\beta})(N-1)}{N}(I\otimes(L^T P_2 L))\Big)\delta_k
\end{aligned}
\tag{12.22}
$$

where

$$
\begin{aligned}
\mathcal{F}_{1,k} &= \begin{bmatrix} 0 & 0 \\ (SM^T\mathcal{N}\tilde{\Lambda}_k\tilde{\mathcal{H}}_k MS^T)\otimes(L_c C) & 0 \end{bmatrix},\\
\mathcal{F}_{2,k} &= \begin{bmatrix} 0 & 0 \\ \bar{\alpha}(SM^T\mathcal{N}\tilde{\Lambda}_k\mathcal{H}MS^T)\otimes(L_c C) & 0 \end{bmatrix},\\
\mathcal{F}_{3,k} &= \begin{bmatrix} 0 & 0 \\ \bar{\beta}(SM^T\mathcal{N}\tilde{\mathcal{H}}_k MS^T)\otimes(L_c C) & 0 \end{bmatrix},\\
\mathcal{D}_{3,k} &= \begin{bmatrix} 0 \\ -(SM^T\mathcal{N}\tilde{\Lambda}_k)\otimes L_c \end{bmatrix}.
\end{aligned}
$$

Taking the condition (12.5) and $||\delta_i(k)||^2 \le \delta_i^2$ into account, one has

$$
\begin{aligned}
\mathbb{E}\{\Delta V_k|\tilde{\eta}_k\} \le\; & \tilde{\eta}_k^T\big((1+\varepsilon)\mathcal{A}_1^T\mathcal{P}\mathcal{A}_1 + \mathcal{A}_2^T\mathcal{P}\mathcal{A}_2 + \mathcal{F} - \mathcal{P}\big)\tilde{\eta}_k \\
& + 2e_k^T\mathcal{D}_1^T\mathcal{P}\mathcal{D}_2\delta_k + e_k^T(1+2\varepsilon^{-1})\mathcal{D}_1^T\mathcal{P}\mathcal{D}_1 e_k \\
& + \delta_k^T\Big((1+2\varepsilon^{-1})\mathcal{D}_2^T\mathcal{P}\mathcal{D}_2 + \frac{\bar{\beta}(1-\bar{\beta})(N-1)}{N}(I\otimes P_2)\Big)\delta_k \\
\le\; & \tilde{\eta}_k\Psi\tilde{\eta}_k^T - \gamma_0\tilde{\eta}_k^T\tilde{\eta}_k + \lambda_{\max}(\Psi)\sum_{i=1}^{N}(\theta_i+\delta_i) \\
\le\; & -\gamma_0\tilde{\eta}_k^T\tilde{\eta}_k + \Theta,
\end{aligned}
\tag{12.23}
$$

which means that

$$
\mathbb{E}\{\mathbb{E}\{\Delta V_k|\tilde{\eta}_k\}\} = \mathbb{E}\{\Delta V_k\} < -\gamma_0\mathbb{E}\{||\tilde{\eta}_k||^2\} + \Theta. \tag{12.24}
$$

According to the above relation, for any $\mu > 1$, it follows that

$$
\begin{aligned}
\mu^{k+1}\mathbb{E}\{V_{k+1}\} - &\mu^k\mathbb{E}\{V_k\} \\
&\le \mu^{k+1}\mathbb{E}\{\Delta V_k\} + \mu^k(\mu-1)\mathbb{E}\{V_k\} \\
&\le \big(\rho_{\max}^P(\mu-1) - \gamma_0\mu\big)\mu^k\mathbb{E}\{||\tilde{\eta}_k||^2\} + \mu^{k+1}\Theta.
\end{aligned}
\tag{12.25}
$$

By selecting a suitable positive scalar μ satisfying $\rho_{\max}(\mu-1) - \gamma_0\mu = 0$, it is computed that

$$
\mu^k\mathbb{E}\{V_k\} - \mathbb{E}\{V_0\} = \sum_{s=1}^{k}\big\{\mu^s\mathbb{E}\{V_s\} - \mu^{s-1}\mathbb{E}\{V_{s-1}\}\big\} \le \sum_{s=1}^{k}\mu^s\Theta, \tag{12.26}
$$

which implies

$$
\begin{aligned}
\mathbb{E}\{V_k\} &\le \mu^{-k}\mathbb{E}\{V_0\} + \sum_{s=1}^{k}\mu^{s-k}\Theta \\
&= \mu^{-k}\mathbb{E}\{V_0\} + \frac{(1-\mu^{-k})\Theta}{1-\mu^{-1}} \\
&= \mu^{-k}\Big(\mathbb{E}\{V_0\} - \frac{\mu\Theta}{\mu-1}\Big) + \frac{\mu\Theta}{\mu-1} \\
&\le \max\Big\{\mathbb{E}\{V_0\}, \frac{\mu\Theta}{\mu-1}\Big\}.
\end{aligned}
\tag{12.27}
$$

Finally, in light of the relationships between (12.7) and (12.13) and also between (12.13) and (12.16), it is not difficult to see that

$$
\begin{aligned}
||x_i(k) - x_j(k)||^2 \le ||\tilde{x}_k||^2 &\le ||\tilde{x}_k||^2 + ||\tilde{\xi}_k||^2 \\
&= ||\tilde{\eta}_{1,k}||^2 + ||\tilde{\eta}_{2,k}||^2 = ||\tilde{\eta}_k||^2
\end{aligned}
\tag{12.28}
$$

with $||\tilde{\eta}_0||^2 = \sum_{i=1}^{N}||x_i(0) - \bar{x}_0||^2 \le \chi_0^2$.

Integrating (12.27) and (12.28) results in

$$U(\chi_0, \delta, \theta) = \max\left\{\frac{2\rho_{\max}\chi_0^2}{\rho_{\min}}, \frac{\mu\Theta}{\mu - 1}\right\},$$

which completes the proof.

Remark 12.4 *In the proof, the elementary inequality $2a^Tb \le \varepsilon a^Ta + \varepsilon^{-1}b^Tb$ has been utilized to deal with the coupling terms among $\tilde{\eta}_k$, e_k and δ_k, which would facilitate the subsequent distributed controller design. It is worth mentioning here that the proposed method is insensitive to the number of agents. Also, the effect from the coupling terms (such as $2\tilde{\eta}_k^T \mathcal{A}_1^T \mathcal{P}\mathcal{D}_1 e_k$ and $2\tilde{\eta}_k^T \mathcal{A}_1^T \mathcal{P}\mathcal{D}_2\delta_k$) is reflected in the solvability of inequality (12.17) involving the parameter ε that can be optimized to reduce the consensus bound. On the other hand, it can be found from (12.18) that the consensus bound is dependent on χ_0^2, which means that a bigger χ_0^2 could lead to a bigger consensus bound.*

12.3 Consensus Controller Design

In the above subsection, the bounded consensus is discussed, and the bound can be calculated via the solution of inequality (12.17). In this subsection, the main task is to propose a suitable approach to obtain the gain matrices. Note that we aim to provide an algorithm whose computational complexity is not affected as the number of agents increases.

Theorem 12.2 *Let the undirected communication graph $\mathscr{G}$ and the scalar $\tilde{\varepsilon}$ be given. Assume that there exist a positive definite matrix $\bar{P}$, matrices Λ, Δ, $\tilde{A}_c$, $\tilde{L}_c$ and $\tilde{K}_c$, and three positive scalars φ_0, φ_1 and γ_0 satisfying*

$$\begin{cases} \Pi_2 = \begin{bmatrix} \Pi_{21} & * & * \\ Q_1 & \bar{\mathcal{P}} - \mathcal{G} - \mathcal{G}^T & * \\ Q_2 & 0 & \Pi_{22} \end{bmatrix} < 0, & (12.29a) \\[2em] \Pi_N = \begin{bmatrix} \Pi_{21} & * & * \\ Q_1 & \bar{\mathcal{P}} - \mathcal{G} - \mathcal{G}^T & * \\ Q_N & 0 & \Pi_{22} \end{bmatrix} < 0 & (12.29b) \end{cases}$$

where

$$\bar{\mathcal{P}} = \begin{bmatrix} P_1 & * \\ P_0 & P_2 \end{bmatrix}, \quad \mathcal{G} = \begin{bmatrix} \Delta W & \Lambda \\ 0 & \Lambda \end{bmatrix},$$

$$\bar{\mathcal{A}}_2 = \begin{bmatrix} D & 0 \\ 0 & 0 \end{bmatrix}, \quad Q_1 = \begin{bmatrix} \lambda_{\max}^{1/2}(\tilde{\mathcal{F}})\tilde{L}_c C & 0 \\ \lambda_{\max}^{1/2}(\tilde{\mathcal{F}})\tilde{L}_c C & 0 \end{bmatrix},$$

$$Q_i = \begin{bmatrix} \tilde{\varepsilon}\Delta W A + \tilde{\varepsilon}\bar{\alpha}\bar{\beta}\lambda_i \tilde{L}_c C & \varepsilon(\tilde{K}_c^* + \tilde{A}_c) \\ \tilde{\varepsilon}\bar{\alpha}\bar{\beta}\lambda_i \tilde{L}_c C & \tilde{\varepsilon}\tilde{A}_c \end{bmatrix},$$

$$\Delta = \begin{bmatrix} \Delta_{11} & \Delta_{12} \\ 0 & \Delta_{22} \end{bmatrix}, \quad \tilde{K}_c^* = \begin{bmatrix} \tilde{K}_c \\ 0 \end{bmatrix},$$

$$\Pi_{21} = -\bar{\mathcal{P}} + \bar{\mathcal{A}}_2^T \bar{\mathcal{P}} \bar{\mathcal{A}}_2 + \gamma_0 I, \quad \Pi_{22} = \bar{\mathcal{P}} - \mathcal{G} - \mathcal{G}^T,$$

$$\mathcal{W} = [\ B(B^T B)^{-1} \quad B^\perp\ ]^T.$$

Then, with the gains given by $K_c = \Delta_{11}^{-1}\tilde{K}_c$, $A_c = \Lambda^{-1}\tilde{A}_c$, *and* $L_c = \Lambda^{-1}\tilde{L}_c$, *the controlled multi-agent system (12.7) is consensus in a mean-square sense with bound*

$$U(\chi_0, \delta, \theta) = \max\left\{ \frac{2\rho_{\max}\chi_0^2}{\rho_{\min}}, \frac{\mu\Theta}{\mu - 1} \right\} \tag{12.30}$$

where $\varepsilon = \tilde{\varepsilon}^2 - 1$ *and other parameters are defined in Theorem 12.1.*

Proof: Firstly, with the aid of the properties of the Kronecker product, $\tilde{\mathcal{F}} \le \lambda_{\max}(\tilde{\mathcal{F}})I$ yields

$$\mathcal{F} \le \text{diag}\{\lambda_{\max}(\tilde{\mathcal{F}})I \otimes (LC)^T P_2(LC), 0\} \triangleq \mathcal{F}_\lambda.$$

Therefore, the inequality (12.17) holds when

$$(1 + \varepsilon)\mathcal{A}_1^T \mathcal{P} \mathcal{A}_1 + \mathcal{A}_2^T \mathcal{P} \mathcal{A}_2 + \mathcal{F}_\lambda - \mathcal{P} + \gamma_0 I < 0. \tag{12.31}$$

Based on the Schur complement lemma, the above inequality is true if and only if

$$\begin{bmatrix} -\mathcal{P} + \mathcal{A}_2^T \mathcal{P} \mathcal{A}_2 + \gamma_0 I & * & * \\ \mathcal{Q}_1 & -\mathcal{P}^{-1} & * \\ \mathcal{Q}_2 & 0 & -\mathcal{P}^{-1} \end{bmatrix} < 0 \tag{12.32}$$

where $\tilde{\varepsilon} = \sqrt{1 + \varepsilon}$ and

$$\mathcal{Q}_1 = \begin{bmatrix} 0 & 0 \\ \lambda_{\max}^{1/2}(\tilde{\mathcal{F}})I \otimes (L_c C) & 0 \end{bmatrix}, \quad \mathcal{Q}_2 = \begin{bmatrix} \tilde{\varepsilon}I \otimes A & \tilde{\varepsilon}I \otimes (BK_c) \\ \tilde{\varepsilon}\bar{\alpha}\bar{\beta}\Xi \otimes (L_c C) & \tilde{\varepsilon}I \otimes A_c \end{bmatrix}.$$

It follows from (12.29b) that $\mathcal{N}$ and Θ_{11} are invertible. Denote

$$\bar{\mathcal{G}} \triangleq \begin{bmatrix} I \otimes (\Delta \mathcal{W}) & I \otimes \Lambda \\ 0 & I \otimes \Lambda \end{bmatrix}.$$

Pre- and post-multiplying inequality (12.32) by $\text{diag}\{I, \bar{\mathcal{G}}, \bar{\mathcal{G}}\}$ and $\text{diag}\{I, \bar{\mathcal{G}}^T, \bar{\mathcal{G}}^T\}$ result in

$$
\begin{bmatrix}
-\mathcal{P} + \mathcal{A}_2^T \mathcal{P} \mathcal{A}_2 + \gamma_0 I & * & * \\
\tilde{\mathcal{Q}}_1 & -\bar{\mathcal{G}}\mathcal{P}^{-1}\bar{\mathcal{G}}^T & * \\
\tilde{\mathcal{Q}}_2 & 0 & -\bar{\mathcal{G}}\mathcal{P}^{-1}\bar{\mathcal{G}}^T
\end{bmatrix} < 0
\tag{12.33}
$$

with

$$
\tilde{\mathcal{Q}}_1 = \begin{bmatrix}
\lambda_{\max}^{1/2}(\tilde{\mathcal{F}}) I \otimes (\Lambda L_c C) & 0 \\
\lambda_{\max}^{1/2}(\tilde{\mathcal{F}}) I \otimes (\Lambda L_c C) & 0
\end{bmatrix},
$$

$$
\tilde{\mathcal{Q}}_2 = \begin{bmatrix}
\tilde{\varepsilon} I \otimes (\Delta W A) + \tilde{\varepsilon}\bar{\alpha}\bar{\beta}\Xi \otimes (\Lambda L_c C) & \tilde{\varepsilon} I \otimes (\Theta W B K_c + \Lambda A_c) \\
\tilde{\varepsilon}\bar{\alpha}\bar{\beta}\Xi \otimes (L_c C) & \tilde{\varepsilon} I \otimes (\Lambda A_c)
\end{bmatrix}.
$$

Due to

$$
\bar{\mathcal{G}} + \bar{\mathcal{G}}^T - \bar{\mathcal{G}}\mathcal{P}^{-1}\bar{\mathcal{G}}^T - \mathcal{P} = -(\bar{\mathcal{G}} - \mathcal{P})\mathcal{P}^{-1}(\bar{\mathcal{G}} - \mathcal{P})^T \le 0,
$$

it follows that the inequality (12.33) is satisfied if

$$
\begin{bmatrix}
-\mathcal{P} + \mathcal{A}_2^T \mathcal{P} \mathcal{A}_2 + \gamma_0 I & * & * \\
\tilde{\mathcal{Q}}_1 & -\bar{\mathcal{N}}_p & * \\
\tilde{\mathcal{Q}}_2 & 0 & -\bar{\mathcal{N}}_p
\end{bmatrix} < 0
\tag{12.34}
$$

where $\bar{\mathcal{N}}_p = \bar{\mathcal{N}} + \bar{\mathcal{N}}^T - \mathcal{P}$.

Denote $\tilde{A}_c \triangleq \Lambda A_c$, $\tilde{L}_c \triangleq \Lambda L_c$, $\tilde{K}_c^* \triangleq \Delta W B K_c$, and $\mathcal{N}_{\bar{p}} = \mathcal{N} + \mathcal{N}^T - \bar{\mathcal{P}}$. By using a series of elementary transformations, the inequality (12.34) can be transformed into a set of matrix inequalities as follows:

$$
\tilde{\Pi}_i = \begin{bmatrix}
-\bar{\mathcal{P}} + \bar{\mathcal{A}}_2^T \bar{\mathcal{P}} \bar{\mathcal{A}}_2 + \gamma_0 I & * & * \\
Q_1 & -\mathcal{N}_{\bar{p}} & * \\
Q_2^i & 0 & -\mathcal{N}_{\bar{p}}
\end{bmatrix} < 0
\tag{12.35}
$$

for $i = 2, 3, \cdots, N$ with

$$
Q_2^i = \begin{bmatrix}
\tilde{\varepsilon}\Delta W A + \tilde{\varepsilon}\bar{\alpha}\bar{\beta}\lambda_i \tilde{L}_c C & \tilde{\varepsilon}(\tilde{K}_c + \tilde{A}_c) \\
\tilde{\varepsilon}\bar{\alpha}\bar{\beta}\lambda_i \tilde{L}_c C & \tilde{\varepsilon}\tilde{A}_c
\end{bmatrix}.
$$

Due to $\lambda_2 \ge \lambda_N$, there is a positive $\Im_i \in (0, 1]$ such as $\lambda_i = \Im_i \lambda_2 + (1 - \Im_i)\lambda_N$. In this case, one has

$$
\tilde{\Pi}_i = \Im_i \tilde{\Pi}_2 + (1 - \Im_i)\tilde{\Pi}_N.
$$

One can easily get $\tilde{\Pi}_i < 0$ if the conditions in Theorem 12.2 are true. It then follows from Theorem 12.1 that the controlled multi-agent system (12.7) with the designed controller is consensus with bound (12.30) in a mean-square sense, which completes the proof.

Remark 12.5 *In Theorem 12.2, a condition for controller design is proposed by exploiting the second and Nth eigenvalues of Laplacian matrix $\mathcal{H}$ as well as the elementary transformation approach. In contrast with that in Theorem 12.1, such a condition can effectively reduce the computational complexity resulting from the increasing number of agents. Furthermore, the parameter $\tilde{\varepsilon}$ can been utilized to optimize the consensus bound via a linear search algorithm outlined as follows.*

Controller design algorithm

Step 1. Denote a positive scalar τ for linearly searching step size, and Ψ_λ, K_λ, A_λ and L_λ for four intermediate variables in order to save the minimal $U(\chi_0, \delta)$ and the controller gains.

Step 2. Let $\tilde{\varepsilon} = 1$, $\Psi_\lambda = 0$, $K_\lambda = 0$, $A_\lambda = 0$ and $L_\lambda = 0$, and then check the solvability of inequalities (12.29a) and (12.29b). If they are solvable, go to the next step, else go to *Step 6.*

Step 3. Let $\tilde{\varepsilon} = \tilde{\varepsilon} + \tau$, and solve the following optimal problem:

$$\mathbf{OP}: \quad \min_{\gamma_0, \tilde{\rho}_{\max}} \tilde{\rho}_{\max} - \gamma_0 \quad \text{s. t. } (12.29a), (12.29b) \text{ and } P < \tilde{\rho}_{\max} I.$$

Step 4. If the optimal problem $\mathbf{OP}$ is solvable, calculate $\rho_{\max}$, $\rho_{\min}$, μ and $\lambda_{\max}(\Psi)$. In what follows, obtain $U(\chi_0, \delta, \theta)$, K_c, A_c and L_c, and then go to the next step. If it is not solvable, go to *Step 6.*

Step 5. If $\Psi_\lambda > U(\chi_0, \delta, \theta)$, denote $\Psi_\lambda = U(\chi_0, \delta, \theta)$, $K_\lambda = K_c$, $A_\lambda = A_c$ and $L_\lambda = L_c$, and then go to *Step 3.*

Step 6. If $\tilde{\varepsilon} > 1 + \tau$, output $U(\chi_0, \delta, \theta)$, K_c, A_c and L_c, else the algorithm is infeasible. Stop.

12.4 Simulation Examples

In this section, a simulation example is presented to illustrate the effectiveness of the proposed event-triggered consensus protocol for discrete-time stochastic multi-agent systems with state-dependent noise.

Consider the system (12.1) with

$$A = \begin{bmatrix} 1 & 0.1 \\ 0 & 1 \end{bmatrix}, \ B = \begin{bmatrix} 1.0 \\ 0.8 \end{bmatrix}, \ C = \begin{bmatrix} -0.2 \\ 0 \end{bmatrix}^T, \ D = \begin{bmatrix} 0 & 0.02 \\ 0 & 0 \end{bmatrix}.$$

Suppose that there are five agents with the undirected communication

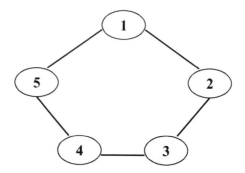

FIGURE 12.2
Communication graph $\mathcal{G}$.

graph $\mathcal{G}$ shown in Fig. 12.2. The associated adjacency matrix $\mathcal{H}$ is selected as follows

$$\mathcal{H} = \begin{bmatrix} -0.2 & 0.1 & 0 & 0 & 0.1 \\ 0.1 & -0.2 & 0.1 & 0 & 0 \\ 0 & 0.1 & -0.2 & 0.1 & 0 \\ 0 & 0 & 0.1 & -0.2 & 0.1 \\ 0.1 & 0 & 0 & 0.1 & -0.2 \end{bmatrix}.$$

For such a Laplacian matrix, the eigenvalues are $\lambda_2 = \lambda_3 = -0.1382$, and $\lambda_4 = \lambda_5 = -0.3618$. Select $\chi_0^2 = 2$, $\theta_i = 0.02$ and $\delta_i = 0.1$ $(i = 1, 2, \cdots, 5)$. The probabilities of cyber-attacks and lossy sensors are, respectively, $\bar{\alpha} = 0.95$ and $\bar{\beta} = 0.9$. By using the MATLAB® software (with YALMIP 3.0), a set of feasible solutions is obtained as follows:

$$P = \begin{bmatrix} 2015 & -1352 & 738 & -200 \\ -1352 & 1661 & -211 & 437 \\ 738 & -211 & 795 & -201 \\ -200 & 437 & -201 & 529 \end{bmatrix},$$

$$\Delta = \begin{bmatrix} 855 & -5749 \\ 0 & 5553 \end{bmatrix}, \quad \Lambda = \begin{bmatrix} 764.96 & -251.67 \\ -220.64 & 503.42 \end{bmatrix},$$

$$\tilde{A}_c = \begin{bmatrix} -177.54 & -67.56 \\ -191.77 & 289.23 \end{bmatrix}, \quad \tilde{L}_c = \begin{bmatrix} 4053 \\ -649 \end{bmatrix},$$

$$\tilde{K}_c = \begin{bmatrix} 757.98 & 63.77 \end{bmatrix}, \quad \tilde{\varepsilon} = 1.001$$

and therefore

$$A_c = \begin{bmatrix} -0.4176 & 0.1177 \\ -0.5640 & 0.6261 \end{bmatrix}, \quad L_c = \begin{bmatrix} 5.6955 \\ 1.2065 \end{bmatrix}, \quad K_c = \begin{bmatrix} 0.8861 \\ 0.0746 \end{bmatrix}^T.$$

In the simulation, all elements of $x_s(0)$ $(s = 1, 2, \cdots, 5)$ are generated that obey the uniform distribution $\mathcal{U}[-0.5,\ 0.5]$ with $\chi_0^2 = 2$. The information sent by attack is $\delta_i(k) = 0.1\sin(\psi_i(k))$ where the stochastic variable $\psi_i(k)$ obeys the Gaussian distribution $\mathcal{N}(2,4)$. Simulation results are shown in Figs. 12.3-12.5. Fig. 12.3 and Fig. 12.4 plot the state trajectories of $x_s(k)$ $(s = 1,2,\cdots,5)$ for the discrete-time stochastic multi-agent system (12.1), respectively, with and without the event-triggered consensus controller (12.6). Obviously, the open-loop multi-agent system cannot reach the desired consensus but the controlled system does so as expected. Furthermore, the update times and deception attack times are shown in Fig. 12.5, from which we can easily find that the proposed consensus control scheme can effectively reduce the update frequency as compared with the traditional time-triggered mechanism.

12.5   Summary

In this chapter, we have investigated the event-triggering consensus control issue for a class of discrete-time stochastic multi-agent systems with lossy sensors and cyber-attacks. A distributed observer has been designed to estimate the relative full states, based on which an event-triggered control protocol has been proposed to guarantee the prescribed consensus, where the state-independent threshold has been adopted to reduce the computational burden. With the aid of eigenvalues and eigenvectors of Laplacian matrix, some sufficient conditions have been derived to guarantee the desired consensus. Furthermore, the controller gains have been designed in terms of a solution for a matrix inequality which is independent of the number of agents. A numerical simulation example has been used to demonstrate the effectiveness of the control technology presented in this chapter.

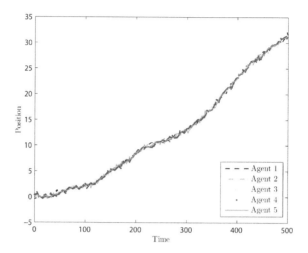

(a) The first element of the state $x_s(k)$.

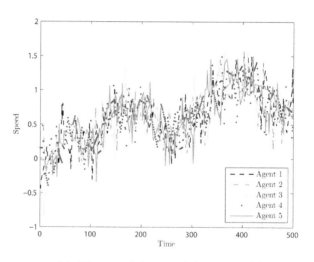

(b) The second element of the state $x_s(k)$.

FIGURE 12.3
The state trajectories of the closed-loop system.

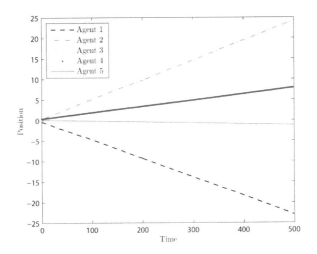

(a) The first element of the state $x_s(k)$.

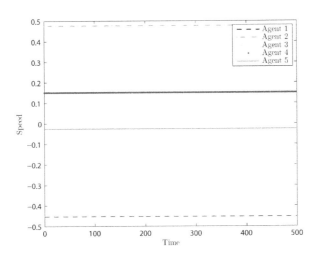

(b) The second element of the state $x_s(k)$.

FIGURE 12.4
The state trajectories of the open-loop system.

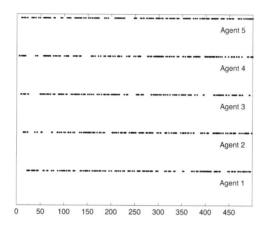

(a) Event-triggering times.

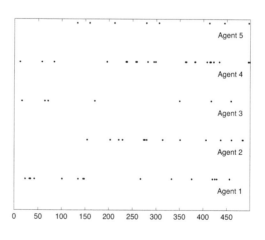

(b) Attack times.

FIGURE 12.5
Event-triggering times and attack times.

Bibliography

[1] F. Abdollahi and K. Khorasani. A decentralized Markovian jump $\mathcal{H}_\infty$ control routing strategy for mobile multi-agent networked systems. *IEEE Transactions on Control Systems Technology*, 19(2):269–283, 2011.

[2] D. Aeyels and F. De Smet. Cluster formation in a time-varying multi-agent system. *Automatica*, 47(11):2481–2487, 2011.

[3] M. Aitrami, X. Chen, and X. Zhou. Discrete-time indefinite LQ control with state and control dependent noises. *Journal of Global Optimization*, 23(3-4):245–265, 2002.

[4] C. Alippi and C. Galperti. An adaptive system for optimal solar energy harvesting in wireless sensor network nodes. *IEEE Transactions on Circuits and Systems-I*, 55(6):1742–1750, 2008.

[5] S. Amin, X. Litrico, S. Sastry, and A. M. Bayen. Cyber security of water SCADA systems-Part I: Analysis and experimentation of stealthy deception attacks. *IEEE Transactions on Control Systems Technology*, 21(5):3–43, 2013.

[6] S. Amin, G. A. Schwartz, and S. S. Sastry. Security of interdependent and identical networked control systems. *Automatica*, 49:186–192, 2013.

[7] D. Angeli. Input-to-state stability of PD-controlled robotic systems. *Automatica*, 35:1285–1290, 1999.

[8] D. Angeli. A Lyapunov approach to incremental stability properties. *IEEE Transactions on Automatic Control*, 47:410–422, 2002.

[9] A. Anta and P. Tabuada. To sample or not to sample self-triggered control for nonlinear systems. *IEEE Transactions on Automatic Control*, 55(9):2030–2042, 2010.

[10] G. Arslan and T. Basar. Disturbance attenuating controller design for strict-feedback systems with structurally unknown dynamics. *Automatica*, 37:1175–1188, 2001.

[11] I. Bashkirtseva and L. Ryashko. Control of equilibria for nonlinear stochastic discrete-time systems. *IEEE Transactions on Automatic Control*, 56(9):2162–2166, 2011.

[12] M. Basin, S. Elvira-Ceja, and E. Sanchez. Mean-square $\mathcal{H}_\infty$ filtering for stochastic systems: Application to a 2DOF helicopter. *Signal Processing*, 92(3):801–806, 2012.

[13] G. K. Befekadu, V. Gupta, and P. J. Antsaklis. Risk-sensitive control under a class of denial-of-service attack models. In *Proceedings of the 2011 American Control Conference, San Francisco, CA, USA*, pages 643–648, 2011.

[14] N. Berman and U. Shaked. $\mathcal{H}_\infty$ control for discrete-time nonlinear stochastic systems. *IEEE Transactions on Automatic Control*, 51(6):1041–1046, 2006.

[15] R. A. Berry and R. G. Gallager. Communication over fading channels with delay constraints. *IEEE Transactions on Information Theory*, 48(5):1135–1149, 2002.

[16] E. Biglieri, J. Proakis, and S. Shamai. Fading channels: Information-theoretic and communications aspects. *IEEE Transactions on Information Theory*, 44(6):2619–2686, 1998.

[17] A. E. Bouhtouri, D. Hinrichsen, and A. Pritchard. $\mathcal{H}_\infty$-type control for discrete-time stochastic systems. *International Journal of Robust Nonlinear Control*, 9(13):923–948, 1999.

[18] R. Caballero-Aguila, A. Hermoso-Carazo, J. D. Jimenez-Lopez, J. Linares-Perez, and S. Nakamori. Signal estimation with multiple delayed sensors using covariance information. *Digital Signal Processing*, 20(2):528–540, 2010.

[19] G. C. Calafiore and F. Abrate. Distributed linear estimation over sensor networks. *International Journal of Control*, 82(5):868–882, 2009.

[20] A. Cantoni, B.-N. Vo, and K. Teo. An introduction to envelope constrained filter design. *Journal of Telecommunications and Information Technology*, pages 3–14, 2001.

[21] Y. Cao, J. Lam, and Y. Sun. Static output feedback stabilization: An ILMI approach. *Automatica*, 34(12):1641–1645, 1998.

[22] A. A. Cárdenas, S. Amin, and S. S. Sastry. Secure control: Towards survivable cyber-physical systems. In *Proceedings of the 1st International Workshop on Cyber-Physical Systems, Beijing, China*, pages 495–500, 2008.

[23] R. Carli, A. Chiuso, L. Schenato, and S. Zampieri. Distributed Kalman filtering based on consensus strategies. *IEEE Journal on Selected Areas in Communications*, 26(4):622–633, 2008.

[24] A. Clark, L. Bushnell, and R. Poovendran. A passivity-based framework for composing attacks on networked control systems. In *Proceedings of the 50th Annual Allerton Conference, Allerton House, UIUC, Illinois, USA*, pages 1814–1821, 2012.

[25] H. Deng and M. Krstić. Output-feedback stabilization of stochastic nonlinear systems driven by noise of unknown covariance. *Systems & Control Letters*, 39(3):173–182, 2000.

[26] D. Ding, Z. Wang, B. Shen, and H. Shu. h_∞ state estimation for discrete-time complex networks with randomly occurring sensor saturations and randomly varying sensor delays. *IEEE Transactions on Neural Networks and Learning Systems*, 23(5):725–736, 2012.

[27] H. Dong, Z. Wang, X. Chen, and H. Gao. A review on analysis and synthesis of nonlinear stochastic systems with randomly occurring incomplete information. *Mathematical Problems in Engineering*, 2012:622–633, 2012.

[28] H. Dong, Z. Wang, and H. Gao. Distributed $\mathcal{H}_\infty$ filtering for a class of Markovian jump nonlinear time-delay systems over lossy sensor networks. *IEEE Transactions on Industrial Electronics*, 60(10):4665–4672, 2013.

[29] H. Dong, Z. Wang, and H. Gao. *Filtering, Control and Fault Detection with Randomly Occurring Incomplete Information*. Wiley, Chichester, UK, 2013.

[30] H. Dong, Z. Wang, D. W. C. Ho, and H. Gao. Variance-constrained $\mathcal{H}_\infty$ filtering for a class of nonlinear time-varying systems with multiple missing measurements: The finite-horizon case. *IEEE Transactions on Signal Processing*, 58(5):2534–2543, 2010.

[31] H. Dong, Z. Wang, D. W. C. Ho, and H. Gao. Robust $\mathcal{H}_\infty$ filtering for Markovian jump systems with randomly occurring nonlinearities and sensor saturation: The finite-horizon case. *IEEE Transactions on Signal Processing*, 59(7):3048–3057, 2011.

[32] J. Dong and G.-H. Yang. Robust static output feedback control for linear discrete-time systems with time-varying uncertainties. *Systems & Control Letters*, 57(2):123–131, 2008.

[33] M. C. F. Donkers and W. P. M. H. Heemels. Output-based event-triggered control with guaranteed $\mathcal{L}_\infty$-gain and improved and decentralized event triggering. *IEEE Transactions on Automatic Control*, 57(6):1362–1367, 2012.

[34] N. Elia. Remote stabilization over fading channels. *Systems & Control Letters*, 54(3):237–249, 2005.

[35] Y. Fan, G. Feng, Y. Wang, and C. Song. Distributed event-triggered control of multi-agent systems with combinational measurements. *Automatica*, 49(2):671–675, 2013.

[36] H. Fang and L. Lin. Stability analysis for linear systems under state constraints. *IEEE Transactions Automatic Control*, 49(6):950–955, 2005.

[37] J. Fax and R. Murray. Information flow and cooperative control of vehicle formations. *IEEE Transactions on Automatic Control*, 49(9):1465–1476, 2004.

[38] H. Foroush and S. Martínez. On event-triggered control of linear systems under periodic denial-of-service jamming attacks. In *Proceedings of the 51st IEEE Conference on Decision and Control, Hawaii, USA*, pages 2551–2256, 2012.

[39] H. Gao and T. Chen. $\mathcal{H}_\infty$ estimation for uncertain systems with limited communication capacity. *IEEE Transactions on Automatic Control*, 52(11):2070–2084, 2007.

[40] E. Garone, B. Sinopoli, A. Goldsmith, and A. Casavola. LQG control for MIMO systems over multiple erasure channels with perfect acknowledgment. *IEEE Transactions on Automatic Control*, 57(2):450–456, 2012.

[41] E. Gershon, U. Shaked, and N. Berman. $\mathcal{H}_\infty$ control and estimation of retarded state-multiplicative stochastic systems. *IEEE Transactions on Automatic Control*, 52(9):1773–1779, 2007.

[42] E. Gershon, U. Shaked, and I. Yaesh. *Control and Estimation of State-Multiplicative Linear Systems*. Springer-Verlag London Limited, London, U.K., 2005.

[43] L. El Ghaoui and G. Calafiore. Robust filtering for discrete-time systems with bounded noise and parametric uncertainty. *IEEE Transactions on Automatic Control*, 46(7):1084–1089, 2001.

[44] D. Gu. A game theory approach to target tracking in sensor networks. *IEEE Transactions on Systems, Man, and Cybernetics-Part B*, 41(1):2–13, 2011.

[45] W. Guan and G. Yang. New controller design method for continuous-time systems with state saturation. *IET Control Theory and Applications*, 4(10):1889–1897, 2010.

[46] X. He, Z. Wang, and D. Zhou. Robust $\mathcal{H}_\infty$ filtering for networked systems with multiple state delays. *International Journal of Control*, 80(8):1217–1232, 2007.

[47] Y. He and Q. Wang. An improved ILMI method for static output feedback control with application to multivariable PID control. *IEEE Transactions on Automatic Control*, 51(10):1678–1683, 2006.

[48] Y. Hong, L. Gao, D. Cheng, and J. Hu. Lyapunov-based approach to multiagent systems with switching jointly connected interconnection. *IEEE Transactions on Automatic Control*, 52(5):943–948, 2007.

[49] T. Hou, W. Zhang, and H. Ma. Finite horizon $\mathcal{H}_2/\mathcal{H}_\infty$ control for discrete-time stochastic systems with Markovian jumps and multiplicative noise. *IEEE Transactions on Automatic Control*, 55(5):1185–1191, 2010.

[50] J. Hu, Z. Wang, and H. Gao. Recursive filtering with random parameter matrices, multiple fading measurements and correlated noises. *Automatica*, 49(11):3440–3448, 2013.

[51] J. Hu, Z. Wang, H. Gao, and L. K. Stergioulas. Extended Kalman filtering with stochastic nonlinearities and multiple missing measurements. *Automatica*, 48(9):2007–2015, 2012.

[52] T. Hu, Z. Lin, and B. M. Chen. An analysis and design method for linear systems subject to actuator saturation and disturbance. *Automatica*, 38(2):351–359, 2002.

[53] L. Huang and X. Mao. On input-to-state stability of stochastic retarded systems with Markovian switching. *IEEE Transactions on Automatic Control*, 54(8):1898–1902, 2009.

[54] M. Huang and J. H. Manton. Stochastic consensus seeking with noisy and directed inter-agent communication: Fixed and randomly varying topologies. *IEEE Transactions on Automatic Control*, 55(1):253–241, 2010.

[55] Y. S. Hung and F. Yang. Robust $\mathcal{H}_\infty$ filtering for discrete time-varying uncertain systems with a known deterministic input. *International Journal of Control*, 75(15):1159–1169, 2002.

[56] Y. S. Hung and F. Yang. Robust $\mathcal{H}_\infty$ filtering with error variance constraints for discrete time-varying systems with uncertainty. *Automatica*, 39(7):1185–1194, 2003.

[57] D. Jacobson. A general result in stochastic optimal control of nonlinear discrete-time systems with quadratic performance criteria. *Journal of Mathematical Analysis and Applications*, 47(1):153–161, 1974.

[58] X. Ji, T. Liu, Y. Sun, and H. Su. Stability analysis and controller synthesis for discrete linear time-delay systems with state saturation nonlinearities. *International Journal of Systems Science*, 42(3):397–406, 2011.

[59] X. Ji, Y. Sun, and T. Liu. Improvement on stability analysis for linear systems under state saturation. *IEEE Transactions Automatic Control*, 53(8):1961–1963, 2008.

[60] V. K. Kandanvli and H. Kar. Robust stability of discrete-time state-delayed systems with saturation nonlinearities: Linear matrix inequality approach. *Signal Processing*, 89(2):161–173, 2009.

[61] H. Kar and V. Singh. Stability analysis of discrete-time systems in a state-space realisation with partial state saturation nonlinearities. *IEE Proceedings-Control Theory Applications*, 150(3):205–208, 2003.

[62] H. Kar and V. Singh. Elimination of overflow oscillations in fixed-point state-space digital filters with saturation arithmetic: An LMI Approach. *IEEE Transactions on Circuits and Systems-II*, 51(1):40–42, 2004.

[63] S. Kar and J. M. F. Moura. Distributed consensus algorithms in sensor networks: quantized data and random link failures. *IEEE Transactions on Signal Processing*, 58(3):1383–1399, 2010.

[64] H. R. Karimi. Robust $\mathcal{H}_\infty$ filter design for uncertain linear systems over network with network-induced delays and output quantization. *Modeling, Identification and Control*, 30(1):27–37, 2009.

[65] S. Kluge, K. Reif, and M. Brokate. Stochastic stability of the extended Kalman filter with intermittent observations. *IEEE Transactions on Automatic control*, 55(2):514–518, 2010.

[66] P. Lancaster and M. T. Tismenetsky. *The Theory of Matrices: With Applications*. Academic, New York, USA, 1985.

[67] T. Li, M. Fu, L. Xie, and J. Zhang. Distributed consensus with limited communication data rate. *IEEE Transactions on Automatic Control*, 56(2):279–292, 2011.

[68] T. Li and J. Zhang. Decentralized tracking-type games for multi-agent systems with coupled ARX models: asymptotic Nash equilibria. *Automatica*, 44(3):713–725, 2008.

[69] J. Liang, Z. Wang, and X. Liu. Distributed state estimation for discrete-time sensor networks with randomly varying nonlinearities and missing measurements. *IEEE Transactions on Neural Networks*, 22(3):66–86, 2011.

[70] J. Liang, Z. Wang, B. Shen, and X. Liu. Distributed state estimation in sensor networks with randomly occurring nonlinearities subject to time-delays. *ACM Transactions on Sensor Networks*, 9(1):Art. No. 4, 2012.

[71] D. Limon, T. Alamo, F. Salas, and E. F. Camacho. Input to state stability of min-max MPC controllers for nonlinear systems with bounded uncertainties. *Automatica*, 42:797–803, 2006.

[72] P. Lin and Y. Jia. Consensus of second-order discrete-time multi-agent systems with nonuniform time-delays and dynamically changing topologies. *Automatica*, 45(9):2154–2158, 2009.

[73] D. Liu and A. N. Michel. Asymptotic stability of discrete-time systems with saturation nonlinearities with application to digital-filters. *IEEE Transactions on Circuits and Systems-I*, 39(10):789–807, 1992.

[74] D. Liu and A. N. Michel. Stability analysis of systems with partial state saturation nonlinearities. *IEEE Transactions on Circuits and Systems-I*, 43(3):230–232, 1996.

[75] L. Liu, Y. Shen, and F. Jiang. The almost sure asymptotic stability and p-th moment asymptotic stability of nonlinear stochastic differential systems with polynomial growth. *IEEE Transactions on Automatic Control*, 56(8):1985–1990, 2011.

[76] S. Liu and J. Zhang. Output-feedback control of a class of stochastic nonlinear systems with linearly bounded unmeasurable states. *International Journal of Robust and Nonlinear Control*, 18(6):665–687, 2008.

[77] S. Liu, J. Zhang, and Z.-P. Jiang. Decentralized adaptive output-feedback stabilization for large-scale stochastic nonlinear systems. *Automatica*, 43:238–251, 2007.

[78] W. Lou and Y. Fang. A multipath routing approach for secure data delivery. In *MILCOM 2001. Communications for Network-Centric Operations: Creating the Information Force, McLean, Virginia, USA*, pages 1467–1473, 2001.

[79] L.-Z. Lu and C. E. MPearce. Some new bounds for singular values and eigenvalues of matrix products. *Annals of Operations Research*, 98(1-4):141–148, 2000.

[80] Y. Luo, Y. Zhu, D. Luo, J. Zhou, E. Song, and D. Wang. Globally optimal multisensor distributed random parameter matrices Kalman filtering fusion with applications. *Sensors*, 8(12):8086–8103, 2008.

[81] C. Ma and J. Zhang. Necessary and sufficient conditions for consensusability of linear multi-agent systems. *IEEE Transactions on Automatic Control*, 55(5):1263–1268, 2010.

[82] Ma09. Decentralized adaptive synchronization of a stochastic discrete-time multiagent dynamic model. *SIAM Journal on Control and Optimization*, 48(2):859–880, 2009.

[83] S. Marano, V. Matta, and L. Tong. Distributed detection in the presence of Byzantine attacks. *IEEE Transactions on Signal Processing*, 57(1):16–29, 2009.

[84] M. Mazo, A. Anta, and Paulo Tabuada. An ISS self-triggered implementation of linear controllers. *Automatica*, 46(8):1310–1314, 2010.

[85] X. Meng and T. Chen. Event based agreement protocols for multi-agent networks. *Automatica*, 49(7):2125–2132, 2013.

[86] Y. Mostofi and R. M. Murray. To drop or not to drop: Design principles for Kalman filtering over wireless fading channels. *IEEE Transactions on Automatic Control*, 54(2):376–381, 2010.

[87] P. Ogren and N. E. Leonard. Obstacle avoidance in formation. In *Proceedings of the IEEE Conference on Robotics and Automation, Taipei, Taiwan*, pages 2492–2497, 2003.

[88] R. Olfati-Saber. Flocking for multi-agent dynamic systems: Algorithms and theory. *IEEE Transactions on Automatic Control*, 51(3):401–420, 2006.

[89] R. Olfati-Saber. Distributed Kalman filtering for sensor networks. In *Proceedings of the 46th IEEE Conference on Decision and Control, New Orleans, LA*, pages 1–7, 2007.

[90] R. Olfati-Saber and R. M. Murray. Consensus problems in networks of agents with switching topology and time-delays. *IEEE Transactions on Automatic Control*, 49(9):1520–1533, 2004.

[91] T. Ooba. Stability of linear discrete dynamics employing state saturation arithmetic. *IEEE Transactions Automatic Control*, 48(4):626–630, 2003.

[92] Z.-H. Pang and G.-P. Liu. Design and implementation of secure networked predictive control systems under deception attacks. *IEEE Transactions on Control Systems Technology*, 20(5):1334–1342, 2010.

[93] S. Patterson, B. Bamieh, and A. Abbadi. Convergence rates of distributed average consensus with stochastic link failures. *IEEE Transactions Automatic Control*, 55(4):880–892, 2010.

[94] L. Patton, S. Frost, and B. Rigling. Efficient design of radar waveforms for optimised detection in coloured noise. *IET Radar, Sonar and Navigation*, 6(1):21–29, 2012.

[95] R. Penrose and J. A. Todd. On best approximate solutions of linear matrix equations. *Mathematical Proceedings of the Cambridge Philosophical Society*, 52(1):17–19, 1956.

[96] D. E. Quevedo, A. Ahlén, A. S. Leong, and S. Dey. On Kalman filtering over fading wireless channels with controlled transmission powers. *Automatica*, 48(7):1306–1316, 2012.

[97] K. Reif, S. Günther, E. Yaz, and R. Unbehauen. Stochastic stability of the discrete-time extended Kalman filter. *IEEE Transactions on Automatic Control*, 44(4):714–728, 1999.

[98] W. Ren, K. Moore, and Y. Chen. High-order consensus algorithms in cooperative vehicle systems. In *Proceedings of the IEEE International Conference on Networking, Sensing and Control, Fort Lauderdale, USA*, pages 457–462, 2006.

[99] A. Ribeiro and G. B. Giannakis. Bandwidth-constrained distributed estimation for wireless sensor networks-part II: Unknown probability density function. *IEEE Transactions on Signal Processing*, 54(7):2784–2796, 2006.

[100] A. Ribeiro and G. B. Giannakis. Bandwidth-constrained distributed estimation for wireless sensor networks-part Part I: Gaussian Case. *IEEE Transactions on Signal Processing*, 54(3):1131–1143, 2006.

[101] A. Ribeiro, I. D. Schizas, S. I. Roumeliotis, and G. B. Giannakis. Kalman filtering in wireless sensor networks. *IEEE Control Systems Magazine*, 30(2):66–86, 2010.

[102] A. J. Rojas and F. Lotero. Signal-to-noise ratio limited output feedback control subject to channel input quantization. *IEEE Transactions on Automatic control*, 60(2):475–479, 2015.

[103] M. Sahebsara, T. Chen, and S. L. Shah. Optimal $\mathcal{H}_2$ filtering in networked control systems with multiple packet dropout. *IEEE Transactions on Automatic Control*, 52(8):1508–1513, 2007.

[104] A. V. Savkin and I. R. Petersen. Optimal $\mathcal{H}_2$ Robust state estimation and model validation for discrete-time uncertain systems with a deterministic description of noise and uncertainty. *Automatica*, 34(2):271–274, 1998.

[105] E. Semsar-Kazerooni and K. Khorasani. Multi-agent team cooperation: a game theory approach. *Automatica*, 45(10):2205–2213, 2009.

[106] G. S. Seyboth, D. V. Dimarogonas, and K. H. Johansson. Event-based broadcasting for multi-agent average consensus. *Automatica*, 49(1):245–252, 2013.

[107] U. Shaked and N. Berman. $\mathcal{H}_\infty$ nonliiear filtering of discrete-time processes. *IEEE Transactions on Signal Processing*, 43(9):2205–2209, 1995.

[108] U. Shaked and V. Suplin. A new bounded real lemma representation for the continuous-time case. *IEEE Transactions on Automatic Control,* 46(9):1420–1426, 2001.

[109] B. Shen, Z. Wang, and Y. S. Hung. Distributed $\mathcal{H}_\infty$-consensus filtering in sensor networks with multiple missing measurements: the finite-horizon case. *Automatica,* 46(10):1682–1688, 2010.

[110] B. Shen, Z. Wang, and X. Liu. A stochastic sampled-data approach to distributed $\mathcal{H}_\infty$ filtering in sensor networks. *IEEE Transactions on Circuits and Systems-Part I,* 58(9):2237–2246, 2011.

[111] B. Shen, Z. Wang, and X. Liu. Bounded $\mathcal{H}_\infty$ synchronization and state estimation for discrete time-varying stochastic complex networks over a finite horizon. *IEEE Transactions on Neural Networks,* 22(1):145–157, 2011.

[112] B. Shen, Z. Wang, H. Shu, and G. Wei. Robust $\mathcal{H}_\infty$ finite-horizon filtering with randomly occurred nonlinearities and quantization effects. *Automatica,* 46(11):1743–1751, 2010.

[113] Z. Shu, J. Lam, and J. Xiong. Non-fragile exponential stability assignment of discrete-time linear systems with missing data in actuators. *IEEE Transactions on Automatic Control,* 54(3):625–630, 2009.

[114] V. Singh. Stability analysis of 2-D discrete systems described by the Fornasini-Marchesini second model with state saturation. *IEEE Transactions Circuits and Systems-II,* 55(8):793–796, 2008.

[115] E. Song, Y. Zhu, J. Zhou, and Z. You. Optimal Kalman filtering fusion with cross-correlated sensor noises. *Automatica,* 43(8):1450–1456, 2007.

[116] E. D. Sontag. Smooth stabilization implies coprime factorization. *IEEE Transactions on Automatic Control,* 34(4):435–443, 1989.

[117] F. Sorrentino, M. Di Bernardo, and F. Garofalo. Synchronizability and synchronization dynamics of weighed and unweighed scale free networks with degree mixing. *International Journal of Bifurcation and Chaos,* 17(7):2419–2434, 2007.

[118] A. Speranzon, C. Fischione, K. H. Johansson, and A. Sangiovanni-Vincentelli. A distributed minimum variance estimator for sensor networks. *IEEE Journal on Selected Areas in Communications,* 26(4):609–621, 2008.

[119] S. Sun, L. Xie, and W. Xiao. Optimal full-order and reduced-order estimators for discrete-time systems with multiple packet dropouts. *IEEE Transactions on Signal Processing,* 56(8):4031–4038, 2008.

[120] Y. Sun, Z. Guan, X. Zhan, and F. Yuan. Consensus of second-order and high-order discrete-time multi-agent systems with random networks. *Nonlinear Analysis: Real World Applications*, 13(5):1979–1990, 2012.

[121] Y. Sun and L. Wang. Consensus of multi-agent systems in directed networks with nonuniform time-varying delays. *IEEE Transactions on Automatic Control*, 54(7):1607–1613, 2009.

[122] M. Syu and G. Tsao. A saturation-type transfer function for backpropagation network modeling of biosystems. In *1994 IEEE International Conference on Neural Networks (ICNN 94)-1st, IEEE World Congress on Computational Intelligence*, pages 3265–3270, 1994.

[123] P. Tabuada. Event-triggered real-time scheduling of stabilizing control tasks. *IEEE Transactions on Automatic Control*, 52(9):1979–1990, 2007.

[124] Z. Tan, Y. C. Soh, and L. Xie. Envelope-constrained IIR filter design: An LMI $\mathcal{H}_\infty$ optimization approach. *Circuits Systems Signal Processing*, 19(3):205–220, 2000.

[125] A. Teixeira, H. Sandberg, and K. H. Johansson. Networked control systems under cyber attacks with applications to power networks. In *Proceedings of the 2010 American Control Conference*, pages 3690–3696, 2010.

[126] Y. Theodor and U. Shaked. Robust discrete-time minimum-variance filtering. *IEEE Transactions on Signal Processing*, 44(2):181–189, 1996.

[127] C. Tseng, K. Teo, and A. Cantoni. Mean square convergence of adaptive envelope-constrained filtering. *IEEE Transactions on Signal Processing*, 50(6):1429–1437, 2002.

[128] C. Tseng, K. Teo, A. Cantoni, and Z. Zang. Envelope-constrained filters: adaptive algorithms. *IEEE Transactions on Signal Processing*, 48(6):1597–1608, 2000.

[129] J. Tsinias. Stochastic input to state stability and application to global feedback stabilization. *International Journal of Control*, 71(5):907–930, 1998.

[130] J. Tsinias. The concept of "Exponential input to state stability" for stochastic systems and applications to feedback stabilization. *Systems and Control Letters*, 36(3):221–229, 1999.

[131] D. J. Tylavsky and G. R. L. Sohie. Generalization of the matrix inversion lemma. *Proceeding of the IEEE*, 74(7):1050–1052, 1986.

[132] V. Ugrinovskii. Distributed robust estimation over randomly switching networks using $\mathcal{H}_\infty$ consensus. *Automatica*, 49(1):160–168, 2013.

[133] A. Vempaty, O. Ozdemir, K. Agrawal, H. Chen, and P. K. Varshney. Localization in wireless sensor networks: Byzantines and mitigation techniques. *IEEE Transactions on Signal Processing*, 61(6):1495–1508, 2013.

[134] A. Vempaty, L. Tong, and P. Varshney. Distributed inference with Byzantine data: State-of-the-art review on data falsification attacks. *IEEE Signal Processing Magazine*, 30(5):65–75, 2013.

[135] B.-N. Vo, A. Cantoni, and K. Teo. Envelope constrained filter with linear interpolator. *IEEE Transactions on Signal Processing*, 45(6):1405–1414, 1997.

[136] T. Wang, L. Chang, and P. Chen. A collaborative sensor-fault detection scheme for robust distributed estimation in sensor networks. *IEEE Transactions on Communications*, 57(10):3045–3058, 2009.

[137] X. Wang and M. D. Lemmon. On event design in event-triggered feedback systems. *Automatica*, 47(10):2319–2322, 2011.

[138] Z. Wang, B. Shen, H. Shu, and G. Wei. Quantized $\mathscr{H}_\infty$ control for nonlinear stochastic time-delay systems with missing measurements. *IEEE Transactions on Automatic Control*, 57(6):1431–1444, 2012.

[139] Z. Wang, F. Yang, D. W. C. Ho, and X. Liu. Robust finite-horizon filtering for stochastic systems with missing measurements. *IEEE Signal Processing Letters*, 12(6):437–440, 2005.

[140] F. Xiao, L. Wang, J. Chen, and Y. Gao. Finite-time formation control for multi-agent systems. *Automatica*, 45(11):2605–2611, 2009.

[141] N. Xiao, L. Xie, and L. Qiu. Feedback stabilization of discrete-time networked systems over fading channels. *IEEE Transactions on Automatic Control*, 57(9):2176–2189, 2012.

[142] X. Xie and J. Tian. State-feedback stabilization for high-order stochastic nonlinear systems with stochastic inverse dynamics. *International Journal of Robust and Nonlinear Control*, 17:1343–1362, 2007.

[143] F. Yang and Y. Li. Set-membership filtering for systems with sensor saturation. *Automatica*, 45(8):1896–1902, 2009.

[144] P. Yang, R. Freeman, and K. Lynch. Multi-agent coordination by decentralized estimation and control. *IEEE Transactions on Automatic Control*, 53(11):2480–2496, 2008.

[145] E. Yaz and R. E. Skelton. Parametrization of all linear compensators for discrete-time stochastic parameter systems. *Automatica*, 30(6):945–955, 1994.

[146] E. Yaz and Y. Yaz. E. Yaz and Y. Yaz, State estimation of uncertain nonlinear stochastic systems with general criteria. *Applied Mathematics Letters*, 14(5):605–610, 2001.

[147] K. You and L. Xie. Coordination of discrete-time multi-agent systems via relative output feedback. *International Journal of Robust and Nonlinear Control*, 21(13):1587–1605, 2011.

[148] K. You and L. Xie. Network topology and communication data rate for consensusability of discrete-time multi-agent systems. *IEEE Transactions on Automatic Control*, 56(10):2262–2275, 2011.

[149] W. Yu, G. Chen Z. Wang, and W. Yang. Distributed consensus filtering in sensor networks. *IEEE Transactions on Systems, Man, and Cybernetics-Part B*, 39(6):1568–1577, 2009.

[150] Z. Zang, A. Cantoni, and K. Teo. Envelope-constrained IIR filter design via $\mathcal{H}_\infty$ optimization methods. *IEEE Transactions on Circuits and Systems-I*, 46(6):649–653, 1999.

[151] H. Zhang, Y. Shi, and A. Mehr. Robust static output feedback control and remote PID design for networked motor systems. *IEEE Transactions on Industrial Electronics*, 58(12):5396–5405, 2011.

[152] J. Zhang, R. S. Blum, X. Lu, and D. Conus. Asymptotically optimum distributed estimation in the presence of attacks. *IEEE Transactions on Signal Processing*, 63(5):1086–1101, 2015.

[153] L. Zhang, Y. Leng, and P. Colaneri. Stability and stabilization of discrete-time semi-Markov jump linear systems via semi-Markov kernel approach. *IEEE Transactions on Automatic Control*, 61(2):503–508, 2016.

[154] W. Zhang and B.-S. Chen. State feedback $\mathcal{H}_\infty$ control for a class of nonlinear stochastic systems. *SIAM Journal on Control and Optimization*, 44(6):1973–1991, 2006.

[155] W. X. Zheng. IEEE Transactions on Circuits and Systems-II. *IEEE Transactions on Circuits and Systems-II*, 50(12):1023–1027, 2003.

[156] J. Zhu, X. Yu, T. Zhang, Z. Cao, Y. Yang, and Y. Yi. The mean-square stability probability of $\mathcal{H}_\infty$ control of continuous Markovian jump systems. *IEEE Transactions on Automatic Control*, 61(7):1918–1924, 2016.

[157] M. Zhu and S. Martinez. On the performance analysis of resilient networked control systems under replay attacks. *IEEE Transactions on Automatic Control*, 59(3):804–808, 2014.

Index

Milton Keynes UK
Ingram Content Group UK Ltd.
UKHW040108071024
449327UK00019B/898

9 780367 570927